DM 120-
U.D
AF769662

Dynamics Reported

Expositions in Dynamical Systems

Board of Editors
H. Amann (Zürich), P. Brunovsky (Bratislava), S. N. Chow (Atlanta), C. K. R.T. Jones (Providence), U. Kirchgraber (Zürich), H. Matano (Tokyo), J. Mawhin (Louvain-la-Neuve), R.D. Nussbaum (NewBrunswick), C. Robinson (Evanston), H. O. Walther (München), E. Zehnder (Zürich)

Advisory Board
L. Arnold, H. Broer, A. Coppel, B. Fiedler, J. Grasman, E. Hairer, J. Henrard, R. A. Johnson, H. Kielhöfer, A. Lasota, M. Misiurewicz, R. Moeckel, R.E. O'Malley, K.J. Palmer, H.O. Peitgen, F. Przytycki, K. Rybakowski, D. G. Saari, J. A. Sanders, A. N. Sharkovsky, J. Scheurle, K. Schmitt, A. Vanderbauwhede, J. Waldvogel, J. A. Yorke

Dynamical Systems are a rapidly developing field with a strong impact on applications. Dynamics Reported is a series of books of a new type. Its principal goal is to make available current topics, new ideas and techniques. Each volume contains about four or five articles of up to 60 pages. Great emphasis is put on an excellent presentation, well suited for advanced courses, seminars etc. such that the material becomes accessible to beginning graduate students. To explain the core of a new method contributions will treat *examples* rather than general theories, they will describe *typical results* rather than the most sophisticated ones. Theorems are accompanied by *carefully written proofs*. The presentation is as *self-contained* as possible.

Authors will receive 5 copies of the volume containing their contributions. These will be split among multiple authors.

Authors are encouraged to prepare their manuscripts in Plain TEX or LATEX. Detailed information and macro packages are available via the Managing Editors.

Manuscripts and correspondence should be addressed to the Managing Editors:

C. K. R. T. Jones
Division of Applied Mathematics
Brown University
Providence, Rhode Island 02912
USA
e-Mail: ckrtj@cfm.brown.edu

U. Kirchgraber
Mathematics
Swiss Federal Institute
of Technology (ETH)
CH-8092 Zürich, Switzerland
e-Mail: kirchgra@math.ethz.ch

H. O. Walther
Mathematics
Ludwig-Maximilians University
W-8000 Munich
Federal Republic of Germany
e-Mail: Hans-Otto. Walther
@mathematik.
uni-muenchen.dbp.de

C.K.R.T. Jones U. Kirchgraber H.O. Walther
(Managing Editors)

Dynamics Reported

Expositions in Dynamical Systems

New Series: Volume 2

With Contributions of
H.S. Dumas Chr. Genecand J. Henrard J. Komorník

Springer-Verlag
Berlin Heidelberg New York
London Paris Tokyo
Hong Kong Barcelona
Budapest

ISBN 3-540-56409-8 Springer-Verlag Berlin Heidelberg New York
ISBN 0-387-56409-8 Springer-Verlag New York Berlin Heidelberg

This work is subject to copyright. All rights are reserved, whether the whole or part of the material is concerned, specifically the rights of translation, reprinting, reuse of illustrations, recitation, broadcasting, reproduction on microfilms or in any other way, and storage in data banks. Duplication of this publication or parts thereof is permitted only under the provisions of the German Copyright Law of September 9, 1965, in its current version, and permission for use must always be obtained from Springer-Verlag. Violations are liable for prosecution under the German Copyright Law.

© Springer-Verlag Berlin Heidelberg 1993
Printed in the United States of America

Typesetting: Camera-ready by authors with Springer TEX in-house system
41/3140 - 5 4 3 2 1 0 - Printed on acid-free paper

Preface

DYNAMICS REPORTED reports on recent developments in dynamical systems.

Dynamical systems of course originated from ordinary differential equations. Today, dynamical systems cover a much larger area, including dynamical processes described by functional and integral equations, by partial and stochastic differential equations, etc. Dynamical systems have involved remarkably in recent years. A wealth of new phenomena, new ideas and new techniques are proving to be of considerable interest to scientists in rather different fields. It is not surprising that thousands of publications on the theory itself and on its various applications are appearing.

DYNAMICS REPORTED presents carefully written articles on major subjects in dynamical systems and their applications, addressed not only to specialists but also to a broader range of readers including graduate students. Topics are advanced, while detailed exposition of ideas, restriction to *typical* results – rather than the *most general* ones – and, last but not least, lucid proofs help to gain the utmost degree of clarity.

It is hoped, that *DYNAMICS REPORTED* will be useful for those entering the field and will stimulate an exchange of ideas among those working in dynamical systems.

Christopher K.R.T Jones
Urs Kirchgraber
Hans-Otto Walther

Managing Editors

Table of Contents

Transversal Homoclinic Orbits near Elliptic Fixed Points of Area-preserving Diffeomorphisms of the Plane

Chr. Genecand

Abstract. *Extending a result due to E. Zehnder [27], we prove that generically, an area-preserving analytic diffeomorphism of the plane has transversal homoclinic orbits and nondegenerate Mather sets in every neighborhood of a stable elliptic fixed point. "Generically" refers to a topology defined by means of the Taylor coefficients of the mapping at the fixed point whose order is higher than a prescribed arbitrary integer, all lower coefficients being held fixed. The proof makes use of the Aubry-Mather theory for monotone twist mappings.*

1. Introduction

1.1 Results and Interests of this Paper. One of the most important purposes of the theory of dynamical systems is to define and check different types of stability for stationary and periodic solutions.

In this paper, we go in the opposite direction: starting from "stable solutions" of 2-dimensional systems – stable in the sense of Lyapunov: nearby starting solutions do not escape –, we prove a general result of so-to-speak "enclosing instability", i.e. of instability in annular regions arbitrarily close to these stable solutions.

To make such a result interesting, we restrict the problem in two ways: first we require *analyticity* of the systems considered; secondly we concentrate on such systems which are "*conservative*", i.e. which preserve some real quantity, called "energy" or "Hamiltonian (function)".

Do these two quite strong restrictions not make the systems particularly "neat" in some sense? Does, for instance, the conjunction of conservativity, analyticity and Lyapunov stability not force such systems to be "locally integrable", in the sense that the stable solutions had to be imbedded in a continuous foliation of invariant cylinders?

We show that the answer is negative: such a "neat", e.g. locally integrable, structure is quite *exceptional*, even for conservative, analytic, stable systems. (Already the presence of isolated orbits prevents integrability, see Siegel [25], Rüssmann [23].)

The general case is not only not integrable, it even exhibits a very "wild" and "chaotic" behaviour in any neighborhood of the stable solution, namely in the so-called "regions of instability" surrounding it.

We show this, in fact, not for the considered 2-dimensional systems themselves, but, for convenience, for their time-one-maps, which are, in this case, *area-preserving analytic diffeomorphisms* defined in the neighborhood of a stable, say elliptic, fixed point ("elliptic" means that the eigenvalues at the fixed point lie on the complex unit circle without the points ± 1).

In this set-up, as it is well-known (see [16]), there are, generally, invariant curves surrounding the fixed point arbitrarily closely. Our result marks an upper limit of this famous property, inasmuch as it shows that, generally, these invariant curves do not form a continuous family as in the integrable case. Rather they alternate with rings containing no invariant curve in their interior, the so-called "rings of instability". The chaotic behaviour in one of these rings can be illustrated by the fact that any two orbits contained in its interior come somewhere arbitrarily close one to the other. In other words, there is a complete mixing of the interior of such rings of instability. Such a non-integrability result for local dynamical systems or diffeomorphisms is not new, but takes place in an already long history beginning 100 years ago with Bruns, Poincaré [20, 21], then Birkhoff [4, 5], and culminating 50 years ago with Siegel [24, 25] and others ([15], ...).

Here the existence of unstable regions is proven by showing existence, generically, of so-called *transversal homoclinic orbits.* These are orbits which are asymptotic in both directions to some hyperbolic periodic orbit (i.e. the eigenvalues are real and $\neq \pm 1$), whose stable and unstable invariant manifolds intersect transversally at it. See Fig. 1.

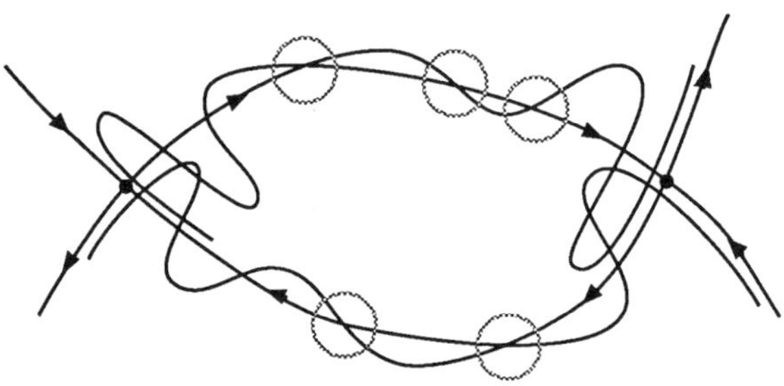

Fig. 1. Transversal Homoclinic Orbits.

In the immediate neighborhood of such orbits, the diffeomorphism actually behaves in a "chaotic" way; in particular, there are subsystems conjugate to so-called "Bernoulli shifts" with infinitely many symbols (see [27], § 8).

We now formulate the result more precisely, except for the analytic topology and for some slight additional condition on the eigenvalue which will be introduced later.

Theorem. *Generically (for some analytic topology), area-preserving, analytic diffeomorphisms defined near an elliptic fixed point have transversal homoclinic orbits in each neighborhood of this point.*

In this form, without specification of the topology, this result is not new, since it has been proven already 20 years ago by Zehnder in [27]. However, his topology was less fine than ours, in the following respect. To obtain existence of the necessitated hyperbolic periodic orbits, Zehnder used a bifurcation method which constrained him to move the eigenvalue λ at the fixed point ($|\lambda| = 1, \lambda^2 \neq 1$) close to some resonant value $\lambda_0 = \mathrm{e}^{2\pi \mathrm{i} p/q}$ with rational p/q. After having done so infinitely many times with an increasing sequence of q's (in order to obtain periodic – and homoclinic – orbits in every neighborhood of the fixed point), it may be that one ends up with an eigenvalue $\lambda_\infty = \mathrm{e}^{2\pi \mathrm{i} a}$ with irrational a of Liouville type, and this could lead one to believe that the arithmetic properties of the eigenvalue are decisive for the result.

Our main contribution consists in showing that this is not the case: to prove the result, we need not change the eigenvalue λ in any way, nor even any coefficient of the Taylor expansion at the fixed point up to a fixed integer ≥ 4: only the Taylor coefficients of higher order are changed, and as little as one wants.

As a consequence, for a given concrete diffeomorphism, the property of having transversal homoclinic orbits arbitrarily close to the fixed point, and thus also the integrability resp. non-integrability of this diffeomorphism, can not be decided if merely a finite number of its Taylor coefficients at the fixed point are known: this information depends always on the knowledge of the whole Taylor expansion. This is the main point cleared up in this paper.

Further interests of this paper lie, besides in the result itself, – and perhaps even more – in the *methods* used to prove it, which are roughly described now. First, the mentioned refinement of Zehnder's result is made possible by the use of a quite recent theory which solves the question of existence of the desired periodic and homoclinic orbits. This is the so-called *theory of minimal states or orbits*, developed simultaneously and independently by Aubry and Mather in the early eighties. It allows us to obtain the desired orbits as "minimal orbits" for a simple variational principle coming from the solid state physics, which will be exposed in Sect. 2.

The difficulty in using this theory is that it is a global theory, while our problem is of local nature. It is therefore necessary to prove some *a priori Lipschitz estimates* for minimal orbits, which allow then to localize the latter into some annuli which are relevant for us. These Lipschitz estimates are derived in Sect. 3, and they are of interest by themselves.

After having obtained, in this way, existence of the desired orbits, we still have to make them nondegenerate, i.e. to make the periodic orbit hyperbolic, and then the homoclinic orbit transversal. This is done essentially by means of elementary perturbation theory, and it necessitates two formulae having their own interest, namely:

- for the first perturbation, which makes the minimal periodic orbit to a hyperbolic minimal orbit: a "*monotone correspondence formula* for generating functions", relating, in a monotone way, perturbations of two different so-called "generating functions" one the other. Generating functions will be defined in 1.2 and in Sects. 2 and 3. See Sect. 4 for this formula.

- for the second perturbation, which makes a minimal homoclinic orbit transversal: a sort of "*discrete Melnikov formula*" measuring the C^1-distance of perturbed invariant manifolds at a point where the unperturbed manifolds intersect nontransversally. This formula consists in a doubly infinite sum – instead of an integral – of the perturbation of the given diffeomorphism, evaluated at all points of the original (nontransversal) homoclinic orbit (Sect. 5). Such a discrete Melnikov formula has also been derived in a more general context by Stoffer (and Kirchgraber) in [32].

Further interesting points are:

1. An immediate consequence of the above theorem, concerning existence of so-called "*nondegenerate Mather sets*". These can be seen as the analogues of transversal homoclinic orbits for *irrational* "rotation numbers", while the latter have *rational* rotation numbers or mean angular speeds of rotation, since they are asymptotic to periodic orbits. See the end of this section and Sect. 6 for more precise definitions.
2. The extension of the above results to some *more special and concrete classes* of area-preserving, analytic diffeomorphisms, namely:
   - the time-one-maps of *potential Hamiltonian systems*, i.e. of systems of the form $\ddot{\xi} = -Q_\xi(t, \xi)$ with an analytic, in t 1-periodic function Q, defined near some stable 1-periodic solution.
   - the *geodesic flows* on analytic Riemannian 2-surfaces near a stable closed goedesic curve.
   - the *analytic autonomous Hamiltonian* systems with 2 degrees of freedom near a stable equilibrium point.

Finally, it should be noticed that in this paper we are only interested for *existence* of transversal homoclinic orbits, but not for their "amount of transversality". Such a more quantitative point of view has been recently the object of several papers, e.g. by Holmes, Marsden and Scheurle ([28, 29]) – see also [30, 31] –, where the splitting distance and the splitting angle for perturbed homoclinic orbits are estimated: these quantities are seen to be of exponentially small order with respect to the perturbation.

1.2 Notation, Normal Form, Generating Functions, Topology, Precise Results. We denote by ϕ ($\phi_0, \psi, \ldots$) analytic, area-preserving diffeomorphisms of the plane defined in some open neighborhood D of the point 0, which has to be an elliptic fixed point. Thus, if λ_1, λ_2 are the eigenvalues of $d\phi(0)$, then $\lambda_1\lambda_2 = 1$, since ϕ preserves areas, i.e. $\lambda_1 = \lambda_2^{-1} =: \lambda$, and $|\lambda| = 1$, $\lambda \neq \pm 1$ by ellipticity, i.e. $\lambda = \mathrm{e}^{2\pi \mathrm{i}\omega_0}$, $2\omega_0 \notin \mathbf{Z}$.

Suppose moreover that $\lambda^k \neq 1$ for $k = 1, \ldots, q$ with some $q \geq 4$, i.e. $k\omega_0 \notin \mathbf{Z}$, $k = 1, \ldots, q$. Then, by a famous result of G. D. Birkhoff [4], there is an area-preserving, analytic change of variables such that, in the new variables $\zeta = \xi + \mathrm{i}\eta = (\xi, \eta)$, the diffeomorphism ϕ takes the following "*normal form*":

$$\phi(\zeta) = \zeta \cdot \exp(2\pi \mathrm{i}\beta(|\zeta|^2)) + P_q(\zeta, \bar{\zeta}) \;, \tag{1.1}$$

where $\beta(X) = \omega_0 + b_1 X + \cdots + b_s X^s$, $s = [q/2] - 1$, and where P_q is a power series in $\zeta, \bar{\zeta}$ beginning with q-th order terms (i.e. 0 is a q-th order zero of P_q).

(In fact, Birkhoff has proven more, namely the existence of a normal form which is a formal power series in $|\zeta|^2$. However, the transformation into this normal form, which is also a power series in $|\zeta|^2$, does not, in general, converge, because of the presence of so-called "small divisors". Our result also implies nonconvergence, since a convergent formal series occurs only in the integrable case).

Suppose, finally, that for some j with $1 \leq j \leq s$, we have $b_j \neq 0$. This assumption, called *twist condition*, is very important for the following, since it allows us to apply the Aubry-Mather theory of minimal states to our situation, and thus to obtain the desired periodic and homoclinic orbits. This is the unique – still quite loose – restriction we have to make on the eigenvalue.

From now on, we assume for definiteness that $q = 4$, $s = 1$, and that $b_1 \neq 0$, say $b_1 > 0$, and we will call a fixed point satisfying these conditions: an elliptic fixed point "*of general stable type*", or a "general stable fixed point".

The other cases with larger q, s and j are treated in exactly the same way, and the Lipschitz estimates of Sect. 3 are even easier to prove in these cases.

As mentioned, we will keep all lower Taylor coefficients of ϕ at 0 fixed up to an arbitrary integer order. Thus we fix $N \geq 4$ and take some area-preserving analytic diffeomorphism ϕ_0 defined in some open neighborhood D_0 of 0, which is a fixed point of general stable type for ϕ_0. We define the class $S = S(\phi_0; N)$ of all area-preserving analytic diffeomorphisms ϕ defined in some open neighborhood D of 0 and having the same $(N-1)$-jet at 0 as ϕ_0. In particular, 0 is a general stable fixed point for every $\phi \in S = S(\phi_0; N)$.

To define the topology we will use in the class S, we need the concept of the "*generating function*" of an area-preserving diffeomorphism. The existence of generating functions is equivalent to the area-preserving character of the diffeomorphism. They can be defined in multiple manners. We choose the following one, well adapted to our situation.

For $\phi \in S = S(\phi_0; N)$, we define the generating function $u : \mathbf{R}^2 \to \mathbf{R}$ with variables ξ, η' by means of the identity

$$\phi_0^{-1} \circ \phi(\xi, \eta) = (\xi', \eta') \quad \Longleftrightarrow \quad \xi' = \xi + u_2(\xi, \eta') \quad \text{and} \quad \eta = \eta' + u_1(\xi, \eta') \,, \tag{1.2}$$

where u_1, u_2 denote the partial derivatives of u.

From the assumptions on ϕ, ϕ_0, it follows at once that u is analytic with a N-th order zero at $\xi = \eta' = 0$, and therefore, in some heighborhood of $\xi = \eta' = 0$, u can be represented by a convergent power series:

$$u(\xi, \eta') = \sum_{k+l \geq N} u_{kl} \xi^k \eta'^l \,. \tag{1.3}$$

Every $\phi \in S = S(\phi_0; N)$ is uniquely determined by ϕ_0, N and by its generating function u, or equivalently by the double sequence of its Taylor coefficients $(u_{kl})_{k+l \geq N}$ at $\xi = \eta' = 0$.

We are now able to define the analytic *topology* on the class $S = S(\phi_0; N)$, with respect to which the "genericity" in the above theorem was meant. An open neighborhood of a $\phi \in S$ with generating function u is given by a sequence $(\varepsilon_{kl})_{k,l\geq 0,\ k+l\geq N}$ of positive real numbers and consists of all $\psi \in S$ whose generating function v satisfies:

$$|u_{kl} - v_{kl}| < \varepsilon_{kl} \quad \text{for all} \quad k,l \geq 0 \quad \text{with} \quad k+l \geq N \; .$$

In other words, the ψ's in this neighborhood have all the same $(N-1)$-jet as ϕ and all higher Taylor coefficients arbitrarily close to those of ϕ. That this topology is very fine, follows from the fact that N can be fixed arbitrarily large and that the $\varepsilon_{kl} > 0$ can be chosen arbitrarily small, for instance as an exponentially decreasing sequence.

Now we can give the exact formulation of the main result.

Theorem 1. *Generically in $S = S(\phi_0; N)$ for the topology just defined, an area-preserving, analytic diffeomorphism $\phi \in S$ has transversal homoclinic orbits arbitrarily close to the elliptic fixed point 0 of general stable type.*

"Generically" means that the subclass of all diffeomorphisms $\phi \in S(\phi_0; N)$ with the property of the theorem is a countable intersection of open dense sets. Such sets are easily found: they can, for instance, be chosen as the subclasses S_n, $n = 1, 2, \ldots$, of all diffeomorphisms $\phi \in S$ having a transversal homoclinic orbit in the open disk $D_{1/n}$ with radius $1/n$ about the fixed point 0. Clearly, the diffeomorphisms lying in the intersection of all S_n, $n = 1, 2, \ldots$ will have the desired property. Furthermore, the S_n are easily seen to be open, by means of elementary perturbation theory. It remains thus to show that the S_n are also dense in S for the above topology. This follows at once from the

Theorem 2. *Given $\phi \in S = S(\phi_0; N)$, an open disk D containing 0 and an open neighborhood B of ϕ in S, there is a $\psi \in B$ with a transversal homoclinic orbit in D.*

Thus the genericity result of Theorem 1 is reduced to the density result of Theorem 2. Using the definition of the topology on S, one can explicit and also strengthen the latter in the following way:

Theorem 2′. *Let $\phi \in S = S(\phi_0; N)$ with generating function u, $R > 0$, and a sequence $(\varepsilon_{kl})_{k+l\geq 0}$, $\varepsilon_{kl} > 0$ for all k, l, be given. Then there are a $M \geq N$ and a polynomial p of the form*

$$p = p(\xi, \eta') = \sum_{N\leq k+l\leq M} p_{kl}\xi^k \eta'^l \; , \tag{1.4}$$

such that the diffeomorphism ψ generated by $u + p$ has a transversal homoclinic orbit in the disk $D_R = \{|\zeta| < R\}$.

Thus, *polynomial* perturbations of the generating function with arbitrarily small coefficients are enough to obtain diffeomorphisms with transversal homoclinic orbits within any given distance of the general stable fixed point. It is this stronger form of Theorem 2 which is proven in the following four sections.

The *proof* has two parts: existence of the desired orbits (Sects. 2 and 3), and their nondegeneracy (Sects. 4 and 5).

As mentioned above, existence is provided by the use of the Aubry-Mather theory of minimal states, whose for us useful results are briefly presented in Sect. 2. Then, Sect. 3 deals with their application to the concrete local situation of Theorem 2′, with the help of a priori Lipschitz estimates for minimal orbits.

To make the so obtained orbits nondegenerate, we construct successively two polynomial perturbations of the form (1.4), the first one making a minimal periodic orbit with a given rotation number to the unique such one, and moreover hyperbolic, the second perturbation making then the obtained homoclinic orbit transversal. The first perturbation is constructed in Sect. 4 with the help of the mentioned monotone correspondence formula between the generating function u of ϕ, and another generating function h, which underlies the variational principle yielding us minimal orbits. The second polynomial perturbation is made up in Sect. 5, where we prove and use the discrete Melnikov formula for perturbations of a nontransversal homoclinic orbit we spoke about. The sum of these two polynomial perturbations gives the polynomial p of Theorem 2′, and the proof of this theorem will then be complete.

Then in Sect. 6, as announced, we turn to the existence of *nondegenerate Mather sets*, which will be defined precisely there, as well as their rotation number, with the help of the concepts of Sects. 2 and 3. Now we only give the following alternative definition. A Mather set is a nonempty closed invariant set lying on a simple Jordan curve surrounding the fixed point and on which the diffeomorphism ϕ induces a map which has to be strictly increasing. This map can thus be extended to a so-called “circle map” with a well-defined rotation number, which has to be irrational for a Mather set. A Mather set either coincides with the underlying curve, which is then invariant (degenerate case), or it is a proper subset of this curve (nondegenerate case), in which case it necessarily has the structure of a Cantor set. It is the latter case which is of interest to us, since it reflects, in the irrational case, the chaotic behaviour of transversal homoclinic orbits we are dealing with. The precise formulation of the result is:

Theorem 3. *A generic $\phi \in S = S(\phi_0; N)$ for the above topology has nondegenerate Mather sets in every open neighborhood of the (general stable) fixed point 0. Moreover the rotation numbers of all nondegenerate Mather sets of ϕ fill a countable union of open real intervals which cluster at ω_0 (= “rotation number” of the fixed point).*

As a result, the neighborhood of a general stable fixed point of an area-preserving, analytic diffeomorphism present the following general picture: invariant curves surrounding the fixed point alternate with rings of instability containing transversal homoclinic orbits and nondegenerate Mather sets. See Fig. 2 below.

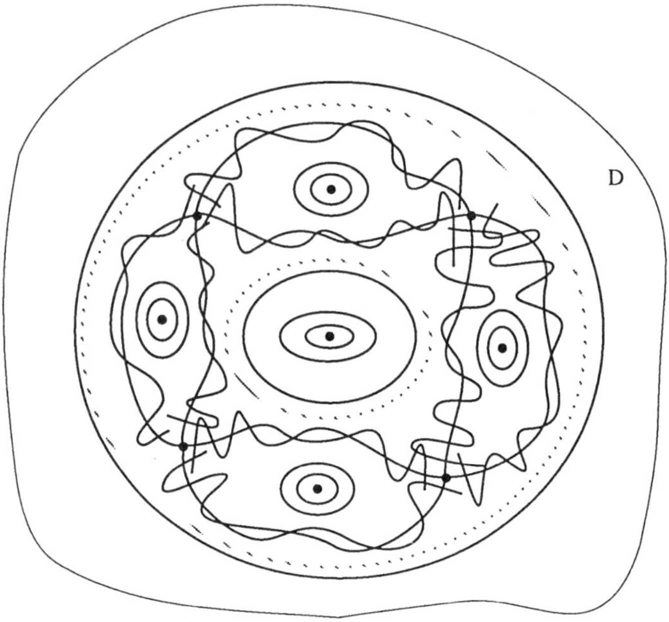

Fig. 2. Neighborhood of a General Stable Fixed Point.

Finally, Sect. 7 deals, as announced, with the extension of the above results (Theorems 1 and 3) to *more special classes of diffeomorphisms.* These are genuine extensions, and not merely applications, because in smaller classes, the density result of Theorem 2 is more difficult to obtain, since there is less freedom to perturb a given diffeomorphism. At the begin of Sect. 7 we shall give a simple criterion allowing one to see more easily if a given special class is still large enough to ensure validity of Theorem 2. This criterion arises from the perturbation part of the proof, i.e. from Sects. 4 and 5. The results proven in Sect. 7 are the following three ones.

- For *potential Hamiltonian systems* $\ddot{\xi} = -Q_\xi(t, \xi)$ with analytic and time-1-periodic potential $Q = Q(t, \xi)$ defined near some "general stable" (in the above sense!) 1-periodic solution, there are, generically for a topology involving only the higher Taylor coefficients of Q along this stable solution, transversal homoclinic solutions as well as nondegenerate Mather sets of solutions, in each C^1-neighborhood of the given stable solution.
- For analytic *geodesic flows* on analytic Riemannian 2-surfaces near a "general stable" closed geodesic curve, there are, generically for a topology concerning only the higher Taylor coefficients of the metric along the closed geodesic curve, arbitrarily close (in the C^1 sense) to the latter, transversal homoclinic geodesic curves as well as nondegenerate Mather sets of geodesic curves.
- In the class of *autonomous analytic Hamiltonian systems* with 2 degrees of freedom near a general stable equilibrium point, there are transversal homoclinic solutions and nondegenerate Mather sets of solutions arbitrarily

close to this equilibrium, generically for an analytic topology defined by means of the higher Taylor coefficients of the Hamiltonian function at the equilibrium point.

Of course, many further concrete examples of special classes could be added to these three ones, in particular also examples of discrete dynamical systems.

2. Elements of the Theory of Minimal States

In this section we briefly present the definitions and results of the theory of minimal states of Aubry and Mather which are of interest to us. For a complete exposition of this theory, we refer to [3], [7], [12] and the references in there. (The original publications are [1], [2] and [13]).

The basic concept is that of "*global (monotone) twist maps*". These are exact symplectic diffeomorphisms $F : \mathbf{R}^2 \to \mathbf{R}^2$ $(x, y) \to (x', y')$ with the periodicity property $F(x+1, y) = F(x, y) + (1, 0)$ and the "*uniform twist property*" $\delta \leq \partial x'/\partial y \leq \delta^{-1}$ for all (x, y) and some $\delta > 0$, and moreover $y'(x, y) \to \pm\infty$ as $y \to \pm\infty$. Such maps possess a "*generating function*" $H = H(x, x')$, defined by

$$F(x,y) = (x',y') \quad \Longleftrightarrow \quad y = -H_1(x,x') \quad \text{and} \quad y' = H_2(x,x')\ . \qquad (2.1)$$

H is defined up to a constant and satisfies $H(x+1, x'+1) = H(x, x')$ and $H_{12}(x, x') \leq -\delta < 0$ for all (x, x'). It follows that $H(x, x+\xi) \to \infty$ as $\xi \to \pm\infty$ and thus that H is bounded from below, such that the following variation problem is meaningful.

Setting $H(x_k \ldots x_j) = \sum_{k \leq i < j} H(x_i, x_{i+1})$, for finite segments $(x_i)_{k \leq i \leq j}$, one calls such a segment "*minimal*" if $H(x_k \ldots x_j) \leq H(x'_k \ldots x'_j)$ for any segment $(x'_i)_{k \leq i \leq j}$ satisfying $x'_k = x_k$ and $x'_j = x_j$. Then a "*state*", i.e. a bi-infinite sequence $(x) = (x_i)_{i \in \mathbf{Z}}$, is said to be "*minimal*" if any finite segment of it is so. Minimal states are stationary and thus define orbits $\{(x_i, y_i)\}_{i \in \mathbf{Z}}$ of F, by means of $y_i = H_2(x_{i-1}, x_i) = -H_1(x_i, x_{i+1})$, which are called "*minimal orbits*". Minimal states and orbits possess a well-defined "*rotation number*" $\omega = \lim_{i \to \infty} x_i/i$. The fundamental existence result is the following:

"For every real number ω, the global twist map F has minimal states and orbits with rotation number ω."

In particular, for rational rotation numbers $\omega = p/q$, we have the following two existence results, which are essential for us:

Lemma 2.1. *i) For every rational p/q, there exist minimal states and orbits which are* "p/q-periodic" *i.e. such that $x_{i+q} = x_i + p$, $i \in \mathbf{Z}$. Moreover, they are exactly the minimal states of the reduced functional $H_{q,p}(x_0 \ldots x_q) = \sum_{0 \leq i \leq q-1} H(x_i, x_{i+1})$ defined on all p/q-periodic states.*

ii) Fix $\omega = p/q$, and let two minimal p/q-periodic states $(x), (x')$ be "neighboring", *i.e. $x_i < x'_i$ for all i and there is no minimal p/q-periodic state between (x) and (x'). Then there exist minimal states $(a), (b)$ such that $0 < a_i - x_i \to 0$*

as $i \to -\infty$, $0 < x_i' - a_i \to 0$ *as* $i \to \infty$, *resp.* $0 < b_i - x_i \to 0$ *as* $i \to \infty$, $0 < x_i' - b_i \to 0$ *as* $i \to -\infty$.

Clearly, the orbits corresponding to the states $(a), (b)$ are $\pm$asymptotic to the periodic orbits corresponding to the states $(x), (x')$.

3. A Priori Lipschitz Estimates for Minimal Orbits

3.1. In this section, we apply the global results of the last section to our particular situation of a local diffeomorphism $\phi \in S = S(\phi_0; N)$. The ideas are prompted by [7], [9]. The methods are, however, somewhat refined and the obtained estimates sharper.

We restrict first the diffeomorphism ϕ to the annulus $A_\rho : \sqrt{1/2}\rho \le |\zeta| \le \sqrt{3/2}\rho$ about the fixed point 0, where $\rho > 0$ is chosen so small that A_ρ is contained in the open disk $D_R = \{|\zeta| < R\}$. Introducing the symplectic polar coordinates x, y defined by

$$2\pi x = \arg\zeta, \quad \rho\sqrt{y+1} = |\zeta| \ , \tag{3.0}$$

ϕ restricted to A_ρ takes the following form:

$$f = f_\rho : (x,y) \to (x',y') = (x + a + by + \alpha(x,y), y + \beta(x,y)) \ , \tag{3.1}$$

defined on $\{x \bmod 1, |y| \le 1\}$, where $a = \omega_0 + b_1\rho^2$, $b = b_1\rho^2$, and where α, β are C^∞-functions 1-periodic in x and with C^1-norm $\varepsilon = |\alpha|_1 + |\beta|_1 = O(\rho^3)$. Thus, for small enough ρ, we have $0 < \varepsilon < b < 1$. Dropping the index ρ we call "*local twist maps*" all maps of the form (3.1) satisfying $0 < \varepsilon < b < 1$. We shall extend them to global twist maps F close to the "*normal form*"

$$F_0 : (x,y) \to (x + a + by, y) \ . \tag{3.2}$$

For these extended maps F, the theory of minimal states will apply, providing periodic and homoclinic orbits for F. But these orbits will in general not be orbits of f, unless they are localized in the domain where F and f coincide. Our main purpose is to prove such a localization by means of a priori Lipschitz estimates for minimal orbits.

3.2 Global Twist Maps. We begin with global twist maps of the form

$$\begin{gathered} F : (x,y) \to (x',y') = (x + a + by + \alpha(x,y), y + \beta(x,y)) \ , \\ 0 < \varepsilon = |\alpha|_1 + |\beta|_1 < b < 1 \ . \end{gathered} \tag{3.3}$$

Instead of minimal orbits, we consider a generalization of them, the so-called "monotone invariant sets". A *monotone invariant set* is a nonempty invariant closed subset M of $\mathbf{R}^2$ which is the graph of a continuous 1-periodic function $w : K \to \mathbf{R}$, (thus $M = \{(x, w(x)),\ x \in K\}$), where K is a closed subset of $\mathbf{R}$, invariant for the translation $x \to x + 1$, and such that the induced map

$\psi : K \to K$ defined by $F(x, w(x)) = F(\psi(x), w(\psi(x)))$ is strictly increasing. Because of the periodicity of F, ψ satisfies $\psi(x+1) = \psi(x) + 1$, and can thus be extended to a circle map $\Psi : \mathbf{R} \to \mathbf{R}$ with a well-defined rotation number ω satisfying $|\Psi(x) - x - \omega| < 1$ for all x.

The following estimate is well-known (see [7]). Let $L := \max(|F|_1, |F^{-1}|_1)$ and recall that $0 < \delta \leq \partial x'/\partial y$ for all (x, y). Then we have

Lemma 3.1. *For every monotone invariant set* $M = \{(x, w(x)),\ x \in K\}$ *of* F, *the function* w *is* Lipschitz-continuous *whose Lipschitz-constant* $\mathrm{Lip}(w) = \sup\{|w(x) - w(x')|/|x - x'|,\ x, x' \in K,\ x \neq x'\}$ *is bounded by* L/δ.

It follows that the induced maps ψ and ψ^{-1} are Lipschitz-continuous as well and that the Lipschitz-constants of w, ψ and ψ^{-1} are uniformly bounded for all monotone invariant sets of the map F. Now for global twist maps F of the form (3.3), we can refine this estimate as follows.

Proposition 3.2. *There are constants* C, k, $k < 1$, *independent of* F, *such that for* $\varepsilon/b < k$, *the following holds:*

i) Let $M = \mathrm{graph}(w)$ *be any monotone invariant set for* F *and let* ω *be its rotation number. Then* $\mathrm{Lip}(w) \leq C\sqrt{\varepsilon/b}$ *and* M *lies in the strip* $D_0 = \{(x, y) \in \mathbf{R}^2,\ |y - y_0| \leq (C+1)\sqrt{\varepsilon/b}\}$, *where* $y_0 = (\omega - a)/b$.

ii) Conversely, in each strip $D_1 = \{(x, y),\ |y - y_1| \leq C'\sqrt{\varepsilon/b}\}$ *with* $C' \leq C + 1$, *there are minimal orbits for every rotation number* ω *close enough to* $\omega_1 = a + by_1$, *more precisely for every* ω *with* $|\omega - \omega_1| \leq (C' - C - 1)\sqrt{\varepsilon b}$.

Proof. 1. The statement ii) is an immediate consequence of i) and of the existence result for minimal orbits of the previous section. Indeed, a minimal orbit is contained in a monotone invariant set M, which is itself contained in the strip D_0 of i). Then for $(x, y) \in M$ we have

$$|y - y_1| \leq |y - y_0| + |y_0 - y_1| \leq (C+1)\sqrt{\varepsilon/b} + |(\omega - a)/b - y_1|$$
$$= (C+1)\sqrt{\varepsilon/b} + |\omega - \omega_1|/b \leq (C+1)\sqrt{\varepsilon/b} + (C' - C - 1)\sqrt{\varepsilon b}/b \leq C'\sqrt{\varepsilon/b}\ .$$

2. The localization in the strip D_0 follows at once from the Lipschitz estimate in i) and from the fact that for the circle map Ψ which extends ψ, the expression $\Psi(x) - x - \omega$ vanishes at some point $x_0 \in [0, 1]$. At this point, we have $|W(x_0) - y_0| \leq \varepsilon/b$, where W is the extension of w corresponding to Ψ. The localization follows from the 1-periodicity of W and from $\varepsilon/b < \sqrt{\varepsilon/b}$ (since $\varepsilon/b < 1$).

3. Proof of the Lipschitz estimate i) *for* $b \geq 1/2$. In this case we have only to show $\mathrm{Lip}(w) = O(\sqrt{\varepsilon})$. Let Ψ denote the affine extension of the induced map ψ to all of $\mathbf{R}$, and let W be the corresponding extension of w, implicitly defined by $\Psi(x) = \pi_1 F(x, W(x))$, $x \in \mathbf{R}$ (π_1 = projection onto the first coordinate). Thus we have, since F is of the form (3.3),

$$\Psi(x) = x + a + bW(x) + \alpha(x, W(x))\ . \tag{3.4}$$

Now by Lemma 3.1, W, Ψ and Ψ^{-1} are Lipschitz-continuous, and thus differentiable at almost all points $x \in \mathbf{R}$. Let x be such a point. Replacing in (3.4) x by $\Psi^{-1}(x)$, and subtracting the so obtained equation from (3.4) again, one obtains

$$\begin{aligned}\Psi(x) + \Psi^{-1}(x) - 2x = b[W(x) - W(\Psi^{-1}(x))] \\ + \alpha(x, W(x)) - \alpha(\Psi^{-1}(x), W(\Psi^{-1}(x))) .\end{aligned}$$

Using now $W(x) - W(\Psi^{-1}(x)) = \beta(\Psi^{-1}(x), W(\Psi^{-1}(x)))$, by (3.3) and the definition of W, and differentiating with respect to x, one obtains

$$\Psi'(x) + (\Psi^{-1})'(x) - 2 = O(\varepsilon) ,$$

where we have used the fact that the derivatives of W, Ψ, Ψ^{-1} are, by Lemma 3.1, bounded, as well as the fact that $|\alpha|_1 + |\beta|_1 = O(\varepsilon)$. Setting now $m = \max(|\psi'|_{L^\infty}, |(\psi^{-1})'|_{L^\infty})$, one has $1 \leq m = O(1)$, and $1/m \leq \min(\psi'(x), (\psi^{-1})'(x))$, and thus, with an appropriate choice of x:

$$m + m^{-1} - 2 = O(\varepsilon), \quad \text{or} \quad (m-1)^2 = O(\varepsilon)m = O(\varepsilon), \quad \text{or} \quad m = 1 + O(\sqrt{\varepsilon}) ,$$

and finally, by (3.4) again, and by the assumption $b \geq 1/2$:

$$\mathrm{Lip}(w) = (\mathrm{Lip}(\psi) - 1 + O(\varepsilon))/(b + O(\varepsilon)) = O(\sqrt{\varepsilon}) + O(\varepsilon) = O(\sqrt{\varepsilon}) .$$

4. Proof of the Lipschitz estimate i) *for* $b < 1/2$. We proceed by iteration. The monotone invariant set M is of course monotone and invariant for any iterate F^n of F as well. Now iterates of global twist maps of the form (3.3) are of the same form, if n is not too large, as the following lemma shows. □

Lemma 3.3 (Iteration Lemma). *Let f be as in (3.3) and $n \leq A/b$, with some fixed constant A. Then*

$$F^n(x, y) = (x + na + b_n y + \alpha_n(x, y), y + \beta_n(x, y)) ,$$

where $b_n = nb$ and $\varepsilon_n = |\alpha_n|_1 + |\beta_n|_1$ is of order $O(n\varepsilon)$.

Thus for $A < 1$ and small enough ε/b, $\varepsilon_n < b_n < 1$ and F^n is of the form (3.3). The proof of this lemma is elementary, and can be seen for instance in [18]. It consists in replacing $F^n - F_0^n$ by the sum of the n differences $F^{j+1}F_0^{n-j-1} - F^j F_0^{n-j}$, $0 \leq j \leq n-1$, estimating then the C^1-norm of the latter by $O(\varepsilon)$.

Now since $b < 1/2$ one can choose n such that $1/2b \leq n < 1/b$ and one obtains $b_n = bn \geq 1/2$, such that, for small enough ε/b, part 3 applies to F^n and yields

$$\mathrm{Lip}(w) = O(\sqrt{\varepsilon_n}) = O(\sqrt{n\varepsilon}) = O(\sqrt{\varepsilon/b}) ,$$

as desired.

3.3 Local Twist Maps. Going out from maps f of the form (3.1), defined in the fixed strip $|y| \leq 1$, we first extend them to the whole plane, obtaining so global twist maps F of the form (3.3), to which we then apply Proposition 3.2.

Lemma 3.4 (Extension Lemma). *For small enough ε/b, there are global twist maps F of the form (3.3) which coincide with f in $\{|y| \leq 1/2\}$ and with the normal form F_0 (3.2) in $\{|y| \geq 1\}$. Moreover the "error" $\varepsilon' = |F - F_0|_1$ for the extended maps F is of the same order as the error ε for the local map, i.e. $\varepsilon' = O(\varepsilon)$ and $\varepsilon = O(\varepsilon')$.*

The proof of this lemma presents no difficulty. To ensure the area-preserving character of the extended map F one has to operate with generating functions of f and of F_0, which are fitted together smoothly in the domain $1/2 \leq |y| \leq 1$ to obtain the generating function of F. An immediate consequence of this lemma and of Proposition 3.2 is the following

Proposition 3.5. *For small enough ε/b, the local twist maps f of the form (3.1) possess, for every rotation number close enough to a, minimal orbits (in fact minimal orbits of the extensions F of Lemma 3.4), which satisfy moreover Lipschitz estimates of order $O(\sqrt{\varepsilon/b})$ and are thus localized in annuli with breadth of the same order.*

Indeed, the extended maps F do have such orbits, by ii) of Proposition 3.2, which, by i) of the same proposition, satisfy Lipschitz estimate of order $O(\sqrt{\varepsilon/b})$, and are localized in strips D_0 with breadth of the same order and with median line $y = y_0 = (\omega - a)/b$. Thus by choosing ε/b small enough and ω close enough to a, one achieves that D_0 lies in the strip $\{|y| \leq 1/2\}$, where the global map F coincides with the local map f. Then the obtained minimal orbits of F are also orbits of f, and we may call them "minimal orbits of the local twist map f".

In particular, for the local twist maps f_ρ obtained by restricting the given diffeomorphism ϕ on the annuli A_ρ, we obtain, for small enough ρ, say for $\rho \leq \rho_1$, minimal orbits of every rotation number close enough to $a = \omega_0 + b_1\rho^2$, which satisfy moreover Lipschitz estimates of order $O(\sqrt{\rho})$. For the diffeomorphism ϕ itself, that means finally:

Proposition 3.6. *The diffeomorphism $\phi \in S$ has minimal orbits of every rotation number ω satisfying $\omega_0 \leq \omega \leq \omega_1 = \omega_0 + b_1\rho_1^2$, and they are localized in annuli with median circle $|\zeta| = \rho$ and breadth of order $O(\rho^{3/2})$, where ρ is defined by $\omega = \omega_0 + b_1\rho^2$.*

Moreover, for every rational $\omega = p/q$ in the above interval, there are minimal p/q-periodic orbits, and also, provided some of the latter are isolated, minimal orbits asymptotic to them.

The last assertion is an application of Lemma 2.1. The minimal asymptotic orbits so obtained will be candidates for the desired homoclinic orbits.

4. First Perturbation: Isolation and Hyperbolicity of Minimal Periodic Orbits

4.1. Lemma 2.1 guarantees the existence of asymptotic orbits, provided there are neighboring minimal p/q-periodic states. One way to guarantee the existence of such states is to make a given minimal p/q-periodic orbit to the unique such one (up to translation of the indices). The corresponding minimal states are then automatically neighboring to each other. The purpose of this section is to construct a polynomial perturbation of the generating function u removing all minimal p/q-periodic orbits except one, which will moreover become hyperbolic. This minimal periodic orbit will then possess, by Lemma 2.1 and by Proposition 3.6, at least two (possibly nontransversal) minimal homoclinic orbits. We formulate the result as follows:

Proposition 4.1. *Given a $\phi \in S$, generated by $u = u(\xi, \eta')$, with a minimal p/q-periodic orbit (Q), then there is a (nonnegative) polynomial*

$$P = P(\xi, \eta') = \sum_{N \leq k+l \leq M} p_{kl} \xi^k \eta'^l \ ,$$

such that for the diffeomorphisms $\phi_s \in S$ generated by $u + sP$ with small positive s, (Q) becomes the unique minimal p/q-periodic orbit (up to translation); moreover, (Q) is hyperbolic for these ϕ_s. Consequently, for small enough $s > 0$, ϕ_s has minimal homoclinic orbits asymptotic to (Q).

4.2 Proof. The difficulty of the proof lies in the fact that we want to perturb the generating function u of the diffeomorphism ϕ, while minimality of orbits is defined in terms of the generating function $h = h^\rho = h(x, x')$ of the local twist map $f = f_\rho = f(x, y)$ defined by restricting ϕ on some annulus A_ρ and using symplectic polar coordinates x, y. It is therefore necessary to know the relation between perturbations of the first and perturbations of the second generating function. Fortunately, as the following Lemma shows, this relation happens to be very simple.

Lemma 4.2. *Given any (C^∞-)perturbation δu of the generating function $u = u(\xi, \eta')$ of ϕ, let ϕ_s be the diffeomorphisms generated by $u + s\delta u$ via formula (1.2), let $f_s = f_{s,\rho}$ be the local twist maps obtained by restricting the ϕ_s to some fixed annulus A_ρ, and let finally h^s $(= h^{s,\rho})$ be their generating functions, as defined by (2.1). Then the relation between the first variation $\delta h = (d/ds)|_{s=0}$ h^s of the family h^s at $s = 0$ and the perturbation δu inducing it is given by*

$$\delta h = \delta h(x, x') = (\det T)^{-1} \delta u \circ T = (\pi \rho^2)^{-1} \delta u \circ T \ , \tag{4.1}$$

where $T : (x, x') \to (\xi, \eta')$ is the change of variables from the variables (x, x') of h to the variables (ξ, η') of u, defined by (1.2), (3.0) and (2.1), and where $\det T = \pi \rho^2$ is its determinant.

In particular, corresponding perturbations of u and of h have the same sign, what we express by saying that the relation (4.1) between first variations of the different generating functions is "*monotone*". Lemma 4.2 is proven below in 4.5. We now use it to prove Proposition 4.1. Since we want to preserve the given p/q-periodic orbit $(Q) = (Q_j)_{0\le j\le q-1}$, we have to choose the polynomial P with (at least) second order zeroes at all q points (ξ_j, η'_j), $0 \le j \le q-1$, which correspond to (Q) in the variables (ξ, η') of u. To preserve also minimality of (Q), it suffices, as we will see on the basis of Lemma 4.2, to take P everywhere nonnegative. Thus we set

$$P(\xi,\eta') = (\xi^2+\eta'^2)^n \prod_{0\le j\le q-1} [(\xi-\xi_j)^2 + (\eta'-\eta'_j)^2], \quad n = [(N+1)/2]\ . \tag{4.2}$$

4.3 Minimality. We have to show that (Q) is the unique minimal p/q-periodic orbit of ϕ_s, the diffeomorphism generated by $u + sP$, for small $s > 0$, where P is the polynomial (4.2). Let $(x) = (x_i)_{i\in\mathbf{Z}}$ be any minimal p/q-periodic state corresponding to the orbit (Q) (thus $x_{i+q} = x_i + p$). Then all other states corresponding to (Q) are translates of (x), i.e. are of the form $T_{a,b}(x) = (x_{i-a}+b)_{i\in\mathbf{Z}}$, with $a, b \in \mathbf{Z}$. They all minimize the functional $h_{q,p}(x_0\ldots x_q) = \sum_{0\le i\le q-1} h(x_i, x_{i+1})$, where h is the generating function $h = h^\rho$ corresponding to u. We show that for $s > 0$, the states $T_{a,b}(x)$ are the unique minimal states for $h^s_{q,p}(x_0\ldots x_q) = \sum_{0\le i\le q-1} h^s(x_i, x_{i+1})$, where h^s is the generating function corresponding to $u^s = u + sP$. To this end, we show that the differences $h^s(x,x') - h(x,x')$, $s > 0$, are everywhere nonnegative, and that they vanish at the points $(x_{i-a}+b, x_{i+1-a}+b)$, $i, a, b \in \mathbf{Z}$ belonging to the state (x) or to its translates, and at these points only. But these differences can be expressed by the following integrals

$$h^s(x,x') - h(x,x') = \int_0^s \frac{d}{d\sigma} h^\sigma(x,x')\, d\sigma = \frac{1}{\pi\rho^2}\int_0^s P\circ T^\sigma(x,x')\, d\sigma\ , \tag{4.3}$$

where we have used formula (4.1) with $\delta u = P$, and where T^σ denotes the transformation from the variables (x,x') of h^σ to the variables of $u^\sigma = u + \sigma P$, $0 \le \sigma \le s$. The dependence of T^s on s appears explicitly in the change (1.2) from the variables (ξ, η') of $u^s = u + sP$ to the variables $(\xi,\eta) \to (\xi', \eta')$ of ϕ_s. However, since the derivatives of P vanish at the orbit points (ξ_j, η'_j), $0 \le j \le q-1$, the inverse of T^s is constant, i.e. independent of s, at these points, and thus T^s maps the state points $(x_{i-a}+b, x_{i+1-a}+b)$, $i, a, b \in \mathbf{Z}$, to orbit points (ξ_j, η'_j), $0 \le j \le q-1$, for all s. Moreover, no other point (x, x') is mapped by T^s to some of the (ξ_j, η'_j), $0 \le j \le q-1$. Since the latter points are the unique zeroes of the polynomial P (besides $(\xi,\eta') = 0$), $P\circ T^\sigma(x,x')$ vanishes at all points $(x_{i-a}+b, x_{i+1-a}+b)$ and only at these points. At all other points (x,x'), $P \circ T^\sigma(x,x')$ is strictly positive. This implies that $h^s(x,x') = h(x,x')$ at the points $(x_{i-a}+b, x_{i+1-a}+b)$ and that $h^s(x,x') > h(x,x')$ elsewhere, for $s > 0$. The same holds of course for the functionals $h^s_{q,p}$ with $s > 0$: they coincide with $h_{q,p}$ at every q-segment of any state $T_{a,b}(x)$ corresponding to (Q), and they are strictly larger than $h_{q,p}$ at any other q-segment. Hence, the states $T_{a,b}(x)$ corresponding to the orbit (Q) become the unique minimal states for $h^s_{q,p}$ with $s > 0$, and thus the orbit (Q) becomes the unique minimal p/q-periodic orbit for ϕ_s, $s > 0$, as desired. In particular, the minimal states $T_{a,b}$, $a, b \in \mathbf{Z}$, are necessarily isolated, and they are thus ordered in a sequence of neighboring minimal p/q-periodic states, between which there are, by Lemma 2.1, $\pm$asymptotic minimal states. The corresponding minimal orbits of ϕ_s, for small enough $s > 0$, are thus $\pm$asymptotic to the minimal p/q-periodic orbit (Q). Moreover, since they are minimal, they are localized, by Proposition 3.6, in annuli with median radius ρ and breadth of order $O(\rho^{3/2})$, which are themselves contained, for small enough ρ, in the disk D_R of Theorem 2$'$. Notice that for small enough $s > 0$, the coefficients of sP satisfy also the smallness conditions of Theorem 2$'$.

4.4 Hyperbolicity. With the above choice (4.2) for the polynomial P, the minimal q-periodic orbit (Q) is automatically hyperbolic for small positive s. This follows from the fact that the zeroes of P at the points (ξ_j, η'_j), $0 \leq j \leq q-1$, corresponding to (Q) are of second, and not higher order. The following lemma, due to MacKay and Stark ([12], 1.8.2), gives the relation between the trace of $d\phi^q$ at the orbit points and the Hessian determinant of the functional $h_{q,p}$ at the corresponding states. The first quantity is known to characterize the type of the periodic orbit (hyperbolic, parabolic or elliptic), while the second one is necessarily nonnegative for a minimal periodic orbit.

Lemma 4.3. *Let (Q) be some p/q-periodic orbit of ϕ and let (x) be a stationary state for $h_{q,p}$ corresponding to (Q). Then one has*

$$\mathrm{tr}(d\phi^q(Q_0)) - 2 = \det(\partial^2 h_{q,p}(x_0 \ldots x_q))/\gamma(x) \ , \tag{4.4}$$

where $\gamma(x) = \prod_{0 \leq i \leq q-1}(-h_{12}(x_i, x_{i+1})) \geq \delta^q > 0$, by the twist property.

See the above reference [12] for the proof. An immediate consequence is

Corollary 4.4. *For periodic minimal orbits, "hyperbolic" and "nondegenerate" cover the same meaning.*

Indeed, if the orbit (Q) is minimal, then the Hessian determinant on the right hand side is nonnegative, and thus the trace of $d\phi^q(Q)$ is ≥ 2, which means that the orbit is hyperbolic or parabolic (without inversion), the latter case corresponding to a degenerate minimal orbit.

Thus we have to show that the minimal orbit (Q) of ϕ_s is nondegenerate for small $s > 0$. Setting $\tau(s) = \mathrm{tr}(d\phi^q_s(Q_0)) - 2$, we show that $\tau(s)$ is a nonconstant analytic function of s. Then it has a smallest positive zero s_0, and for s between 0 and s_0, $\tau(s)$ has to be positive since $\tau(s) \geq 0$ for $s \geq 0$, by minimality of (Q) for ϕ_s with $s \geq 0$. Hence for $0 < s < s_0$, (Q) is a nondegenerate minimal orbit of ϕ_s, and thus is hyperbolic.

In fact, $\tau(s)$ is even a *polynomial* of degree $2q$ in s, as is seen by writing out the differential $d\phi^q_s(Q_0)$ as a product of the q differentials $d\phi_s(Q_j)$, $0 \leq$

$j \leq q-1$, which one can then express in terms of the generating functions $u^s = u + sP$, by differentiating the equations (1.2). In each of the q matrices $d\phi_s(Q_j)$, s appears then in the first and second powers with derivatives of P as factors. More precisely, we have the nonvanishing term $P_{11}(\xi_j, \eta_j')s^2$ in the first column of $d\phi_s(Q_j)$, and s appears only linearly in the second column. Thus the trace of the product of the q matrices $d\phi_s(Q_j)$, $0 \leq j \leq q-1$, is a polynomial in s with largest exponent $2q$, as claimed.

4.5 Proof of Lemma 4.2. We use the concept of "*variational Hamiltonian*" of a one-parameter C^1-family of symplectic diffeomorphisms ϕ_s, $|s| \leq s_0$. First, the "variational vector field" X is defined by $X = (d/ds)|_{s=0}\, \phi^{-1} \circ \phi_s = d\phi^{-1}\delta\phi$, where $\phi = \phi_s|_{s=0}$. Then for symplectic ϕ_s, the field X is Hamiltonian, i.e. $X = X_H = JH_\zeta$: the function H is called the variational Hamiltonian of the ϕ_s (at $s = 0$).

Let u^s be the generating function of ϕ_s, as defined by (1.2). We claim that the relation between the first variation $\delta u = (d/ds)|_{s=0}\, u^s$ and the variational Hamiltonian H is very simple, namely $H = H(\zeta) = \delta u \circ T_u^{-1}$, where T_u is the transformation $(\xi, \eta') \to (\xi, \eta)$ from the variables of u to the variables of ϕ. To prove this, we need the following elementary lemma about the transformation of Hamiltonian vector fields.

Lemma 4.5 (Transformation Lemma). *Given a Hamiltonian vector field $X = X_H = X(\zeta)$ and a change of variables $T : z \to \zeta = \zeta(z)$, then the transformed vector field $Y(z) = (dT(z))^{-1}X \circ T(z)$ satisfies*

$$Y = \alpha^{-1}JG_z, \quad \alpha(z) = \det(dT(z)), \quad G(z) = H \circ T(z)\ . \tag{4.5}$$

If, moreover, $\alpha(z) = \alpha$ is constant, then Y is Hamiltonian again with Hamiltonian $H' = \alpha H \circ T$, thus $Y = Y_{H'} = JH'_z$.

The proof is a simple computation which is left to the reader.

Expressing now the variational vector field $X = X(\zeta) = (\delta\xi, \delta\eta)$ in the new variables $\omega = (\xi, \eta')$ of u, we obtain a field $Y = Y(\omega) = (\delta\xi, \delta\eta')$ satisfying, by Lemma 4.5,

$$Y = \det(T_u)^{-1}JG_\omega = (1 + u_{12})^{-1}JG_\omega, \quad G = G(\omega) = H \circ T_u(\omega)\ ,$$

where T_u is the change of variables $\omega = (\xi, \eta') \to (\xi, \eta) = \zeta$. But the field Y can also be computed directly by differentiating formulae (1.2) for $u^s = u + s\delta u = u^s(\xi_s, \eta_s')$ with respect to s at $s = 0$. This gives

$$Y = (\delta\xi, \delta\eta') = (1 + u_{12})^{-1}J(\delta u_1, \delta u_2) = (1 + u_{12})^{-1}J\delta u_\omega\ .$$

Comparing both formulae for Y, we see that $\delta u(+\text{const}) = G = H \circ T_u$, or

$$H = H(\zeta) = \delta u \circ T_u^{-1}\ , \tag{4.6}$$

which is precisely the above formula. Thus, the variational Hamiltonian H is the same as the first variation δu of the generating function, up to the evident change of variables. This fact can be seen as a variant of the well-known Hamilton-Jacobi equations.

The same argument, applied to the local twist map $f_s = f_{\rho,s}$ with the variational Hamiltonian $H' = H'(z)$ defined by $df(z)^{-1}\delta f(z) = JH'_z(z)$, and with the generating function $h^s = h^{\rho,s} = h^s(x,x')$, yields

$$H' = H'(z) = -\delta h \circ T_h^{-1}, \quad \text{or} \quad \delta h = -H' \circ T_h \ , \tag{4.7}$$

where T_h is the change of variables $(x,x') \to (x,y)$. The minus-sign comes from the fact that $(\delta x, \delta x') = h_{12} J \delta h_{(x,x')} = -\det T_h^{-1} J \delta h_{(x,x')}$.

Finally, the relation between both variational Hamiltonians $H = H(\zeta)$ and $H' = H'(z)$ is obtained by another application of Lemma 4.5:

$$H' = \det(T_1^{-1}) H \circ T_1 = -(\pi\rho^2)^{-1} H \circ T_1 \ , \tag{4.8}$$

where T_1 is the change of variables $z = (x,y) \to (\xi,\eta) = \zeta$. Combining now the three formulae (4.6)–(4.8), we obtain the desired relation between δh and δu: $\delta h = -H' \circ T_h = (\pi\rho^2)^{-1} H \circ T_1 \circ T_h = (\pi\rho^2)^{-1} \delta u \circ T_u^{-1} \circ T_1 \circ T_h$, thus

$$\delta h = (\pi\rho^2)^{-1} \delta u \circ T = (\det T)^{-1} \delta u \circ T \ ,$$

where $T = T_u^{-1} \circ T_1 \circ T_h$ is the change of variables $(x,x') \to (\xi,\eta')$. This is the formula (4.1) of Lemma 4.2, whose proof is now complete. □

5. Second Perturbation: Nondegeneracy of Homoclinic Orbits

5.1. In Proposition 4.1, we have obtained diffeomorphisms ϕ_s generated by $u^s = u + sP$, having, for small enough positive s, a unique hyperbolic minimal q-periodic orbit $(Q) = (Q_j)_{0 \le j \le q-1}$ with at least two – possibly degenerate – minimal homoclinic orbits $(H) = (H_i)_{i \in \mathbf{Z}}$. From now on, we shall denote by ϕ, u these modified ϕ_s, u^s, and assume that the obtained homoclinic orbits (H) are nontransversal – otherwise we are done. In order to prove Theorem 2′, it therefore remains to show that these orbits can be made nondegenerate by means of a second polynomial perturbation of u. More precisely, the orbits (H) will be replaced by nearby transversal homoclinic orbits (H'), asymptotic to the same hyperbolic periodic orbit (Q), which will remain minimal.

In a C^∞-context, it would suffice to apply a C^∞-perturbation of u with small compact support to the region of, say, the homoclinic point H_0, changing the direction of one of the invariant manifolds at this point and letting the other unchanged. Therefore, in a first step, we will work with C^∞-perturbation of u with small compact support, approximating them in a second step with a polynomial perturbation of the desired form. The difficulty comes from the smallness conditions of Theorem 2′ on the coefficients of these polynomials, which forces us to find a *continuous family of polynomials* with decreasing coefficients. This will be achieved by an application of the Implicit Function Theorem to a certain map Ω characterizing the C^1-distance of the invariant manifolds at the point H_0, and which has thus to be differentiable. This method is inspired from [27]. The precise result is

Proposition 5.1. *If $\phi \in S$, generated by $u = u(\xi, \eta')$, has a* nontransversal *homoclinic orbit (H) asymptotic to a minimal hyperbolic periodic orbit (Q), then there is a one-parameter C^1-family of polynomials*

$$P^s = P^s(\xi, \eta') = \sum_{N \le k+l \le M} p^s_{kl} \xi^k \eta'^l, \quad p^s_{kl} = O(s), \quad |s| \le s_0 ,$$

such that for all $s \neq 0$, the diffeomorphism ϕ_s generated by $u + P^s$ has a transversal homoclinic orbit $(H^s) = (H^s_i)_{i \in \mathbf{Z}}$, $\pm$asymptotic to the same hyperbolic minimal p/q-periodic orbit (Q), and satisfying $|H_i - H^s_i| = O(s)$ uniformly in i.

5.2 Definition of the Map Ω. For C^∞-diffeomorphisms ψ C^1-close to ϕ, the hyperbolic periodic orbit (Q) continues to exist as a nearby hyperbolic periodic orbit (Q_ψ) of ψ, whose invariant manifolds $\gamma^\pm(\psi)$ are C^1-close to those, $\gamma^\pm$, of (Q) for ϕ. By assumption, the latter intersect tangentially at H_0. Let e_0 be the common unit tangent vector to $\gamma^\pm$ at H_0 pointing to $H_1 = \phi(H_0)$. Draw, at both ends of e_0, two straight lines l_1, l_2 perpendicular to it. For ψ C^1-close enough to ϕ, the line l_1 through H_0 cuts each curve $\gamma^\pm(\psi)$ at exactly one point $H_0^\pm(\psi)$. Let $f_0^\pm(\psi)$ be the tangent vector to $\gamma^\pm(\psi)$ at $H_0^\pm(\psi)$ which ends up at the line l_2 perpendicular to e_0 (see Fig. 3). Thus the vectors $f_0^\pm(\psi)$ are normalized so that the differences $f_0^\pm(\psi) - e_0$ are orthogonal to e_0.

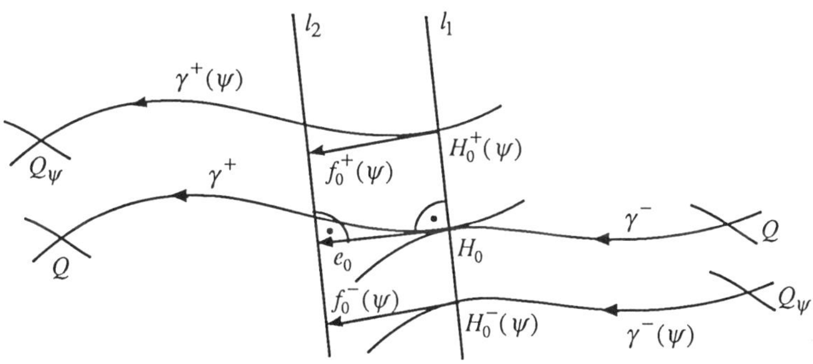

Fig. 3. Definition of Ω.

The map Ω is defined, for C^∞-diffeomorphisms ψ which are sufficiently C^1-close to ϕ, by

$$\Omega(\psi) = (e_0 \wedge (II_0^-(\psi) - H_0^+(\psi)), c_0 \wedge (f_0^-(\psi) - f_0^+(\psi))) \in \mathbf{R}^2 , \tag{5.1}$$

where $\wedge$ denotes the outer product in $\mathbf{R}^2$. Thus $\Omega(\psi)$ is the orthogonal C^1-distance of the curves $\gamma^\pm(\psi)$ at H_0. If in particular $\Omega(\psi) = (0, a)$, $a \neq 0$, then the perturbed invariant manifolds $\gamma^\pm(\psi)$ intersect transversally at $H_0^+(\psi) = H_0^-(\psi)$, which thus defines a transversal homoclinic orbit for ψ.

We have to remark that the components of the map Ω are nothing other than, respectively, the splitting distance and (in first order) the splitting angle between the perturbed manifolds at the point H_0. As already mentioned in the introduction (p. 4), these quantities have been estimated in recent papers (references [28, 29], see also [30, 31]), where they appear to be exponentially small with respect to the perturbation $\psi - \phi$.

We first show in 5.3 that Ω is differentiable for symplectic ψ. Then in 5.4 we prove the proposition under a certain surjectivity assumption on $d\Omega(\phi)$, by use of the Implicit Function Theorem. In 5.5, finally, we verify this surjectivity assumption for $d\Omega(\phi)$ in the present case.

5.3 Differentiability of Ω. In [27], Zehnder proved that the map Ω is Fréchet-differentiable on some C^1-neighborhood of ϕ with the C^2-topology, using functional analytic methods and tricky estimates. Here we give an alternative proof of the differentiability of Ω, but only for *symplectic* C^∞-diffeomorphisms ψ C^1-close to ϕ, using an explicit formula for the derivative of Ω at such ψ, which we will need later anyhow.

Let ψ_s, $|s| \leq s_0$, be a one-parameter C^1-family of C^∞-diffeomorphisms with $\psi_0 = \psi$, and set $\pi = \delta\psi_0 = (d/ds)|_{s=0}\psi_s$ and $d\Omega(\psi)\pi = (d/ds)|_{s=0}\Omega(\psi_s) \in \mathbf{R}^2$. We have to show that $d\Omega(\psi)\pi$ does not depend on the family ψ_s but only on π and on ψ, and that this dependence is continuous. Now the components of $d\Omega(\psi)\pi$ will appear as bi-infinite sums of π, resp. $d\pi$, at all points of the nontransversal homoclinic orbit (H). These sums can be viewed as a discrete version of the Melnikov formulae, and they will actually be continuous in π and in ψ. The method in this subsection is inspired from [8].

We consider the stable and the unstable manifolds separately, and define thus the partial maps $\Omega^\pm$ by means of $\Omega^\pm(\psi) = (e_0 \wedge (H_0^\pm(\psi) - H_0), e_0 \wedge (f_0^\pm(\psi) - e_0))$, such that $\Omega = \Omega^- - \Omega^+$. Using the fact that $H_0^\pm(\psi) - H_0$ and $f_0^\pm(\psi) - e_0$ are orthogonal to e_0, one can also write

$$\Omega^\pm(\psi) = (f_0^\pm(\psi) \wedge (H_0^\pm(\psi) - H_0), f_0^\pm(\psi) \wedge (f_0^\pm(\psi) - e_0)) \ .$$

Set $H_0^\pm(s) = H_0^\pm(\psi_s)$, $H_i^\pm(s) = \psi_s^i(H_0^\pm(s))$, $i \in \mathbf{Z}$, and $f_0^\pm(s) = f_0^\pm(\psi_s)$, $f_i^\pm(s) = d\psi_s^i(H_0^\pm(s)) \cdot f_0^\pm(s)$, $i \in \mathbf{Z}$. By a standard result of perturbation theory for invariant manifolds (see [8], Lemma 4.5.2), the $H_i^\pm(s)$ and the $f_i^\pm(s)$, $i \in \mathbf{Z}$, are differentiable with respect to s at $s = 0$ and the directional derivatives $\delta H_i^+(\psi) = (d/ds)|_{s=0}H_i^+(s)$ and $\delta f_i^+(\psi)$ are uniformly bounded for positive i, as well as the $\delta H_i^-(\psi)$ and $\delta f_i^-(\psi)$ for negative i.

Dropping the argument ψ, we see that the quantities $d_i^+ = f_i^+ \wedge \delta H_i^+$ tend to 0 exponentially as $i \to \infty$, since the $f_i^+ = d\psi^i(H_0^+) \cdot f_0^+$ do so and the δH_i^+ are bounded for $i > 0$. Therefore

$$d_0^+ = d_0^+ - \lim_{i \to \infty} d_i^+ = \sum_{i \leq 0} d_i^+ - d_{i+1}^+ \ .$$

Now $\delta H_{i+1}^+ = \delta(\psi(H_i^+(\psi))) = \delta\psi(H_i^+) + d\psi(H_i^+)\delta H_i^+$ and $f_{i+1}^+ = d\psi(H_i^+)f_i^+$, such that

$$\begin{aligned} d_{i+1}^+ - d_i^+ &= f_{i+1}^+ \wedge \delta H_{i+1}^+ - f_i^+ \wedge \delta H_i^+ \\ &= f_{i+1}^+ \wedge \delta\psi(H_i^+) + d\psi(H_i^+) f_i^+ \wedge d\psi(H_i^+)\delta H_i^+ - f_i^+ \wedge \delta H_i^+ \\ &= f_{i+1}^+ \wedge \delta\psi(H_i^+) + (\det d\psi(H_i^+) - 1) f_i^+ \wedge \delta H_i^+ \\ &= f_{i+1}^+ \wedge \delta\psi(H_i^+) \ , \end{aligned}$$

where we have used the formula $(Ax) \wedge (Ay) = (\det A) x \wedge y$ for $A = d\psi(H_i^+)$, so that $\det(A) = 1$. Thus

$$d_0^+ = \sum_{i \geq 0} -f_{i+1}^+(\psi) \wedge \delta\psi(H_i^+) = -\sum_{i \geq 0} f_{i+1}^+(\psi) \wedge \pi(H_i^+(\psi)) \ ,$$

where the sum converges absolutely. Hence the directional derivative d_0^+ of the first component of $\Omega^+(\psi)$ does not depend on the particular family ψ_s with $\psi_0 = \psi$ and $\delta\psi_0 = \pi$, but only on π and on ψ. Moreover, it is clearly linear in π, and also continuous for the C^0-topology, since the sum converges uniformly for bounded π. By a standard argument, it follows that the first component Ω_1^+ of Ω^+ is Gâteaux-differentiable at ψ for the C^1-topology, with the above sum for d_0^+ as Gâteaux-derivative $d\Omega_1^+(\psi)\pi$. Finally, since the $P_i^+(\psi)$ and $f_{i+1}^+(\psi)$ depend continuously on ψ, so does the sum, what shows that Ω_1^+ is in fact Fréchet-differentiable in some symplectic C^1-neighborhood of ϕ with the C^1-topology.

The same argument shows that also the first component Ω_1^- of Ω^- is Fréchet-differentiable with Fréchet-derivative

$$d\Omega_1^-(\psi)\pi = d_0^- = \sum_{i<0} f_{i+1}^-(\psi) \wedge \pi(H_i^-) \ .$$

From this it follows that the maps $\psi \to H_0^\pm(\psi)$, and thus also all maps $\psi \to H_i^\pm(\psi)$, $i \in \mathbf{Z}$, are Fréchet-differentiable, because the points $H_0^\pm(\psi)$ varies orthogonally to e_0 by definition, and thus their derivatives with respect to ψ are essentially given by $d_0^\pm$.

The differentiability of the second components of the maps $\Omega^\pm$ is proven in the same way, but requires the C^2-topology on the symplectic diffeomorphisms ψ, and the sum formulae we obtain are a bit more complicated, namely they are given by

$$\begin{aligned} d\Omega_2^+(\psi)\pi &= -\sum_{i \geq 0} f_{i+1}^+ \wedge [d\pi(H_i^+) f_i^+ + d^2\psi(H_i^+)(\delta H_i^+, f_i^+)] \ , \\ d\Omega_2^-(\psi)\pi &= \sum_{i<0} f_{i+1}^- \wedge [d\pi(H_i^-) f_i^- + d^2\psi(H_i^-)(\delta H_i^-, f_i^-)] \ , \end{aligned}$$

the argument ψ being dropped everywhere. The expressions $\delta H_i^\pm(\psi)$ appearing in these sums are the derivatives of the maps $\psi \to H_i^\pm(\psi)$, $i \in \mathbf{Z}$, which we have just seen to exist as Fréchet-derivatives.

Thus in the formulae for $d\Omega^\pm(\psi)\pi$, the first variations of an invariant manifold and of its tangent vector at a point appear as sums of the perturbation

$\pi = \delta\psi$ and of $d\pi$ evaluated at all image points for the stable manifold, respectively at all preimage points for the unstable manifold.

The map $\Omega = \Omega^- - \Omega^+$ is differentiable as well for the C^2-topology in some symplectic C^1-neighborhood of ϕ, and $d\Omega(\psi)\pi = d\Omega^-(\psi)\pi - d\Omega^+(\psi)\pi$. In particular, for the derivative at the diffeomorphism ϕ, we obtain

$$d\Omega(\phi)\pi = \left(\sum_{i\in\mathbf{Z}} f_{i+1} \wedge \pi(H_i), \sum_{i\in\mathbf{Z}} f_{i+1} \wedge [d\pi(H_i)\cdot f_i + d^2\phi(H_i)(\delta H_i^\sigma, f_i)]\right), \tag{5.2}$$

where $\sigma = +$ for $i \geq 0$ and $\sigma = -$ for $i < 0$. This is a discrete analogon of the well-known Melnikov formula for the perturbation of a nontransversal homoclinic orbit (see [8]). Such a discrete Melnikov formula has also been derived in a more general context by Stoffer (and Kirchgraber) in [32].

5.4 Proof of the Proposition for Surjective $d\Omega(\phi)$**.** In 5.3, we have restricted the map Ω to symplectic C^∞-diffeomorphisms C^1-close to ϕ. So we may express it in terms of the generating functions $v = v(\xi, \eta')$ of ψ. Ω is now defined for C^∞-functions v C^2-close to u and is differentiable for the C^3-topology on the functions v. We assume now that the differential $d\Omega(u)$ is surjective when restricted to C^∞-perturbations w of u with "*small*" compact support, "small" meaning that the support of the corresponding perturbation of ϕ contains only the point H_0 of the homoclinic orbit (H). In particular, it contains no point of the periodic orbit (Q), which therefore is preserved by the perturbation.

Let thus w^1, w^2 be two C^∞-perturbations of u with small support such that $d\Omega(u)w^{1,2}$ form a basis of $\mathbf{R}^2$. By continuity of $d\Omega(u)$, and by the Weierstraß-Stone theorem, there are two polynomials P^1, P^2 approximating w^1, w^2 such that $d\Omega(u)P^{1,2}$ is still a basis of $\mathbf{R}^2$. Moreover, one can choose these polynomials with N-th order zeroes at 0 and second order zeroes at the q points (ξ_j, η'_j), $0 \leq j \leq q-1$, corresponding to the periodic orbit (Q), since by assumption these points do not lie in the support of $w^{1,2}$. Thus, every perturbation of u defined by $c \cdot P = c_1 P^1 + c_2 P^2$ for small $c = (c_1, c_2) \in \mathbf{R}^2$ will preserve the hyperbolic periodic orbit (Q).

The auxiliary map $F : \mathbf{R}^3 \to \mathbf{R}^2$, $(c, s) \to \Omega(u + c\cdot p) - (0, s)$ is C^1 by 5.3, and it satisfies $F(0) = 0$ since the orbit (H) is nontransversal, and $F_c(0) = (d\Omega(u)P^1, d\Omega(u)P^2)$ is regular. By the Implicit Function Theorem, there is a C^1-function $s \to c(s) \in \mathbf{R}^2$ defined for small $|s|$ such that for these s, $F(c(s), s) = 0$, i.e. $\Omega(u + c(s)\cdot P) = (0, s)$. Thus for small $|s| \neq 0$, the diffeomorphism ϕ_s generated by $u + c(s)\cdot P$ has a transversal homoclinic orbit (H^s) asymptotic to (Q), which, by continuity, is close to the original orbit (H) of order $O(s)$.

Since (Q) is hyperbolic and minimal for ϕ, it is nondegenerate minimal by Corollary 4.4, and thus it is continued in a unique manner to a hyperbolic locally minimal periodic orbit for diffeomorphisms ψ C^1-close to ϕ. Moreover, since (Q) was the *unique* minimal p/q-periodic orbit of ϕ, its continuation remains absolutely minimal, for ψ C^1-close enough of ϕ. Thus in the case of the just defined perturbation ϕ_s of ϕ, the orbit (Q) remains also hyperbolic and *minimal*, for small enough $|s|$.

5.5 Surjectivity of $d\Omega(u)$. It remains to show that $d\Omega(u)$ is surjective when restricted to C^∞-perturbations w of u with "small" compact support. "Small" means in this case that the support of w contains the single point $N_0 = (\xi_0, \eta'_0)$ corresponding to the nontransversal homoclinic orbit (H). For such a function w, let ϕ_s be the diffeomorphism generated by $u + sw$, with small $|s|$. Then $\phi_s \equiv \phi$ at all points H_i with $i \neq 0$, such that $\pi(H_i) = \delta\phi(H_i) = 0$ and $d\pi(H_i) = \delta d\phi(H_i) = 0$ for $i \neq 0$, and thus the sums (5.2) for $d\Omega(u)w = d\Omega(\phi)\pi$ contain only the term with $i = 0$. Expressing, in this term, π with the help of the perturbation w of u, we obtain

$$\pi(H_0) = d\phi(H_0) \cdot [w_2 a + w_1 b] \; ,$$
$$d\pi(H_0) = d\phi(H_0)[w_{11}A + w_{12}B + w_{22}C] + L(w_1, w_2) \; ,$$

where $w_1, w_2, w_{11}, w_{12}, w_{22}$ are the partial derivatives of w, evaluated at the point N_0, L is some linear operator, irrelevant for our purpose, and where

$$a = \begin{pmatrix} 0 \\ -1 \end{pmatrix}, \quad b = \frac{1}{1+u_{12}} \begin{pmatrix} 1 \\ u_{12} \end{pmatrix},$$
$$A = \begin{pmatrix} 0 & 0 \\ -1 & 0 \end{pmatrix}, \quad B = \frac{1}{1+u_{12}} \begin{pmatrix} 1 & 0 \\ 2u_{11} & -1 \end{pmatrix}, \quad C = \frac{1}{(1+u_{12})^2} \begin{pmatrix} -u_{11} & 1 \\ -u_{11}^2 & u_{11} \end{pmatrix},$$

the partial derivatives u_{11} and u_{12} of u being also evaluated at the point N_0.

Now the surjectivity follows at once from the fact that the vectors a, b are linearly independent in $\mathbf{R}^2$, as well as the matrices A, B, C in $\mathrm{sl}(2; \mathbf{R})$. With an appropriate choice of the derivatives w_1, w_2 and w_{11}, w_{12}, w_{22} of w, one can give $\pi(H_0)$ and $d\phi(H_0)^{-1} d\pi(H_0)$ any value in $\mathbf{R}^2$, resp. in $\mathrm{sl}(2; \mathbf{R})$. Thus $d\Omega(u)w$ can be given any arbitrary value in $\mathbf{R}^2$ by an appropriate choice of the partial derivatives of w at N_0. This completes the proof of the Proposition 5.1. □

6. Application to Mather Sets

Before proving Theorem 3, we give a stronger version of Theorem 1, which we have in fact proven in the previous sections:

Theorem 1′. *Generically, a diffeomorphism $\phi \in S$ has in every neighborhood of the elliptic fixed point 0 of general stable type, hyperbolic minimal p/q-periodic orbits with transversally intersecting invariant manifolds. These orbits are the unique minimal p/q-periodic orbits of ϕ (up to translation of the indices).*

We now give a precise definition of a Mather set, using the concept of "monotone invariant set" introduced at the beginning of Sect. 3. A *Mather set* can be defined as a monotone invariant set with irrational rotation number, which is "minimal" in the sense that it contains no other such set. Hence a Mather set is perfect, and every orbit it contains is recurrent and dense in it. For a diffeomorphism $\phi \in S$, there are Mather sets for every not too large rotation number $\omega > \omega_0$, in view of the general existence result for minimal states. Some of them are invariant curves: they are called degenerate. To prove Theorem 3, we have to show that generically, the Mather sets of a $\phi \in S$ are nondegenerate. Then they are Cantor sets on curves surrounding the fixed point. Theorem 3 is a consequence of Theorem 1′ above, and of the following proposition, which says that near a minimal p/q-periodic orbit with transversally intersecting invariant manifolds, there are no invariant curves.

Proposition 6.1. *Let* (Q) *be a minimal* p/q*-periodic orbit with transversally intersecting invariant manifolds. Then there is an* $\varepsilon > 0$, *such that there is no invariant curve with rotation number in the interval* $(p/q - \varepsilon, p/q + \varepsilon)$.

Proof. Assume the contrary. Then there is a sequence of invariant curves whose rotation numbers ω_n converge to p/q. By a result of Birkhoff ([4]), such curves are, in some lift, graphs of Lipschitz-continuous functions with a uniform Lipschitz-constant. Therefore, a subsequence converges, in the sup norm, to an invariant curve Γ, whose rotation number is p/q by continuity ([10]).

Now Mather showed ([14], Prop. 2.8) that an invariant curve which is a graph necessarily consists of minimal orbits. It has the rotation number p/q in the present case. But since the p/q-periodic (Q) is itself minimal, and since minimal states with the same rotation number can not cross (see [3]), the orbit (Q) must lie on the invariant curve Γ. Since (Q) is hyperbolic, parts of the invariant manifolds of (Q) are contained in Γ, and thus overlap smoothly each other. This contradicts the assumption that the invariant manifolds of (Q) intersect transversally. □

7. Special Classes of Diffeomorphisms

7.1. In this last section, we show how to generalize the results obtained so far to more special and more concrete classes of diffeomorphisms. In general, the existence results of Sect. 3 hold without change in a special class, and we only have to show that the nondegeneracy of the obtained orbits can be achieved by means of two perturbations similar to those of Sects. 4 and 5. The following criterion allows us to see if a given special class is large enough to admit such perturbations. It uses the concept of the "generating functions of a special class". These are functions characterizing the class in a natural way, and in terms of which an analytic topology in the class can be defined, exactly as the functions $u = u(\xi, \eta')$ are the generating functions for the whole class S. For instance, the "generating functions" in this sense for the special class of potential systems are the potential functions $Q = Q(t, \xi)$.

Criterion 7.1. *In a given special class, the general results hold if*

1) there is a monotone correspondence *between perturbations of the "generating functions" of the class and the variational Hamiltonians induced by these perturbations;*

2) for C^∞-perturbations of the generating function with "small" compact support, the expression

$$(e_0 \wedge \delta\phi(H_0), e_0 \wedge d\delta\phi(H_0)e_0) , \tag{7.1}$$

evaluated at some nontransversal homoclinic point H_0, can take any value in $\mathbf{R}^2$, by an appropriate choice of the perturbation $\delta\phi$.

The "variational Hamiltonian" was defined in 4.5, and the perturbations with "small" support in 5.4 (the support of the corresponding perturbation of ϕ contains the point H_0, but no other point of the orbit of H_0). Notice that the expression (7.1) is precisely the term to which the sums in (5.2) for $d\Omega(\phi)\pi$ reduce, when the perturbation π has "small" support. The sufficiency of these two conditions appears clearly in the proofs of Proposition 4.1 and 5.1. We now use this criterion to prove the general results for three special classes.

7.2 Potential Case. We consider first nonautonomous 1-dimensional potential systems

$$\ddot{\xi} = -Q_\xi(t,\xi), \quad Q(t+1,\xi) = Q(t,\xi) , \tag{7.2}$$

where the continuous potential function Q is analytic in ξ. We assume that $\xi \equiv 0$ is a stationary solution of (7.2), such that Q admits the Taylor expansion $Q = Q(t,\xi) = \sum_{n\geq 2} Q_n(t)\xi^n$, $Q_n(t+1) = Q_n(t)$. We assume moreover that $\xi \equiv 0$ is "of general stable type" in the sense that 0 is a general stable fixed point of the time-one map $\phi : (\xi(0), \dot{\xi}(0)) \to (\xi(1), \dot{\xi}(1))$. This map ϕ is area-preserving, since the above system is Hamiltonian. These maps form a special class whose generating functions are the potentials $Q = Q(t,\xi)$. The topology is defined by means of the coefficient functions $Q_n = Q_n(t)$ with $n \geq N$, the lower Q_n remaining unchanged, where $N \geq 5$ is an arbitrary integer. The *result* is:

"Generically in this class, in every C^1-neighborhood of the stationary solution $\xi \equiv 0$, there are transversal homoclinic solutions and nondegenerate Mather sets of solutions."

To verify the criterion 7.1 in this case, we need three integral formulae relating the variational Hamiltonian and the first variations of ϕ and $d\phi$ on the one hand, to perturbations δQ of Q and to their derivatives on the other hand. Let $\delta Q = \delta Q(t,\xi) = \delta Q(t+1,\xi)$ be some C^∞-perturbation of Q with compact support on the cylinder, and let ϕ_s denote the time-one maps of the equation

$$\ddot{\xi} = -Q_\xi(t,\xi) - s\delta Q_\xi(t,\xi) , \tag{7.2$_s$}$$

with small enough $|s|$. Then the variational Hamiltonian $H = H(\zeta)$ of the vector field $X = d\phi(\zeta)^{-1}\delta\phi(\zeta) = JH_\zeta(\zeta)$ ($\phi = \phi_s|_{s=0}$) is given by

$$H(\zeta_0) = \int_0^1 \delta Q(t,\xi(t))\, dt , \tag{7.3}$$

where $\xi = \xi(t)$ is the solution of (7.2) with $(\xi(0), \dot{\xi}(0)) = \zeta_0$. From this formula it appears that the relation between the perturbation δQ and the variational

Hamiltonian H is "monotone", in the sense that nonnegative δQ induce nonnegative H, which shows condition 1) of the criterion.

Let now (H) be some nontransversal homoclinic orbit of ϕ. Then the first variations $\pi(H_0) = \delta\phi(H_0)$ and $d\pi(H_0) = \delta d\phi(H_0)$ are given by

$$\begin{aligned} \pi(H_0) &= d\phi(H_0) \int_0^1 \begin{pmatrix} x_2 \\ -x_1 \end{pmatrix} \delta Q_\xi(t, \xi(t))\, dt \quad \text{and} \\ d\pi(H_0) &= d\phi(H_0) \int_0^1 \begin{pmatrix} x_1 x_2 & x_2^2 \\ -x_1^2 & -x_1 x_2 \end{pmatrix} \delta Q_{\xi\xi}(t, \xi(t))\, dt + \cdots \,, \end{aligned} \tag{7.4}$$

where $\xi = \xi(t)$ is the solution of (7.2) with $(\xi(0), \dot{\xi}(0)) = H_0$, and where the points in the second formula stand for terms depending on the first derivative δQ_ξ of δQ only. The functions $x_1 = x_1(t)$ and $x_2 = x_2(t)$ are the fundamental solution of the equation (7.2) linearized at $\xi = \xi(t)$. In other words $\ddot{x}_i = -Q_{\xi\xi}(t, \xi(t))x_i$, $i = 1, 2$, and $(x_i(0), \dot{x}_i(0)) = (1, 0)$ for $i = 1$, resp. $= (0, 1)$ for $i = 2$.

To verify condition 2) of the criterion 7.1, we have to show first that the C^∞-perturbation $\delta Q = \delta Q(t, \xi)$ can be chosen with "small" support, i.e. such that the corresponding perturbation of ϕ contains the point H_0, and no other point of the orbit of H_0, in its support. To do so, we localize the support of $\delta Q = \delta Q(t, \xi)$ to some open (periodically extended) neighborhood U of $\{(t, \xi(t)),\ t_1 \le t \le t_2\}$, where $(t_1, t_2) \subset (0, 1)$ is so chosen that for all $t \in (t_1, t_2)$ and for every solution ξ_n of (7.2) with $(\xi_n(0), \dot{\xi}_n(0)) = H_n = \phi^n(H_0)$, $n \in \mathbf{Z} \setminus \{0\}$, then $(t, \xi_n(t)) \notin U$. Or equivalently, since $H_n = (\xi(n), \dot{\xi}(n))$, there is no point in U of the form $(t, \xi(t+n))$ with $t_1 \le t \le t_2$ and $n \in \mathbf{Z}\setminus\{0\}$. Then the corresponding perturbation of ϕ will affect only the point H_0 of the orbit (H), as desired. The existence of such an interval $(t_1, t_2) \subset (0, 1)$ follows from the identity principle for solutions of differential equations and from the fact that the homoclinic solution $\xi = \xi(t)$ is by definition $\pm$-asymptotic to some periodic solution, which implies that one has only a finite numbers of points $(t, \xi(t + n))$ to exclude from U.

The surjectivity requirement of condition 2) above follows then at once from formulae (7.4) above. It suffices to choose the derivatives $\delta Q_\xi, \delta Q_{\xi\xi}$ of δQ along the curve $\{(t, \xi(t)),\ t_1 \le t \le t_2\}$ appropriately, in order to give $\pi(H_0)$ and $d\pi(H_0)$ any value in $\mathbf{R}^2$, resp. in $\mathrm{sl}(\mathbf{R}; 2)$, since the fundamental solutions x_1, x_2, as well as their products x_1^2, x_1x_2, x_2^2, are linearly independent in $L^1(t_1, t_2)$. The criterion 7.1 is verified, and thus also the validity of the general results in the potential case.

7.3 Geodesic Curves on a 2-surface. Let Γ be a closed geodesic curve on an analytic 2-dimensional Riemann surface. In some C^1-neighborhood of Γ, one can define so-called Gaussian coordinates u, v, such that Γ becomes the line $v \equiv 0$ and the metric takes the form $ds^2 = E(u, v)du^2 + dv^2$, with a positive analytic function $E = E(u, v)$, which is 1-periodic in u and satisfies $E(u, v) = 1 + \sum_{n \ge 2} E_n(u)v^n$.

We look for curves $v = v(u)$ which are locally geodesic, i.e. extremals of the variational problem

$$I(\gamma) = \int_\gamma ds = \int \sqrt{E(u, v(u)) + \dot{v}(u)^2}\, du, \quad \text{defined for } C^1\text{-curves } \gamma\colon v = v(u) \,.$$

They are solutions of the Euler-Lagrange equation

$$\frac{d}{du}\frac{\partial}{\partial \dot{v}}\sqrt{E(u,v)+\dot{v}^2} = \frac{\partial}{\partial v}\sqrt{E(u,v)+\dot{v}^2}\ . \tag{7.5}$$

The time-one-maps of such systems are area-preserving analytic diffeomorphisms having a fixed point corresponding to the closed geodesic Γ. We assume that this fixed point is of "general stable type". The "generating functions" for this class are the "metrics" $E = E(u,v)$, and the topology is defined by means of the coefficient functions $E_n = E_n(u) = E_n(u+1)$ of $E = E(u,v)$. The *result* is then:

"In every C^1-neighborhood of the stable geodesic Γ, there are, generically, minimal hyperbolic p/q-periodic geodesic curves, transversal homoclinic geodesic curves asymptotic to them and also nondegenerate Mather sets of geodesic curves."

To verify the criterion 7.1 in this case, we use integral formulae of the same form as for the potential case. In particular, C^∞-perturbations $\delta E = \delta E(u,v)$ of E with "small" supports are constructed in the same way as there. We do not go into more details here, but we rather give an alternative formulation of the above result. Instead of changing the metric on a fixed surface, one can consider the metric as induced by the Euclidean metric of $\mathbf{R}^3$, and vary the surface near the given geodesic curve Γ. Then the above result becomes:

"Given an analytic 2-surface in $\mathbf{R}^3$ with a closed geodesic Γ of general stable type, there is a nearby surface osculating the given one of arbitrary high order along Γ, such that on the new surface, the geodesic Γ is arbitrarily well approached in C^1-norm by transversal homoclinic geodesic curves and by nondegenerate Mather sets of geodesic curves."

7.4 Hamiltonian Systems with 2 Degrees of Freedom. We consider analytic autonomous Hamiltonian systems with 2 degrees of freedom near a stable equilibrium which we take to be 0. Thus we look at systems of the form

$$\begin{aligned} &\dot{z} = JG_z, \quad G = G(z), \quad z = (x,y) = (x_1,x_2,y_1,y_2) \in \mathbf{R}^4, \\ &G(z) = \sum_{k\geq 2} G_k(x,y), \quad G_k = \text{homogeneous polynomial of degree } k\ . \end{aligned} \tag{7.6}$$

We also consider potential systems $\ddot{x} = -Q_x(x)$, $x \in \mathbf{R}^2$, in which case $G(x,y) = y^2/2 + Q(x)$. Such systems have been examined by Kozlov in [11].

After a linear transformation, the quadratic part of G takes the form $2G_2 = \alpha_1(x_1^2+y_1^2) + \alpha_2(x_2^2+y_2^2)$. If $|\alpha_1/\alpha_2| \neq p/q$, for $p,q = 1,2,3,4$, there is another change of variables such that (Birkhoff normal form)

$$G(x,y) = \frac{1}{2}\sum_{\nu=1,2}\alpha_\nu(x_\nu^2+y_\nu^2) + \frac{1}{4}\sum_{\nu,\mu=1,2}\beta_{\nu\mu}(x_\nu^2+y_\nu^2)(x_\mu^2+y_\mu^2) + \sum_{k\geq 5} G_k\ . \tag{7.7}$$

We assume the equilibrium 0 to be "of general stable type" in the sense that

$$\alpha_1,\alpha_2 > 0 \quad \text{and} \quad \Delta = \alpha_1^2\beta_{22} - 2\alpha_1\alpha_2\beta_{12} + \alpha_2^2\beta_{11} \neq 0, \quad \text{say} > 0\ . \tag{7.8}$$

The diffeomorphisms belonging to such systems will be defined below. The "generating functions" of the so obtained special class are the Hamiltonians $G = G(x, y)$, resp. the potentials $Q = Q(x)$, and the analytic topology is defined in terms of the higher Taylor coefficients of G at $x = y = 0$, resp. of Q at $x = 0$. The *result* is then:

"In every neighborhood of the equilibrium 0 in $\mathbf{R}^4$, the systems (7.6) have, generically, transversal homoclinic solutions and nondegenerate Mather sets of solutions."

First we explain how diffeomorphisms ϕ of the general form are defined in this case. The systems (7.6) are restricted first to fixed energy surfaces $G = \varepsilon^2$, on which they can be reduced to nonautonomous systems of one degree of freedom, for which the diffeomorphisms $\phi = \phi_\varepsilon$ are section maps. But in general, the so obtained diffeomorphisms have no fixed point, such that Proposition 3.6 does not apply to them. However, local twist maps $f = f_\varepsilon$ of the form (3.1) can be defined, such that the existence of the desired orbits will follow directly from Proposition 3.5 (instead of Prop. 3.6). We describe briefly these successive transformations.

Going first to "action-angle variables" $I_n = (x_n^2 + y_n^2)/2$, $\theta_n = \arg(x_n + \mathrm{i}y_n)$, $n = 1, 2$, one gets $G = G(\theta, I) = \alpha_1 I_1 + \alpha_2 I_2 + \sum_{\nu,\mu=1,2} \beta_{\nu\mu} I_\nu I_\mu + \cdots$. Rescaling then by $I = \varepsilon^2 J$, $G = \varepsilon^2 E$, one obtains

$$E = E(\theta, J) = \alpha_1 J_1 + \alpha_2 J_2 + \varepsilon^2 \sum_{\nu,\mu=1,2} \beta_{\nu\mu} J_\nu J_\mu + O(\varepsilon^3) .$$

Restricting the system (7.6) to the energy surface $G = \varepsilon^2$, thus to $E = 1$, one can reduce it to one degree of freedom by eliminating the time t as follows. Since $(\partial/\partial J_1)E = \alpha_1 + O(\varepsilon^2) \neq 0$ for small ε, one can solve $E = 1$ with respect to J_1, obtaining a function $J_1 = J_1(\theta_1, \theta_2, J_2, \varepsilon)$. Taking then $\tau = \theta_1/2\pi$ as the new time, the conjugate variable $\Phi_\varepsilon = \Phi_\varepsilon(\tau, \theta_2, J_2) = 2\pi J_1(\theta_1, \theta_2, J_2, \varepsilon)$ becomes the new Hamiltonian, which is nonautonomous and 1-periodic in τ. The diffeomorphism $\phi = \phi_\varepsilon$ is then defined as the time-one map of the corresponding (reduced) system.

Now to define the corresponding local twist maps $f = f_\varepsilon$, we restrict first J_2 to the interval $1/4\alpha_2 \leq J_2 \leq 3/4\alpha_2$ (notice that on the energy surface $E = 1$, J_2 varies from 0 to $1/\alpha_2 + O(\varepsilon^2)$), and we set $y = 4\alpha_2 J_2 - 2$ and $x = \theta_2/2\pi$. The diffeomorphisms ϕ_ε give rise to symplectic maps $f_\varepsilon : (x, y) \to (x', y')$, defined in $\{(x, y) \in \mathbf{R}^2,\ x \bmod 1,\ |y| \leq 1\}$, from which we have to show that they have the form (3.1) (here ε plays the rôle of a small parameter). To do so, we compute the terms of order $O(\varepsilon^2)$ in the reduced system. This gives (see [17]):

$$\frac{\partial x}{\partial \tau} = \frac{\dot\theta_2}{\dot\theta_1} = \frac{E_{J_2}}{E_{J_1}} = \frac{\alpha_2 + 2\varepsilon^2(\beta_{12} J_1 + \beta_{22} J_2) + O(\varepsilon^3)}{\alpha_1 + 2\varepsilon^2(\beta_{11} J_1 + \beta_{12} J_2) + O(\varepsilon^3)} ,$$

$$\frac{\partial y}{\partial \tau} = \frac{4\alpha_2 \dot J_2}{\dot\theta_1} = \frac{-4\alpha_2 E_{\theta_2}}{E_{J_1}} = O(\varepsilon^3) ,$$

and thus, by using $J_1 = \frac{1-\alpha_2 J_2}{\alpha_1} + O(\varepsilon^2)$ on the energy surface $E = 1$,

$$\frac{\partial x}{\partial \tau} = \frac{\alpha_2}{\alpha_1} + \frac{2\varepsilon^2}{\alpha_1^3}(\alpha_1\beta_{12} - \alpha_2\beta_{11}) + \frac{2\varepsilon^2}{\alpha_1^3}(\alpha_1^2\beta_{22} - 2\alpha_1\alpha_2\beta_{12} + \alpha_2^2\beta_{11})J_2 + O(\varepsilon^3)$$
$$= a + by + O(\varepsilon^3) ,$$
$$\text{with} \quad b = \frac{\Delta}{2\alpha_1^3\alpha_2}\varepsilon^2 \quad \text{and} \quad a = \frac{\alpha_2}{\alpha_1} + 2\frac{\alpha_1\beta_{12} - \alpha_2\beta_{11}}{\alpha_1^3}\varepsilon^2 + 2b ,$$

where Δ is the positive quantity in (7.8). It follows, since $(\partial/\partial\tau)y = O(\varepsilon^3)$, that the time-one-map f_ε of this system is of the form (3.1) with a and b as defined above. Thus Proposition 3.5 applies to the maps f_ε, for small $\varepsilon > 0$, and yields the desired minimal periodic and homoclinic orbits, which correspond to periodic and homoclinic solutions of the reduced systems, and thus also to solutions of (7.6) lying on the surface $G = \varepsilon^2$, i.e. arbitrarily close to the equilibrium 0, as desired.

We still had to verify the criterion 7.1 in this case. We do not do this here, but we can say that the integral formulae used are very similar again to those (7.3), (7.4) for the potential case, and also the localization of the support is very similar.

Many other examples of special classes could be given, either continuous ones as the three cases treated here, or also discrete special classes, such as the class of the iterates of standard twist maps.

Note. This paper is an abbreviated version of the dissertation I did at the ETH Zürich under the direction of Prof. J. Moser, to whom I am very thankful for his constant help and many useful advices. The details omitted here are contained in this dissertation.

References

[1] S. Aubry. The twist map, the extended Frenkel-Kontorova model and the devil's staircase. Order in Chaos. Physica D **7** (1983), No. 1–3, 240–258

[2] S. Aubry and P.-Y. Le Daeron. The discrete Frenkel-Kontorova model and its extensions I: exact results for the ground-states. Physica D **8** (1983), No. 3, 381–422

[3] V. Bangert. Mather sets for twist maps and geodesics on tori. Dynamics Reported **1** (1988), 1–56

[4] G.D. Birkhoff. Surface transformations and their dynamical applications. Acta Math. **43** (1922), 1–119, § 44–47

[5] G.D. Birkhoff. Dynamical Systems. American Mathematical Society **8**. New York (1927). 295 p.

[6] H.W. Broer and F. Takens. Formally symmetric normal forms and genericity. Dynamics Reported **2** (1989), 39–59

[7] A. Chenciner. La dynamique au voisinage d'un point fixe elliptique conservatif: de Poincaré et Birkhoff à Aubry et Mather. Sém. Bourbaki, Exposé 622, Vol. 1983/84. Astérisque **121/122** (1984), 147–170

[8] J. Guckenheimer, Ph. Holmes. Nonlinear oscillations, dynamical systems and bifurcations of vector fields. Applied Math. Sciences **42**. Springer, New York (1983), 184–190 (§ 4.5)

[9] R. Herman. Introduction à l'étude des courbes invariantes par les difféomorphismes de l'anneau. Astérisque **103/104** (1983)

[10] A. Katok. Some remarks on the Birkhoff and Mather twist theorems. Erg. Th. Dynam. Sys. **2** (1982), 183–194

[11] V.V. Kozlov. Integrability and nonintegrability in Hamiltonian mechanics. Russ. Math. Surv. **38** (1983), No. 1, 1–76

[12] R.S. MacKay and J. Stark. Lectures on orbits of minimal action for area-preserving maps. Preprint. University of Warwick, May 1985

[13] J.N. Mather. Existence of quasi-periodic orbits for twist homeomorphisms of the annulus. Topology **21** (1982), 457–467

[14] J.N. Mather. Variational construction of orbits of twist diffeomorphisms. Preprint. ETH Zürich, FIM. March 1990

[15] J. Moser. Nonexistence of integrals for canonical systems of differential equations. Comm. on Pure and Applied Math. **8** (1955), 409–436

[16] J. Moser. On invariant curves of area-preserving mappings of an annulus. Nachr. Akad. Wiss. Göttingen, Math.-Phys. Kl. IIa **1** (1962)

[17] J. Moser. Stable and random motions in dynamical systems. Ann. of Math. Studies 77. Princeton NJ: Princeton Univ. Press, 1973

[18] J. Moser. Proof of a generalized form of a fixed point theorem due to G.D. Birkhoff. LNM 597. Springer (1976), 464–494

[19] J. Moser. Monotone twist mappings and the calculus of variations. Ergod. Th. and Dyn. Syst. **6** (1986), 401–413

[20] H. Poincaré. Sur le problème des trois corps et les équations de la dynamique. Acta Math. **13** (1890), 1–270, spec. § 22

[21] H. Poincaré. Les méthodes nouvelles de la mécanique céleste, Vol. I–III. Paris, 1892, 1893, 1899

[22] R.C. Robinson. Generic properties of conservative systems. Amer. J. Math. **92** (1970), 562–603

[23] H. Rüssmann. Über die Existenz einer Normalform inhaltstreuer elliptischer Transformationen. Math. Ann. **137** (1959), 64–77

[24] C.L. Siegel. On the integrals of canonical systems. Ann. of Math., II. Ser. **42** (1941), No. 3, 806–822

[25] C.L. Siegel. Über die Existenz einer Normalform analytischer Hamiltonscher Differentialgleichungen in der Nähe einer Gleichgewichtslösung. Math. Ann. **128** (1954), 144–170

[26] C.L. Siegel and J. Moser. Lectures on celestial mechanics. Springer (1971), 290 p.

[27] E. Zehnder. Homoclinic points near elliptic fixed points. Commun. on Pure and Applied Math. **26** (1973), 131–182

a) References for splitting distance and angle

[28] P.J. Holmes, J.E. Marsden, J. Scheurle. Exponentially small splittings of separatrices. Proc. Nat. Acad. Sci. (to appear)

[29] P.J. Holmes, J.E. Marsden, J. Scheurle. Exponentially small splittings of separatrices with applications to KAM theory and degenerate bifurcations. Cont. Math. **81** (1988), 213–244

[30] A.I. Neisthadt. The separation of motions in systems with rapidly rotating phase. P.M.M. USSR **48** (1984), 133–139

[31] V.F. Lazutkin, I.G. Schachmannski, M.B. Tabanov. Splitting of separatrices for standard and semistandard mappings. Physica D **40** (1989), 235–248

b) Reference for the discrete Melnikov function

[32] D. Stoffer. Transversal homoclinic points and hyperbolic sets for non-autonomous maps, I and II. Journ. of Appl. Math. and Phys. (ZAMP) **39** (1988), 518–549, resp. 783–812. (See § 1.2 of II, 792–798, in particular Theorem 1.2.2, p. 297)

Asymptotic Periodicity of Markov and Related Operators

J. Komorník

Introduction

The aim of this paper is to provide a unified exposition of some results in the theory of asymptotic behaviour of Markov (and related) operators. These results followed the invited address of A. Lasota at ICM '82. He introduced the notion of constrictive operators saying that an operator P is weakly (strongly) constrictive if there exists a weakly (strongly) compact set F such that all trajectories of densities converge in L^1 norm to F. Lasota et al. investigated the asymptotic periodicity of Markov operators and proved that it holds for

(i) strongly constrictive Markov operators [14]
(ii) weakly constrictive Frobenius-Perron operators [19].

Lasota conjectured that the weak constrictiveness of Markov operators is equivalent to the strong one.

The positive solution of this problem, published in [7], has some practical consequences. Namely, for many operators appearing in applied problems the weak constrictiveness is much easier to verify. The reader interested in applications of constrictive operators can find them in the book [16] by Lasota and Mackey.

The proof in [7] suggested that a weaker sufficient condition for asymptotic periodicity than weak constrictiveness could still be found. This was really done in [10] where the so-called quasi-constrictive (or smoothing) operators were introduced. The importance of this new condition becomes transparent when we realize that it can be considered as a generalization of the famous Doeblin condition that has been utilized in the theory of Markov processes with transition function (cf. [4], [24]).

The quasi-constrictiveness of Markov operators is a substantial refinement of their quasi-compactness. (The later property was considered in [24] as well as in [6] where it was applied to the important class of Harris operators.) Namely, if P is a quasi-compact Markov operator then the image $P(D)$ of the set D of densities must be a proper subset of D. On the other hand, the Frobenius-Perron operator P induced by the diadic transformation of the unit interval is quasi-constrictive since it is asymptotically stable [16] but $P(D) = D$.

Furthermore, quasi-constrictiveness of a positive power bounded operator clearly follows from the existence of a nontrivial lower-bound function for this operator (that is a well-known criterion of asymptotic stability in Markovian case).

The approach of Lasota et al. to the study of asymptotic properties of Markov operators provides a common generalization to the analogical results achieved in ergodic theory (where Frobenius-Perron operators were studied) and in the theory of Harris operators (that included the practically important classes of kernel operators as well as clasical Markov chains).

The results of Lasota, Lee and Yorke were generalized to the class of linear contractions [1], [25] and to positive power bounded operators [20], where strong constrictiveness was proved to be a sufficient condition of asymptotic periodicity. The latter result was generalized in [8] and [9] where the asymptotic periodicity of constrictive and quasi-constrictive positive linear power bounded operators on L^1 was proved.

Most of the results mentioned above can be extended to the class of operators defined on bands of signed measures using the technique developed in [11].

1. Basic Notions and Results

The principal objects of study of this paper are linear continuous positive operators on Banach lattices. Throughout the paper, by an operator we shall always understand one having the properties listed above. Except the very last chapter we shall deal with operators on $L^1 = L^1(\mu)$, where μ is a σ-finite measure defined on a σ-algebra $\mathcal{B}$ of subsets of basic set X.

We shall denote the set of nonnegative elements of L^1 by

$$L^1_+ = \{f \in L^1 : f(x) \geq 0 \text{ a.e.}\} \tag{1.1}$$

and densities by

$$D = \{f \in L^1_+ : \|f\| = 1\} . \tag{1.2}$$

For any $f \in L^1$, the support of f is defined by

$$\operatorname{supp}(f) = \{x \in X : f(x) \neq 0\} . \tag{1.3}$$

Recall that an operator P is called Markov if $P(D) \subset D$ [16].

An important class of Markov operators is represented by Frobenius-Perron operators induced by measurable transformations of the basic space X that are nonsingular (i.e. $\mu(S^{-1}(A)) > 0$ if $\mu(A) > 0$). For any $f \in L^1$ the image $P_S f$ is uniquely determined by

$$\int_B P_S f \, d\mu = \int_{S^{-1}(B)} f \, d\mu \quad \text{for } B \in \mathcal{B} . \tag{1.4}$$

Definition 1.1. An operator P is called

(i) power bounded if there exists an $M > 0$ such that

$$\|P^n\| \le M \quad \text{for every } n \in N \tag{1.5}$$

(ii) trivial if $\lim_{n\to\infty} \|P^n f\| = 0$ for every $f \in L^1$
(iii) strictly nontrivial if

$$\liminf \|P^n f\| > 0 \quad \text{for } f \in D \tag{1.6}$$

(Note that Markov operators are strictly nontrivial.)
(iv) weakly almost periodic if for every $f \in L^1$ the trajectory $\{P^n f\}$ is weakly precompact
(v) constrictive if there exists a weakly compact subset $F \subset L^1$ such that

$$\lim_{n\to\infty} d(P^n f, F) = 0 \quad \text{for } f \in D \tag{1.7}$$

where $d(g, F)$ is the infimum of $\|g - h\|$, $h \in F$. (Note that a constrictive operator is power bounded.)
(vi) asymptotically periodic if it is either trivial or if there exist finitely many distinct functions $g_1, \dots, g_a \in L^1_+$, a permutation α of the set $\{1, \dots, a\}$ and positive continuous linear functionals $\lambda_1, \dots, \lambda_a$ on L^1 such that

$$\lim_{n\to\infty} \left\| P^n \left(f - \sum_{i=1}^{a} \lambda_i(f).g_i \right) \right\| = 0 \quad \text{and} \quad Pg_i = g_{\alpha(i)},\ i = 1, \dots, a \tag{1.8}$$

(vii) asymptotically stable if it is either trivial or satisfies (1.8) for $a = 1$.

All operators considered henceforth will be supposed to be power bounded and nontrivial.

Note that if P is an asymptotically periodic operator then there exists $q \le a!$ such that

$$\alpha^q(i) = i \quad \text{and} \quad P^q g_i = g_i \quad \text{for } i = 1, \dots, a\ . \tag{1.8a}$$

Moreover the function

$$Q(f) = \sum_{i=1}^{a} \lambda_i(f).g_i \tag{1.8b}$$

satisfies

$$P^q(Q(f)) = Q(f), \quad \lim_{n\to\infty} P^n(f - Q(f)) = 0\ . \tag{1.8c}$$

Remark 1.1. The set $F \subset L^1$ satisfying (1.7) is called a constrictor for P (cf. [25]).

Theorem 1. *A constrictive operator is asymptotically periodic.*

The proof for Markov operators is in Chap. 3, for power bounded operators in Chap. 5.

Let us recall two kinds of necessary and sufficient conditions for weak precompactness of a bounded subset $F \subset L^1$ (cf. [5]).

(C1) If $\{B_m\} \in \mathcal{B}$ is any decreasing sequence with empty intersection then

$$\lim_{m\to\infty}\left[\sup\left\{\int_{B_m} f\,d\mu : f \in F\right\}\right] = 0 \tag{1.9}$$

(C2) For every $\varepsilon > 0$ there exist a set $K \in \mathcal{B}$, $\mu(K) < \infty$ and a positive number δ such that

$$\int_{B\cup(X-K)} f\,d\mu < \varepsilon \quad \text{for } f \in F \quad \text{and} \quad B \in \mathcal{B},\ \mu(B) < \delta\ . \tag{1.10}$$

Note that if $\mu(X) < \infty$ than the condition (C2) can be replaced by
(C2′) For every $\varepsilon > 0$ there exists $\delta > 0$ such that

$$\int_{B} f\,d\mu < \varepsilon \quad \text{for } f \in F \quad \text{and} \quad B \in \mathcal{B},\ \mu(B) < \delta\ . \tag{1.10a}$$

The above conditions inspired the following definition.

Definition 1.2. A Markov operator P is called
(i) quasi-constrictive if it has a constrictor $F \subset L^1_+$ satisfying
(C3) There exist a $K \in \mathcal{B}$, $\mu(K) < \infty$ and numbers $\kappa < 1$, $\delta > 0$ such that

$$\int_{B\cup(X-K)} f\,d\mu \le \kappa \quad \text{for } f \in F \quad \text{and} \quad B \in \mathcal{B},\ \mu(B) < \delta\ . \tag{1.11}$$

Remark 1.2. It is straightforward to observe that if P is a quasi-constrictive Markov operator and $\mu(X) < \infty$, then the following condition is satisfied.
(C3′) There exist $\kappa < 1$ and $\delta > 0$ such that for every $f \in L^1_+$ the inequality

$$\limsup_{n\to\infty}\left[\sup\left\{\int_{B} Pf^n\,d\mu : \mu(B) < \delta\right\}\right] \le \kappa.\|f\| \tag{1.11a}$$

holds.

As will be shown in Chap. 2, the condition (C3) for (a constrictor) $F \subset D$, is equivalent to the following conditions (C4) and (C5).

(C4) There exists $\kappa < 1$ and a weakly compact set F_0 such that the neighbourhood $U(F,\kappa) = \{f \in L^1 : d(f, F_0) \le \kappa\}$ is a constrictor for P.

(C5) There exists a $\kappa < 1$ such that for any decreasing sequence $\{B_m\}$ with empty intersection the inequality

$$\lim_{m\to\infty}\left[\sup\left\{\int_{B_m} f\,d\mu : f \in F\right\}\right] < \kappa \tag{1.12}$$

holds (for a constrictor F).

Remark 1.3. The conditions (C3) and/or (C5) for (a constrictor) $F \subset L^1_+$ can be equivalently expressed in the form

(C6) There exists a $K \in \mathcal{B}$, $\mu(K) < \infty$ and numbers $\varepsilon, \delta > 0$ such that

$$\int_{K-B} f\,d\mu \geq \varepsilon \quad \text{for } f \in F,\ B \in \mathcal{B},\ \mu(B) < \delta \tag{1.13}$$

and/or

(C7) There exists $\varepsilon > 0$ such that

$$\lim_{m\to\infty}\left[\inf\left\{\int_{X-B_m} d\mu : f \in F\right\}\right] > \varepsilon \quad \text{for all} \quad \{B_m\} \to \emptyset\ . \tag{1.14}$$

The advantage of these conditions is that they are directly extendable to positive operators on L^1. (The latter makes sense on the spaces of signed measures as well.)

The existence of a constrictor F (for P) satisfying (C6) yields that the following condition is satisfied

(C6′) There exist a $K \in \mathcal{B}$, $\mu(K) < \infty$, real numbers $\varepsilon, \delta > 0$ and a function $n_0 : D \to N$ such that

$$\int_{K-B} P^n f\,d\mu \geq \varepsilon \quad \text{for } B \in \mathcal{B},\ \mu(B) < \delta,\ f \in D,\ n \geq n_0(f)\ . \tag{1.13a}$$

On the other hand, the condition (C6′) obviously implies that the set $F = \{P^n f : f \in D,\ n \geq n_0(f)\}$ is a constrictor for P satisfying (1.13).

Similarly, the existence of a constrictor F satisfying (C7) implies that the following condition holds

(C7′) There exists $\varepsilon > 0$ and a function $n_0 : D \to N$ such that

$$\lim_{m\to\infty}\left[\inf\left\{\int_{X-B_m} P^n f\,d\mu : f \in D,\ n \geq n_0(f)\right\}\right] > \varepsilon\ . \tag{1.14a}$$

On the other hand, the condition (C7′) implies that the set

$$F = \{P^n f : f \in D,\ n \geq n_0(f)\}$$

is a constrictor for P satisfying (1.14).

Definition 1.3. A Markov operator P is called almost constrictive if it satisfied

(C8) There exists $\kappa < 1$ such that for any decreasing sequence $\{B_m\}$ with empty intersection

$$\lim_{m\to\infty}\left[\limsup_{n} \int_{B_m} P^n f\,d\mu\right] < \kappa \quad \text{for } f \in D \tag{1.15}$$

holds.

Note that the condition (1.15) can be equivalently expressed in the form

(C8′)
$$\lim_{m\to\infty}\left[\liminf_{n} \int_{X-B_m} P^n f\,d\mu\right] > \varepsilon \tag{1.15a}$$

for some $\varepsilon > 0$ and all $f \in D$, $\{B_m\} \to \emptyset$.

Definition 1.4. An operator P is called
(i) quasi-constrictive if it satisfied (1.13.a) (equivalent to (1.14.a))
(ii) almost constrictive if it satisfied (1.15.a).

Note that, unlike the situation in the class of Markov operators, constrictive operators need not be quasi-constrictive. (See Example 1.2 below.)

The next theorem is a generalization of the main result of [10].

Theorem 2. *If P is a quasi-constrictive operator then it is asymptotically periodic.*

The proof for Markov operators is in Chap. 4, for power bounded operators in Chap. 5.

A more detailed characterization of asymptotic periodicity of operators is given by the following result.

Proposition 1.1. *Let P be an asymptotically periodic operator.*
(i) The functions $g_1, \dots, g_a \in L^1_+$ satisfying (1.8) are linearly independent.
(ii) For every $i, j \in \{1, \dots, a\}$, $i \neq j$ and $f \in L^1_+$

$$\operatorname{supp}(f) \subset \operatorname{supp}(g_i) \cap \operatorname{supp}(g_j) \Rightarrow \lim_{n\to\infty} \|P^n f\| = 0 \ . \tag{1.16}$$

(iii) If P is a strictly nontrivial operator, then the functions $g_1, \dots, g_a$ satisfying (1.8) have disjoint supports.
(iv) If P is a Markov operator then there exist densities $g_1, \dots, g_a$ with disjoint supports satisfying (1.8).

The proof is in Chap. 5 (part (iv) is proved in Chap. 3).

The part (iv) of the above result inspired a simple criterion of asymptotic stability of Markov operators that can be extended to power bounded operators as follows.

Definition 1.5. A set $B \in \mathcal{B}$, $\mu(B) > 0$ is called a lower set for an operator P if

$$\lim_{n\to\infty} \mu(B - \operatorname{supp}(P^n f)) = 0 \quad \text{for all} \quad f \in D \ . \tag{1.17}$$

Proposition 1.2. *If P is an asymptotically periodic strictly nontrivial operator having a lower set, then P is asymptotically stable.*

This result (that is very well known for Markov operators [16]) is an easy consequence of the following one, that may be of independent interest.

Proposition 1.3. *If P is a weakly almost periodic strictly nontrivial operator having a lower set, then it is weakly asymptotically stable i.e. there exists a P-invariant density g and a positive linear functional λ on L^1 such that for every $f \in L^1$ the differences*

$$P^n f - \lambda(f) \cdot g$$

converge weakly to 0.

The proof for Markov operators is in Chap. 3, for power bounded operators in Chap. 5.

Another "classical" criterion of asymptotic stability of a Markov operator P is the existence of nontrivial lower-bound function. Recall that $h \in L^1_+$, $\|h\| > 0$ is a nontrivial lower-bound function for an operator P if the sequence of negative parts $\{[h - P^n f]^+\}_{n\in N}$ converges strongly to 0 for all $f \in D$. It is obvious that if an operator P has a nontrivial lower-bound function h, then it is quasi-constrictive, hence it is asymptotically periodic. Moreover, it is strictly nontrivial and has a lower set $\mathrm{supp}(h)$, hence P is asymptotically stable. On the other hand, an asymptotically stable Markov operator has a unique invariant density that is the maximal lower-bound function for P.

A formally symmetric criterion of asymptotic periodicity of Markov operators is that of the existence of an upper-bound function. A function $h \in L^1_+$ is said to be an upper-bound function for P if the sequence $\{[P^n f - h]^+\}_{n\in N}$ converges strongly to 0 for every $f \in L^1_+$. It is obvious that if h is an upper-bound function for an operator P, then the set

$$F^h = \{g \in L^1_+ : g \leq h\}$$

is a weakly compact constrictor for P, hence P is asymptotically periodic.

Example 1.1. Consider the unity preserving positive operator P defined on the space R^3 (of column vectors) by

$$Px = \mathbb{P}.x, \quad \text{where}$$

$$\mathbb{P} = \begin{bmatrix} 0 & 1 & 0 \\ 1 & 0 & 0 \\ 1/3 & 1/6 & 1/2 \end{bmatrix}.$$

Obviously, the restriction of P to the 1-dimensional subspace X_0 spanned by $g_3 = (0,0,1)'$ is trivial. The operator P^2 is represented by the matrix

$$\mathbb{P}^2 = \begin{bmatrix} 1 & 0 & 0 \\ 0 & 1 & 0 \\ 1/3 & 5/12 & 1/4 \end{bmatrix},$$

thus it has linearly independent eigenvectors $g_1 = (1,0,4/9)'$, $g_2 = (0,1,5/9)'$. Since the vectors g_1, g_2, g_3 form the basis of R^3, the convex envelope $F = \mathrm{co}(g_1, g_2)$ is the smallest constrictor for P. Note that the functions g_1 and g_2, that are permuted by P, do not have disjoint supports. Moreover, they have different norms, thus they cannot be simultaneously renormalized and replaced by densities.

In the next example we show that an asymptotically stable positive operator need not be quasi-constrictive (hence it need not have a nontrivial lower-bound function).

Example 1.2. Let $X = N$, $\mu(\{m\}) = 2^{-m}$ and the operator P be given by

$$P(\{m\}) = 2.3^{-m}.(1,1,\ldots,1,\ldots) .$$

Obviously, P is a power bounded ($M = 4/3$), unity preserving and asymptotically stable operator. Consider the sequence of densities $f_m = 2^m.1_{\{m\}}$. We have

$$P^n f_m = (2^{m+1}/3^m).(1,1,1,\ldots,1,..) \quad \text{for } m \in N .$$

Therefore

$$\lim_{n\to\infty} \|P^n f_m\| = 2(2/3)^m$$

and thus

$$\inf_{f\in D} \liminf \|P^n f\| = 0 .$$

Therefore P does not satisfy (1.13.a), hence it is not quasi-constrictive but constrictive.

Chapters 2, 3, 4 and 5 of this article are mainly devoted to the proofs of the results listed above. They also contain auxiliary results that are of independent interest. The main result of Chap. 2 is following proposition.

Proposition 1.4. *Let P be an operator.*

(i) For a given $f \in L^1_+$ the trajectory $\{P^n f\}$ is weakly precompact if and only if the sequence

$$A_n(f) = n^{-1}.\sum_{i=0}^{n-1} P^i f \tag{1.18}$$

converges in L^1.

(ii) P is weakly almost periodic if and only if there exists a $f_0 \in D$ such that $f_0 > 0$ a.e. and the sequence $\{A_n(f_0)\}$ converges in L^1 to a P-invariant density g. Moreover, the set $G = \mathrm{supp}(g)$ and its characteristic function 1_G satisfy

$$\lim_{n\to\infty} \|P^n f - 1_G.P^n f\| = 0 \quad \textit{for every} \quad f \in L^1 . \tag{1.19}$$

Remark 1.4. The above results (except for the part "only if" in (i)) are not direct consequences of the Mean Ergodic Theorem. The positivity of the operator P is essential.

This result enables us to follow the approach of [14] and restrict our attention to the case when the measure μ is finite and the characteristic function 1_X of the whole space X is P-invariant, i.e. P is a unity preserving operator.

Chapter 3 is devoted to unity preserving Markov operators and plays a crucial role in the whole article. It follows (in a simplified way) the author's proof of asymptotic periodicity of constrictive Markov operators published in [7]. First it goes parallel to Chap. VIII of [6] and provides us with a weak version of Theorem 2. It characterizes all cluster points of trajectories $P^n f$, $f \in L^1$, in the weak topology, making a substantial use of the notion of nice sets. Recall that a set $B \in \mathcal{B}$ is a nice set if all images $P^n(1_B)$, $n \in N$, are characteristic functions (cf. [14]).

The most difficult part of Chap. 3 is connected with mixing operators. Recall that a unity preserving Markov operator P is mixing if for every $f \in L^1_+$ the sequence $\{P^n f\}$ converges weakly to $[\|f\|/\mu(X)].1_X$. We say that P is an exact operator if the above convergence is strong (cf. [16]).

Proposition 1.5. *Let P be a Markov operator with invariant unity. Then for every $f \in L^1$ the trajectory $\{P^n f\}$ is weakly precompact and all its weak cluster points are measurable with respect to the σ-algebra $\mathcal{B}_0$ of nice sets.*

If P is a quasi-constrictive operator, then $\mathcal{B}_0$ is finite. Moreover, there exists a finite decomposition of the space X formed by disjoint atoms $C_1, \ldots, C_a$ of $\mathcal{B}_0$ and a number $q < a!$ such that the power $R = P^q$ preserves the subspaces

$$L_i = \{f \in L^1 : \mathrm{supp}(f) \subset C_i\}, \ i = 1, \ldots, a$$

and the restrictions $R_{|L_i}$ are mixing and quasi-constrictive i Markov operators.

Asymptotic periodicity of the operator P follows directly from the next interesting result.

Proposition 1.6. *A mixing and quasi-constrictive Markov operator is exact.*

This fact is a direct consequence of a more general result that characterizes the relation between mixing and exact operators using the following simple notion.

Definition 1.6. Let $\rho \geq 0$ be a given number.

(i) We say that function $f_1, \ldots, f_k \in L^1_+ - \{0\}$ are ρ-orthogonal if there exist disjoint sets $B_1, \ldots, B_k \in \mathcal{B}$ such that

$$\int_{X-B_i} f_i \, d\mu \leq \rho.\|f_i\| \quad \text{for } i = 1, \ldots, k \ . \tag{1.20}$$

(ii) We say that trajectories of the functions $f_1, \ldots, f_m$ are ρ-orthogonal if for every $n \in N$ the functions $P^n f_1, \ldots, P^n f_m$ are ρ-orthogonal.

Theorem 3. *Let P be a mixing Markov operator that is not exact. Then for every $\rho > 0$ and $k \in N$ there exist functions $f_1, \ldots, f_{2^k} \in L^1_+ - \{0\}$ with ρ-orthogonal trajectories.*

Chapter 4 contains an interesting criterion of existence of invariant density for a given strictly nontrivial operator. It resembles the analogical results for Markov operators from [26] and [22].

Theorem 4. *A strictly nontrivial operator P has an invariant density if and only if there exists $f \in D$ such that for every decreasing sequence $\{B_m\} \subset \mathcal{B}$ with empty intersection the inequality*

$$\lim_{m\to\infty}\left[\liminf_n \int_{X-B_m} A_n(f)\,d\mu\right] > 0 \tag{1.21}$$

holds.

This result enables us to prove (in Chap. 4) the following generalization of the main result of [18].

Theorem 5. *An almost constrictive operator is weakly almost periodic.*

Combining this theorem with the results of Chap. 2 and Chap. 3 we obtain the proof of Theorem 2 for Markov operators. Finally the proof of Theorems 1 and 2 for power bounded operators is completed in Chap. 5 where we apply Sucheston decomposition of power bounded operators to trivial and strictly nontrivial parts.

In Chap. 6 we generalize Theorems 1, 2, 4 and 5 for power bounded operators defined on a band L of signed measures on a σ-algebra $\mathcal{B}$ (i.e. a Banach lattice L such that $\mu \in L$, $\nu \ll \mu \Rightarrow \nu \in L$).

This generalization is based on the fact that for every operator P on L and measure $\mu \in L$ there exists a nonnegative measure $\mu_0 \in L$ such that $\mu \leq \mu_0$ and the subband

$$L_0 = \{\nu \in L : \nu \ll \mu_0\} \approx L^1(\mu_0) \text{ is } P\text{-invariant} .$$

We say that an operator P on a band L is

(i) Markov if it preserves the set D of nonnegative normalized elements (called here probabilities)

(ii) constrictive if it has a weakly compact constrictor F (i.e. $\lim_{n\to\infty} d(P^n\mu, F) = 0$ for $\mu \in D$)

(iii) quasi-constrictive if there exist $\varepsilon > 0$ and the function $n_0 : D \to N$ such that

$$\lim_{m\to\infty}[\inf\{P^n\mu(X - B_m) : \mu \in D,\ n \geq n_0(\mu)\}] > \varepsilon \tag{1.22}$$

(iv) asymptotically periodic if it has a compact constrictor contained in a P-invariant band $L_0 \approx L^1(\mu_0)$ for some $\mu_0 \in L^+$.

2. The Reduction Procedure

The aim of this chapter is to prove Proposition 1.4 and to show that it enables us to reduce our attention to unity preserving operators. In order to achieve this goal we first prove some auxiliary results.

Note that there exist two kinds of natural partial orderings on L^1. The first one is given by the pointwise ordering

$$f \leq g \Leftrightarrow g - f \in L^1_+ , \tag{2.1}$$

which induces on L^1 the structure of Banach lattice with the lattice operations $f \vee g$ and $f \wedge g$ defined pointwise.

Note that every $f \in L^1$ can be expressed as

$$f = f^+ - f^-, \quad \text{where} \quad f^+, f^- \in L^1_+ \ . \tag{2.2}$$

The second partial ordering on L^1 that we consider is given by

$$f \ll g \Leftrightarrow \mu[\text{supp}(f) - \text{supp}(g)] = 0 \ . \tag{2.3}$$

The following very simple results will be frequently used in the sequel.

Proposition 2.1. *Let P be an operator on L^1.*
(i) The following inequalities hold for any $f, g \in L^1$.

$$P(f \vee g) \geq Pf \vee Pg, \quad P(f \wedge g) \leq Pf \wedge Pg \tag{2.4}$$

and

$$P(f^+) \geq (Pf)^+, \quad P(f^-) \geq (Pf)^-, \quad P(|f|) \geq |Pf| \ . \tag{2.5}$$

(ii) If $f, g \in L^1_+$ and $G = \text{supp}(g)$, then we have

$$1_{X-G}.f \leq [f - g]^+ \tag{2.6}$$

$$1_G.f = \lim_{k\to\infty} [f \wedge k.g] \tag{2.7}$$

and

$$f \ll g \Leftrightarrow f = \lim_{k\to\infty} [f \wedge k.g] \Leftrightarrow \lim_{k\to\infty} \|[f - k.g]^+\| = 0 \ . \tag{2.8}$$

Corollary 2.1. *Let P be an operator and $f, g \in L^1_+$, $f \ll g$.*
(i) For every $n \in N$ we have $P^n f \ll P^n g$.
(ii) If $\lim_{n\to\infty} \|P^n g\| = 0$, then $\lim_{n\to\infty} \|P^n f\| = 0$.
(iii) If $\{P^n g\}$ is weakly precompact then the same holds for $\{P^n f\}$.

Proof. Let M be a constant satisfying (1.5). Let k and $n \in N$ be given. From (2.5) (applied to P^n) we get that

$$[P^n f - k.P^n g]^+ = [P^n(f - k.g)]^+ \leq P^n([f - k.g]^+) \ .$$

(i) Let $n \in N$ be given. From (1.5) and (2.8) we directly get

$$\lim_{k\to\infty} \|[P^n f - k.P^n g]^+\| \leq M. \lim_{k\to\infty} \|[f - k.g]^+\| = 0 \ .$$

Hence $P^n f \ll P^n g$.

(ii) For any $\varepsilon > 0$ there exist a $k \in N$ such that

$$\|[f - k.g]^+\| < \varepsilon/M$$

and thus

$$\sup_n \|(P^n f - k.P^n g)^+\| \leq \varepsilon \ . \tag{2.9}$$

Therefore

$$\limsup_n \|P^n f\| \leq k. \lim_{n\to\infty} \|P^n g\| + \sup_n \|(P^n f - k.P^n g)^+\| \leq \varepsilon .$$

Since $\varepsilon > 0$ was arbitrary, we have

$$\lim_{n\to\infty} \|P^n f\| = 0 .$$

(iii) Since $\{P^n g\}$ is weakly precompact it satisfies (1.9) for any given sequence $\{B_m\} \to \emptyset$. Let $\varepsilon > 0$ be given. Let $k \in N$ be such that (2.9) holds. We have

$$\lim_{m\to\infty} \left[\sup_n \left\{\int_{B_m} P^n f \, d\mu\right\}\right] \leq k. \lim_{m\to\infty} \left[\sup_n \left\{\int_{B_m} P^n g \, d\mu\right\}\right] + \sup_n \|(P^n f - k.P^n g)^+\| \leq \varepsilon .$$

Since $\varepsilon > 0$ and $\{B_m\} \to \emptyset$, $\{P^n f\}$ is weakly precompact. □

Proof of Proposition 1.4. (i) Suppose that $f \in L^1$ and that the trajectory $\{P^n f\}$ is weakly precompact. We get directly from the condition (C2) of weak precompactness that the sequence $\{A_n f\}$, given by (1.18) is weakly precompact.

According to the Mean Ergodic Theorem (cf. [5]) the sequence $\{A_n f\}$ converges strongly to a P-invariant function g.

Suppose now that $f_0 \in L^1_+$ be such that the sequence $\{A_n f_0\}$ converges strongly to a P-invariant function g, satisfying

$$\lim_{n\to\infty} \|A_n f_0 - g\| = 0 . \tag{2.10}$$

Put $G = \text{supp}(g)$. From Corollary 2.1 (i) we get that for any $n \in N$ and $f \in L^1_+$ we have

$$P(1_G.P^{n-1} f) \ll Pg = g .$$

Therefore the equalities

$$1_{X-G}.P(1_G.P^{n-1} f) = 0$$

and

$$1_{X-G}.P^n f = 1_{X-G}.P(1_{X-G}.P^{n-1} f) \tag{2.11}$$

hold for every $n \in N$.

Consider the operator R defined on L^1 by

$$Rf = 1_{X-G}.Pf , \tag{2.12}$$

(R is obviously linear and positive).

From (2.11) we get that for every $f \in L^1$ and $n \in N$ we have

$$R^n f = 1_{X-G}.P^n f , \tag{2.13}$$

which implies that R is a power bounded operator and the constant M, satisfying (1.5) for R, can be taken the same as for P. Therefore, the inequality

$$\limsup \|R^n f\| \leq M. \liminf \|R^n f\| \tag{2.14}$$

obviously holds.

From (2.10) we get

$$0 = \lim_{n \to \infty} \int 1_{X-G}.A_n f_0 \, d\mu \ ,$$

which together with (2.14) implies that

$$\lim_{n \to \infty} \|R^n f_0\| = 0 \ .$$

Now we show that for every $f \in L^1_+$, $f \ll f_0$ the trajectory $\{P^n f\}$ is weakly precompact. From Corollary 2.1 (ii) we get that

$$\lim_{n \to \infty} \|R^n f\| = 0 \ .$$

Therefore, for every $\varepsilon > 0$ there exists $r \in N$ such that the inequality

$$\|P^r f - 1_G.P^r f\| < \varepsilon/3M \tag{2.15}$$

holds (where M is the constant satisfying (1.5)).

This implies that for any $n > r$ we have

$$\|P^n f - P^{n-r}(1_G.P^r f)\| < \varepsilon/3 \ . \tag{2.16}$$

From Proposition 2.1 (ii) we get that there exists $k \in N$ such that

$$\|[1_G.P^r f - k.g]^+\| < \varepsilon/3M \ ,$$

which, combined with (2.16) gives

$$\|[P^n f - k.g]^+\| < 2\varepsilon/3 \quad \text{for } n > r \ . \tag{2.17}$$

Consider any nonincreasing sequence $\{B_m\} \subset \mathcal{B}$ with empty intersection. There exists $m_0 \in N$ such that for every $m \geq m_0$ we have

$$\int_{B_m} g \, d\mu < \varepsilon/3k \ . \tag{2.18}$$

Combining (2.18) and (2.17) we get

$$\int_{B_m} P^n f \, d\mu < \varepsilon \quad \text{for } n \geq r \quad \text{and} \quad m > m_0 \ . \tag{2.19}$$

Hence the trajectory $\{P^n f\}$ is weakly precompact.

ii) If there exists $f_0 \in L^1_+$ such that $\operatorname{supp}(f_0) = X$ and $\{A_n f_0\}$ converges strongly to $g = Pg$, then for every $f \in L^1_+$ we have $f \ll f_0$, thus $\{P^n f\}$ is weakly precompact and (1.19) holds.

The rest of the proof follows from the decomposition (2.2) that holds for every $f \in L^1$. □

Corollary 2.2. *If* $\operatorname{supp}(f_0) = X$, $\lim \|A_n f_0 - g\| = 0$ *and* $f \in L^1_+$ *is P-invariant, then* $\operatorname{supp}(f) \subset \operatorname{supp}(g)$.

Proof. From (1.19) we obtain that $f = 1_G.f$. □

Now we are going to prove the equivalence of conditions (C3) and (C4) for quasi-constrictiveness of Markov operators.

The implication (C4) $\Rightarrow$ (C3) follows from (1.10).

Suppose that (C3) holds for a constrictor F. Let $K \in \mathcal{B}$, $\delta > 0$ and $\kappa < 1$ satisfy (1.11). Put $g_0 = \delta^{-1}.1_K$, $F_0 = \{g \in L^1_+ : g \leq g_0\}$. Obviously F_0 is a weakly precompact subset of L^1_+. Put $\lambda = (\kappa + 1)/2$. We show that the neighbourhood $U(F, \lambda)$ is a constrictor for P.

The properties of the constrictor F yield that there exists a function $n_0 : D \to N$ such that

$$d(P^n f, F) < (1-\kappa)/4 = (\lambda - \kappa)/2 \quad \text{for } f \in D \quad \text{and} \quad n \geq n_0(f) .$$

We show, that for any fixed $f \in D$ and $n \geq n_0(f)$ the iterate $P^n f \in U(F_0, \lambda)$.

There exists $h \in F$ such that $\|P^n f - h\| < \lambda - \kappa$.

Put $B = \{x : P^n f(x) > \delta^{-1}\}$. Since $P^n f \in D$ we have $\mu(B) < \delta$. From (1.11) we get

$$\int_{B \cup (X-K)} h \, d\mu \leq \kappa .$$

Therefore

$$\int_{B \cup (X-K)} P^n f \, d\mu < \lambda .$$

Put $g = g_0 \wedge P^n f$. Obviously, $g \in F_0$ and

$$\|P^n f - g\| = \int (P^n f - g)^+ d\mu = \int_{B \cup (X-K)} (P^n f - g_0)^+ d\mu$$
$$\leq \int_{B \cup (X-K)} P^n f \leq \lambda .$$

The fact that the conditions (C3) and (C5) for constrictors of Markov operators are equivalent is a direct consequence of the equivalence of the conditions (C6) and (C7) for constrictors of positive operators that we are going to prove now.

Let a set $F \subset L^1_+$ satisfies the condition (C6) with $K \in \mathcal{B}$, $\mu(K) < \infty$, δ and $\varepsilon > 0$. We show that it satisfies (C7) with the same ε.

Let $\{B_n\} \to \emptyset$. There obviously exists $m_0 \in N$ such that $\mu(K \cap B_m) < \delta$ for $m \geq m_0$. Therefore

$$\int_{X-B_m} f \, d\mu \geq \int_{K-B_m} f \, d\mu \geq \varepsilon \quad \text{for } f \in F \quad \text{and} \quad m \geq m_0 . \tag{2.20}$$

Now suppose that set $F \subset L^1_+$ satisfies (C7) with the given $\varepsilon > 0$. We show that F satisfies (C6) with the same ε and some $K \in \mathcal{B}$ and $\delta > 0$. Suppose that it is not the case.

For any $K \in \mathcal{B}$, $\mu(K) < \infty$ there exists $B_m \subset K$, $\mu(B_m) < 2^{-m}$ and $f_m \in D$ such that

$$\int_{K-B_m} f_m \, d\mu < \varepsilon \,. \tag{2.21}$$

Let $\{K_m\} \subset X$, $\mu(K_m) < \infty$, be a increasing sequence such that $\bigcup_{m=1}^{\infty} K_m = X$. For every $m \in N$ choose $B_m \subset K_m$, $\mu(B_m) < 2^{-m}$ and $f_m \in F$ such that (2.21) holds for $K = K_m$.

Put

$$E_m = (X - K_m) \cup \bigcup_{k=m}^{\infty} B_k - \bigcap_{n=1}^{\infty} B_n \quad \text{for } m \in N \,.$$

Obviously, $\mu(\bigcap_{n=1}^{\infty} B_n) = 0$ and $\{E_m\} \to \emptyset$.

Moreover, $\mu([X - E_m] - [K_m - B_m]) = 0$ for $m \in N$. Therefore,

$$\inf \left\{ \int_{X-E_m} f \, d\mu : f \in F \right\} \leq \int_{K_m - B_m} f_m \, d\mu \leq \varepsilon \,,$$

which contradicts (C7). □

Proposition 2.2. *Let P be a weakly almost periodic operator on L^1. Let $f_0 \in D$ be such that* $\operatorname{supp}(f_0) = X$, $\lim_{n\to\infty} A_n(f_0) = g$ *and* $\operatorname{supp}(g) = G$.

(i) The subspace $L^1_G = \{f \in L^1 : \operatorname{supp}(f) \subset G\} = L^1(\mu_G)$ (where μ_G is the restriction of μ to $\mathcal{B}_G = \{B \in \mathcal{B} : B \subset G\}$) is P-invariant.

(ii) The restriction P_G of P to L^1_G is a weakly almost periodic operator.

(iii) For every $f \in L^1$ the sequence $\{A_n f\}$ given by (1.18) converges strongly to a limit Af. This way we obtain a positive linear operator on L^1 that preserves the subspace L^1_G.

(iv) The mapping

$$\Phi_g(f) = f/g \tag{2.22}$$

is an isomorphism of Banach lattices $L^1(\mu_G)$ and $L^1(\mu_g)$, where μ_g is given by $d\mu_g/d\mu = g$.

(v) The operator

$$P_g = \Phi_g . P_G . \Phi_g^{-1} \tag{2.23}$$

is weakly almost periodic.

(vi) If an operator P is Markov or constrictive or quasi-constrictive, then the same is true about P_G and P_g.

(vii) If P_g is asymptotically periodic or asymptotically stable or weakly asymptotically stable, then the same is true about P_G and P.

Proof. Part (i) and (ii) follow from Corollary 2.1 (i) and (iii), part (iii) follows from Proposition 1.4.

(iv) It is obvious that the mapping Φ_g is surjective, linear and positivity preserving. We show that it is a isometry. For every $f \in L^1_G$ we have

$$\int |\Phi_g(f)| \, d\mu_g = \int |f/g| . g \, d\mu = \|f\| \,.$$

Part (v) follows directly from (iv) and (ii), parts (vi) and (vii) follow from (iv) and from (1.19), that implies that the operators P and P_G have the same constrictors. □

3. Asymptotic Periodicity of Constrictive Markov Operators

Throughout this chapter we suppose that P is a unity preserving Markov operator i.e. it preserves the characteristic function 1_X of the whole space X and that $\mu(X) < \infty$.

Proposition 3.1. *The class $\mathcal{B}_0$ of nice sets is a σ-algebra.*

Proof. Let $B \in \mathcal{B}_0$. Then for any given $n \in N$ there exists an $E \in \mathcal{B}_0$ such that $P^n 1_B = 1$. This yields $P^n 1_{X-B} = P^n(1_X - 1_B) = 1_{X-E}$, thus $X - B \in \mathcal{B}_0$. Hence $\mathcal{B}_0$ is closed with respect to complements. We show that it is closed with respect to finite unions. Let $B_1, B_2 \in \mathcal{B}_0$. Let $n \in N$ be fixed and $P^n 1_{B_i} = 1_{E_i}$, $i = 1, 2$. Applying (2.4) we get

$$\begin{aligned} 1_{E_1 \cup E_2} &\leq P^n(1_{B_1 \cup B_2}) = P^n(1_{B_1} \wedge 1_{B_2} + 1_{B_1} \wedge 1_{X-B_2} + 1_{X-B_1} \wedge 1_{B_2}) \\ &\leq 1_{E_1} \wedge 1_{E_2} + 1_{E_1} \wedge 1_{X-E_2} + 1_{X-E_1} \wedge 1_{E_2} = 1_{E_1 \cup E_2} \,. \end{aligned}$$

Therefore $\mathcal{B}_0$ is an algebra of sets. Continuity of P on L^1 implies that if $\{B_m\}$ is an increasing sequence of elements of $\mathcal{B}_0$, $B = \bigcup_{m=1}^{\infty} B_m$, $n \in N$ is fixed, $P^n(1_{B_m}) = 1_{E_m}$ for $m \in N$ and $E = \bigcup_{m=1}^{\infty} E_m$, then $P^n(1_B) = 1_E$. Hence $B \in \mathcal{B}_0$. □

Proof of Proposition 1.5. Weak precompactness of all trajectories $\{P^n f\}$, $f \in L^1$ is a direct consequence of Proposition 1.4 (ii). From the fact that the operator P is unity preserving we conclude that L^∞ is a P-invariant subset of L. Applying the Riesz convexity theorem we obtain that L^2 is a P-invariant subspace of L^1 and that the restriction of P to L^2 is a contraction with respect to the norm $\|.\|_2$ of L^2. Let U be the dual operator to P (defined on L^∞). From the fact that P is a unity preserving operator we conclude that U preserves the L^1 norm of the functions from L^∞ and can be extended to a Markov operator on L^1. Moreover, the sequence $\{U^n P^n\}$ restricted to L^2 is nonincreasing in the natural ordering of positive semidefinite operators on L^2. Therefore, it converges in the weak operator topology to a symmetric positive definite operator W satisfying

$$\|f\|_2^2 \geq (Wf, f) = \lim_{n\to\infty} \|P^n f\|_2^2 = \|W^{1/2} f\|_2^2 \geq \|Wf\|_2^2 \tag{3.1}$$

where $W^{1/2}$ is a symmetric positive semidefinite square root of W (cf. [27]).

From (3.1) we easily obtain

$$W = U^n W P^n \quad \text{for any} \quad n \in N \tag{3.2}$$

and

$$\|(I-W)P^n f\|_2^2 = \|P^n f\|_2^2 - 2.(Wf, f) + \|Wf\|_2^2 \tag{3.3}$$

for $f \in L^2$ and $n \in N$. This implies that

$$\lim_{n\to\infty} \|(I-W)P^n f\|_2 = 0 \quad \text{for every } f \in L^2 \ . \tag{3.4}$$

For every $f \in L^2$ the trajectory $\{P^n f\}$ is weakly precompact in L^2 (cf. [28]). If g is its given cluster point then it n is a weak limit of some subsequence $\{P^{n_k} f\}$. The image $(I-W)g$ is a weak limit of the sequence $\{(I-W)P^{n_k} f\}_{k\in N}$ which converges strongly to 0. Therefore $g \in \mathrm{Ker}(I-W)$.

Next we show that the subspace $\mathrm{Ker}(I-W)$ consists of functions that are measurable with respect to $\mathcal{B}_0$. Since the unity 1_X belongs to $\mathrm{Ker}(I-W)$, for every $f \in \mathrm{Ker}(I-W)$ and $c \in R$ we have $f - c.1_X \in \mathrm{Ker}(I-W)$.

Consider any fixed $f \in \mathrm{Ker}(I-W)$, $f = f^+ - f^-$, where $f^+, f^- \in L^2_+$. For any given $n \in N$ we have

$$\begin{aligned}\|f\|_2^2 &= \|P^n f\|_2^2 = \|P^n f^+\|_2^2 + \|P^n f^-\|_2^2 - 2(P^n f^+, P^n f^-) \\ &\le \|f^+\|_2^2 + \|f^-\|_2^2 - 2(P^n f^+, P^n f^-) = \|f\|_2^2 - 2.\int P^n(f^+).P^n(f^-)\, d\mu \ .\end{aligned}$$

Hence $P^n(f^+).P^n(f^-) = 0$ a.e., which means that the functions $P^n(f^+)$ and $P^n(f^-)$ have disjoint support for every $n \in N$. Using the same arguments we obtain that for every fixed $c \in R$ and $n \in N$ the functions $P^n[(f-c)^+]$ and $P^n[(f-c)^-]$ have disjoint supports.

Put $N_f = \{c \in R : \mu[f^{-1}(c)] = 0\}$.

For any given $c \in N_f$ put $h_1 = 1_{f^{-1}(-\infty,c)}$, $h_2 = 1_X - h_1$. We have

$$\mathrm{supp}(h_1) = \mathrm{supp}[(f-c)^-] \quad \text{and} \quad \mathrm{supp}(h_2) = \mathrm{supp}[(f-c)^+] \ .$$

According to Corollary 2.1 (i) for every $n \in N$ the functions $P^n h_1, P^n h_2$ have disjoint supports. However, $h_1 + h_2 = 1_X$. Therefore $P^n h_1$ is a characteristic function for every $n \in N$. Hence $f^{-1}(-\infty, c)$ is a nice set for every $c \in N_f$. It is evident that the set N_f is dense in R, which implies that f is measurable with respect to $\mathcal{B}_0$.

The fact that L^2 is dense in L^1 and P is a contraction on L^1 implies that all cluster points of the trajectories $\{P^n f\}$, $f \in L^1$ are also contained in the closed subspace of functions from L^1 that are measurable with respect to $\mathcal{B}_0$.

Let P be a quasi-constrictive Markov operator. Let the constants $\delta > 0$ and $\kappa < 1$ satisfy (1.11.a). It is easy to show that the following inequality holds

$$\mu(B) \ge \delta \quad \text{for } B \in \mathcal{B}_0 \ . \tag{3.5}$$

Suppose that $B \in \mathcal{B}_0$ is such that $\mu(B) < \delta$. For every $n \in N$ there exists $E_n \in \mathcal{B}_0$ such that $P^n 1_B = 1_{E_n}$, $\mu(E_n) = \mu(B) < \delta$ for $n \in N$. On the other hand $f = 1_B/\mu(B) \in D$ and

$$\limsup_n \left[\sup\left\{\int_E P^n f\, d\mu : \mu(E) \le \delta\right\}\right] \ge \limsup_n \int_{E_n} P^n f\, d\mu = 1 \ ,$$

which obviously contradicts (1.11.a). Therefore, $\mathcal{B}_0$ is atomic and the number of its atoms is not greater than $\delta^{-1}.\mu(X)$.

Let $C_1, \ldots, C_a$ be atoms of $\mathcal{B}_0$. There exist $B_1, \ldots, B_a \in \mathcal{B}_0$ such that $P(1_{C_i}) = 1_{B_i}$ for $i = 1, \ldots, a$. Moreover,

$$\sum_{i=1}^{a} 1_{B_i} = P\left(\sum_{i=1}^{a} 1_{C_i}\right) = 1_X ,$$

which means that the sets B_i, $i = 1, \ldots, a$, are pairwise disjoint. On the other hand, they all belong to $\mathcal{B}_0$, hence they can be expressed as unions of atoms $C_1, \ldots, C_a$. Therefore, in every B_i there exists exactly one atom $C_{\alpha(i)}$ equal to B_i. In this way we obtain a permutation α of the set $\{1, \ldots, a\}$ and an integer $q \leq a!$ such that

$$P1_{C_i} = 1_{C_{\alpha(i)}} \quad \text{and} \quad \alpha^q(i) = i,\ i = 1, \ldots, a . \tag{3.6}$$

According to Corollary 2.1 (i) the operator $R = P^q$ preserves the subspaces

$$L_i = \{f \in L^1 : \operatorname{supp}(f) \subset C_i\} . \tag{3.7}$$

We show that the restrictions of R to the subspaces L_i, $i = 1, \ldots, a$ are mixing.

Let $i \in \{1, \ldots, a\}$ be fixed. Let $f \in L_i \cap D$. All weak cluster points of $\{R_i^n\}_{n \in N}$ are $\mathcal{B}_0$-measurable and in L_i. As C_i is an atom of $\mathcal{B}_0$, they are multiples of C_i. Moreover, they belong to D. Therefore, $R^n f$ converges weakly to $1_{C_i}/\mu(C_i)$. This implies, that for every $f \in L$ the sequence $\{R^n f\}$ converges weakly to the multiple $\gamma_i(f).1_{C_i}$ of the function 1_{C_i}, where

$$\gamma_i(f) = \int f\, d\mu/\mu(C_i) . \tag{3.8}$$

Hence the operators R_i are mixing. □

Proof of Proposition 1.3 for Markov Operators. Suppose that P is a unity preserving Markov operator. We show that if P has a lower set then the σ-algebra $\mathcal{B}_0$ of nice sets is trivial. Let there exist $C \in \mathcal{B}_0$ such that $0 < \mu(C) < \mu(X)$. Let B be a lower set for P. From (1.17) we obtain

$$0 = \lim_{n \to \infty} \mu[B - \operatorname{supp}(P^n(1_C)] = \lim_{n \to \infty} \mu[B \cap \operatorname{supp}(P^n 1_{X-C})]$$

as well as

$$0 = \lim_{n \to \infty} \mu[B - \operatorname{supp}(P^n 1_{X-C})] ,$$

which implies $\mu(B) = 0$. But this is a contradiction. Hence the operator P is mixing i.e. weakly asymptotically stable. □

Proof of Theorem 3. First we present two simple auxiliary results. For every $f_1, \ldots, f_k$, $h_1, \ldots, h_k$ the inequality

$$[f_1 \wedge \cdots \wedge f_k - h_1 \wedge \cdots \wedge h_k]^+ \leq \sum_{i=1}^{k} [f_i - h_i]^+ \tag{3.9}$$

evidently holds. Moreover, if $f_i \in L^\infty$, $\|f_i\|_\infty \leq 1$ for $i = 1, \ldots, k$, then

$$f_1 \wedge \cdots \wedge f_k \geq \prod_{i=1}^{k} f_i \ . \tag{3.10}$$

The fact that L^∞ is dense in L^1 evidently yields that there exists an $f \in L^\infty$ such that $\int f \, d\mu = 0$, $\|f\|_\infty < 1$ and $\{P^n f\}$ converges to 0 weakly but not strongly (otherwise P is exact). Put

$$M_1 = \lim_{n \to \infty} \|P^n f\|/2 \ .$$

For every $n \in N$ we have

$$\|(P^n f)^+\| = \|(P^n f)^-\| = \|P^n f\|/2 \geq M_1 = \lim_{n \to \infty} \|(P^n f)^+\| \leq 1 \ .$$

For given numbers $k \in N$ and $\rho > 0$ put

$$b = \rho . M_1^{k-1}/2k, \quad c = 2^{1/(k-1)} \ .$$

The sequence $\{\|P^n f\|\}$ is nonincreasing, thus there exists n_0 such that for all $n \in N$ the inequality

$$M_1 \leq \|(P^{n_0+n} f)^+\| = \|(P^{n_0+n} f)^-\| \leq M_1.(1+b) \tag{3.11}$$

holds for all $n \in N$.

Put

$$h = P^{n_0} f \ . \tag{3.12}$$

From (2.5) and (3.11) we obtain that for every $s \in S = \{+, -\}$ and $n \in N$ we have

$$P^n(h^s) \geq (P^{n_0+n} f)^s \tag{3.13}$$

and

$$\|P^n(h^s) - (P^{n_0+n} f)^s\| = \|P^n(h^s)\| - \|(P^{n_0+n} f)^s\| \leq b.M_1 \ . \tag{3.14}$$

The fact that P is a mixing operator on L^1 implies that for every $s_0, s_1 \in S$ we have

$$\lim_{n \to \infty} (P^n h^{s_1}, h^{s_0}) = \|h^{s_0}\|.\|h^{s_1}\| \geq M_1^2 \ .$$

Hence there exists an $n_1 \in N$, such that for every $s = (s_0, s_1) \in S^2$ the inequalities

$$\int (P^{n_1} h^{s_1}).h^{s_0} \, d\mu \geq M_1^2/c \tag{3.15}$$

hold. Combining this with (3.10) we obtain

$$\|(P^{n_1} h^{s_1}) \wedge h^{s_0}\| \geq M_1^2/c \ . \tag{3.16}$$

Repeating this process inductively, we get that for every $j = 2, \ldots, k-1$ there exists an $n_j \in N$ such that for every $s \in S^{j+1}$ the inequalities

$$M_1^{j+1}/c^j \leq \|h^{s_0} \wedge P^{n_1} h^{s_1} \wedge \cdots \wedge P^{n_j} h^{s_j}\| \tag{3.17}$$

hold.

If we now put

$$g_s = h^{s_0} \wedge P^{n_1} h^{s_1} \wedge \cdots \wedge P^{n_{k-1}} h^{s_{k-1}} \quad \text{for } s = (s_0, \ldots, s_{k-1}) \in S^k \tag{3.18}$$

we get from (3.17) that

$$\|g_s\| \geq M_1^k/2 \tag{3.19}$$

holds for all $s \in S^k$.

Now we show that the trajectories of the functions $g_s : s \in S^k$ are ρ-orthogonal. For any $n \in N$ and $s \in S^k$ we put

$$\sigma_{n,s} = (P^{n_0+n} f)^{s_0} \wedge (P^{m_1+n} f)^{s_1} \wedge \cdots \wedge (P^{m_{k-1}+n} f)^{s_{k-1}} \tag{3.20}$$

where $m_i = n_0 + n_i$ for $i = 1, \ldots, k-1$.

For any fixed n $\in$ N the sets

$$B_{n,s} = \operatorname{supp}(\sigma_{n,s}) : s \in S^k \tag{3.21}$$

are obviously disjoint.

Moreover, applying (2.4) we conclude that the inequalities

$$P^n(g_s) \leq P^n h^{s_0} \wedge P^{n+n_1} h^{s_1} \wedge \cdots \wedge P^{n+n_{k-1}} h^{s_{k-1}} \tag{3.22}$$

hold for all $n \in N$ and $s \in S^k$.

Applying (3.22), (3.9) and (3.14) we get

$$\begin{aligned}\int_{X-B_{n,s}} P^n g_s \, d\mu &\leq \int_{X-B_{n,s}} [P^n h^{s_0} \wedge P^{n+n_1} h^{s_1} \wedge \cdots \wedge P^{n+n_{k-1}} h^{s_{k-1}} - \sigma_{n,s}]^+ d\mu \\ &\leq \sum_{i=0}^{k-1} \int [P^{n+n_i} h^{s_i} - (P^{n+m_i} f)^{s_i}]^+ d\mu \leq k.b.M_1 .\end{aligned}$$

Combining this with (3.19) we get that $g_s : s \in S^k$ satisfy the inequalities

$$\int_{X-B_{n,s}} P^n g_s \, d\mu \leq \rho.\|g_s\| = \rho.\|P^n g_s\| \quad \text{for } n \in N . \tag{3.23}$$

Hence they are ρ-orthogonal. □

Proof of Proposition 1.6. Let P be a mixing and quasi-constrictive Markov operator that is not exact. Let the constants $\kappa < 1$ and $\delta > 0$ satisfy (1.11.a). Put

$$\rho = (1-\kappa)/2 \quad \text{and} \quad k = \log_2[\mu(X)/\delta] + 1 .$$

Let $S = \{+, -\}$. Let the functions $g_s : s \in S^k$ and the sets $B_{n,s} : n \in N$, $s \in S^k$ be defined as in the previous proof. Since for any given $n \in N$ the sets

$B_{n,s} : s \in S^k$ are disjoint, there exists $s \in S^k$ such that $\mu(B_{n,s}) < \delta$ for infinitely many S. Applying (1.11.a) we get

$$\liminf \int_{B_{n,s}} P^n g_s \, d\mu < \kappa . \|g_s\| \ . \tag{3.24}$$

Combining this with (3.23) we get

$$\begin{aligned} \liminf \|P^n g_s\| &= \liminf \left[\int_{B_{n,s}} P^n g_s \, d\mu + \int_{X - B_{n,s}} P^n g_s \, d\mu \right] \\ &\leq (\kappa + \rho) . \|g_s\| < \|g_s\| \ , \end{aligned}$$

which contradicts the fact that P is a Markov operator. □

Corollary 3.1. *A constrictive Markov operator is asymptotically periodic (i.e. Theorem 1 is true for Markov operators). Moreover, there exist densities $g_1, \dots, g_a$ with disjoint supports, satifying (1.8). The convex envelope*

$$F_0 = \mathrm{co}\{g_1, \dots, g_a\} = \left\{ \sum_{i=1}^{a} \lambda_i g_i : 0 \leq \lambda_i, \ \sum_{i=1}^{a} \lambda_i = 1 \right\} \tag{3.25}$$

is the smallest constrictor for P.

Proof. Let P be a constrictive unity preserving Markov operator and $C_1, \dots, C_a$ be atoms of $\mathcal{B}_0$. For any $f \in L^1$ we have

$$f = \sum_{i=1}^{a} 1_{C_i} . f \ .$$

According to Theorem 3 there exist strong limits

$$\lim_{n \to \infty} R^n (1_{C_i} . f) = \gamma_i(f) . 1_{C_i} \quad (i = 1, \dots, a)$$

where $q \in N$ satisfies (3.6), $R = P^q$ and positive linear functionals γ_i are defined by (3.8). For $i = 1, \dots, a$ put

$$\bar{g}_i(f) = 1_{C_i} / \mu(C_i), \quad \lambda_i(f) = \gamma_i(f) . \mu(C_i) = \int_{C_i} f \, d\mu \tag{3.26}$$

and

$$\bar{Q}(f) = \sum_{i=1}^{a} \gamma_i(f) . 1_{C_i} = \sum_{i=1}^{a} \lambda_i(f) . \bar{g}_i \ . \tag{3.27}$$

Obviously

$$\sum_{i=1}^{a} \lambda_i(f) = 1 \quad \text{holds for } f \in D \ . \tag{3.28}$$

For any $n \in N$ we have

$$\|R^n f - \bar{Q}f\| = \sum_{i=1}^{a} \|R^n(1_{C_i}.f) - \gamma_i(f).1_{C_i}\|$$

hence

$$\lim_{n\to\infty} \|R^n f - \bar{Q}f\| = 0 \ , \tag{3.29}$$

and

$$\lim_{n\to\infty} \|P^n(f - \bar{Q}f)\| \le M. \lim_{n\to\infty} \|R^n(f - \bar{Q}f)\| = 0 \ . \tag{3.30}$$

Therefore P has a constrictor formed by the convex envelope of invariant densities.

The proof for arbitrary Markov operator follows from Proposition 2.2 and from the fact that the isomorphism Φ_g given by (2.22) satisfies

$$\mathrm{supp}(\Phi_g(f)) = \mathrm{supp}(f) \ . \tag{3.31}$$

□

4. Weakly Almost Periodic Operators

The aim of this chapter is to prove Theorems 4 and 5 that provide very weak sufficient conditions for weak almost periodicity of positive operators. We follow the ideas of Straube [26] and Socala [22], where the canonical immersion of L^1 to its second dual $Z = (L^\infty)'$ was utilized.

Recall that L^1 is dense in Z [28], thus the second dual P'' of P can be obtained as the unique extension of P to Z. Obviously, Z is a Banach lattice containing L^1 (with the partial ordering $\le$) as a sublattice and P'' is a lattice homomorphism.

We shall make substantial use of the fact that the unit ball of Z is compact in the w^*-topology which is the smallest topology that makes all mapping $z \to z(y)$ for $y \in L^\infty$ continuous.

Proof of Theorem 4. For any fixed $f \in D$ the sequence $\{A_n(f)\}$, given by (1.18) is precompact in the w^*-topology. Any cluster point z of this sequence obviously satisfies

$$z \ge 0, \quad P''z = z \ . \tag{4.1}$$

Moreover, z determines a finitely additive measure z on $\mathcal{B}$ given by

$$\mu_z(B) = z(1_B) \ , \tag{4.2}$$

that evidently satisfies

$$\liminf \int_B A_n(f)\, d\mu \le \mu_z(B) \le \limsup \int_B A_n(f)\, d\mu \tag{4.3}$$

and thus also

$$\mu_z(B) = 0 \quad \text{if} \quad \mu(B) = 0 \ . \tag{4.4}$$

Using the Yosida-Hewett decomposition we can uniquely express the measure μ_z as a sum

$$\mu_z = \mu_0 + \mu_a \tag{4.5}$$

where μ_0 is σ-additive and μ_a is purely additive [29].

Moreover, (4.4) implies that μ_0 is absolutely continuous with respect to μ, hence $\mu_0 = \mu_g$ for some $g \in L^1$.

Recall, that μ_0 is the maximum of all σ-additive measures that are majorized by μ_z in Z [29]. Hence the implication

$$f \in L^1, \quad f \leq z \Rightarrow f \leq g \tag{4.6}$$

holds true.

We have

$$Pg = P''g \leq P''z = z\ , \tag{4.7}$$

which yields

$$Pg \leq g\ . \tag{4.8}$$

Now we prove that if f satisfies (1.21) then $\|g\| > 0$, hence $\|g\|^{-1}.g$ is a P-invariant density. It is well known [25] that any purely additive measure (hence also μ_a) has the following strange property. There exists a decreasing sequence $\{B_m\}$ such that

$$\lim_{m\to\infty} \mu(B_m) = 0 \quad \text{and} \quad \mu_a(B_m) = \mu_a(X) \quad \text{for all } m \in N\ . \tag{4.9}$$

Put $B_0 = \bigcap_{m=1}^{\infty} B_m$. From (4.4) we get that $\mu_a(B_0) = 0$.

Replacing B_m by $B_m - B_0$ we can supplement (4.9) by

$$\bigcap_{m=1}^{\infty} B_m = \emptyset\ . \tag{4.10}$$

From (4.5) and (4.9) we obtain

$$\mu_g(X) = \mu_z(X) - \mu_a(X) = \mu_z(X) - \mu_a(B_m) \geq \mu_z(X - B_m) \quad \text{for all } m \in N\ . \tag{4.11}$$

Combining this with (4.3) and (1.21) we obtain

$$\|g\| = \mu_g(X) \geq \lim_{m\to\infty} \left[\liminf_{n} \int_{X-B_m} A_n(f)\, d\mu\right] > 0\ . \tag{4.12}$$

From (4.8) we get that the sequence $\{P^n g\}$ is nonincreasing and has a P-invariant limit $g_0 \leq g$. If $c = \|g - g_0\| > 0$, then the density $h = (g - g_0)/c$ satisfies $\lim_{n\to\infty} \|P^n h\| = 0$, which contradicts (1.6). Therefore $Pg = g$ and $g/\|g\|$ is a P-invariant density. □

Remark 4.1. Let $Pg = g \in D$ and $G = \text{supp}(g)$. According to Proposition 1.4 and 2.2 the restriction P_G of P to the subspace $L^1_G = \{f \in L^1 : \text{supp}(f) \subset G\}$ is weakly almost periodic. Moreover, the sequences $\{A_n(f)\}$ converge strongly for $f \in L^1_G$ and determine an operator A on L^1_G.

Proposition 4.1. *Let P be an operator that has an invariant density.*

(i) There exists a P-invariant density g_0 with maximal support. That means that for any P-invariant density g the inclusion

$$\mathrm{supp}(g) \subset G = \mathrm{supp}(g_0) \tag{4.13}$$

holds.

(ii) For any $f \in L^1$ the sequence $\{A(1_G.P^n f)\}$ converges strongly and determines a P-invariant function $A(f)$ satisfying

$$A(f) = \lim_{n\to\infty} A(1_G.P^n f) = \lim_{n\to\infty} A(1_G.A_n(f)) \ . \tag{4.14}$$

(This implies that $A(f) = A(P^m f) = A(A_m(f))$ for $m \in N$.)

(iii) If $f \in L^1_+$, z is a cluster point of the sequence $\{A_n(f)\}$ (in w^*-topology of Z) and $\mu_z = \mu_g + \mu_a$ is the Yosida-Hewitt decomposition of the additive measure μ_z to σ-additive part μ_g and purely additive part μ_a, then the equality

$$g = A(f) \tag{4.15}$$

holds. Hence g does not depend on the choice of z.

Proof. (i) Let $f_0 \in D$ be such that $\mathrm{supp}(f_0) = X$. Let the measure μ_0 be determined by

$$d\mu_0/d\mu = f_0 \ . \tag{4.16}$$

Consider the class

$$\mathcal{B}_1 = \{B \in \mathcal{B} : B = \mathrm{supp}(h),\ h \in D,\ Ph = h\}$$

and put

$$M_0 = \sup\{\mu_0(B) : B \in \mathcal{B}_1\} \ .$$

Choose a sequence $h_n \in D$ such that $Ph_n = h_n$ and

$$\lim_{n\to\infty} \mu_0(\mathrm{supp}(h_n)) = M_0 \ .$$

Put

$$g_0 = \sum_{n=1}^{\infty} 2^{-n}.h_n, \quad G = \mathrm{supp}(g_0) \ .$$

Obviously $\mu_0(G) = M_0$. If $g \in D$, $Pg = g$ and $\mu(\mathrm{supp}(g) - G) > 0$ then for

$$g_1 = (g + g_0)/2$$

we have

$$Pg_1 = g_1, \quad \mathrm{supp}(g_1) = G \cup \mathrm{supp}(g) \in \mathcal{B}_1, \quad \mu_0(\mathrm{supp}(g_1)) > \mu_0(G) = M_0 \ ,$$

which is a contradiction. Therefore (4.13) holds.

(ii) According to Corollary 2.1 (i) the subspace

$$L^1_G = \{f \in L^1 : \mathrm{supp}(f) \subset G\} \tag{4.17}$$

is P-invariant. Therefore, for every fixed $f \in L^1_+$ and $m \in N$ we have

$$P(1_G.P^m f) = 1_G.P(1_G.P^m f)$$

and thus also

$$P(1_G.P^m f) \leq 1_G.P^{m+1} f \ . \tag{4.18}$$

According to Remark 4.1 the restriction of P to L^1_G is weakly almost periodic. Hence for every fixed $f \in L^1_+$ and $m \in N$ the sequence $\{A_n(1_G.P^m f)\}$ has a strong limit $A(1_G.P^m f)$.

Furthermore, the inequality (4.18) obviously implies that $A(1_G.P^m f)$ is a bounded nondecreasing sequence of P-invariant functions. Therefore, it has a strong limit Af satisfying

$$\lim_{m\to\infty} \|A(f) - A(1_G.P^m f)\| = 0 \ . \tag{4.19}$$

The equalities (4.14.a) follow directly from (4.14).

(iii) The equality $Pg = g$ and the inclusion $\text{supp}(g) \subset G$ imply that $Ag = g$. Let A' be the dual operator to the operator A. A' is a positive linear operator on L^∞, hence for any given $h \in L^\infty_+$ we have

$$(g, h) = (Ag, h) = (g, A'h) \leq (z, A'h) \ .$$

The properties of w^*-topology imply that there exists a subsequence $\{m_k\}$ such that

$$(z, A'h) = \lim_{k\to\infty} (A_{m_k} f, A'h) = \lim_{k\to\infty} (A(A_{m_k} f), h) = (Af, h) \ .$$

This yields the inequality $g \leq Af$.

On the other hand, for any $h \in L^\infty_+$ there exists a subsequence $\{n_k\}$ such that $(z, h) = \lim_{k\to\infty}(A_{n_k} f, h)$. For any $m \in N$ we have

$$\|A_{n_k} f - A_{n_k}(P^m f)\| \leq \frac{2.m.M}{n_k} \ .$$

Hence

$$(A(1_G.P^m f), h) = \lim_{k\to\infty} A_{n_k}(1_G P f, h) \leq \lim_{k\to\infty} \left[(A_{n_k} f, h) + \frac{2.m.M}{n_k} \right] = (z, h) \ .$$

From (4.14) we get that $(Af, h) \leq (z, h)$ for $h \in L^\infty_+$. This, together with (4.6), yields the inequality $Af \leq g$. □

Proof of Theorem 5. Let P be an almost constrictive operator and let $\varepsilon < 1$ satisfy (1.15.a). Combining Proposition 4.1 (iii) with (4.12) and (1.15.a) we obtain the inequality

$$\|Af\| \geq \varepsilon.\|f\| \quad \text{for } f \in L^1_+ \ . \tag{4.20}$$

From (4.14) we get that for any $f \in L^1_+$ and $n \in N$ we have

$$\|Af - AP^n f\| = \lim_{m\to\infty} \|Af - A(1_G.P^{n+m} f)\| = 0 \ .$$

Therefore

$$\lim_{n\to\infty} \|AP^n f - A(1_G.P^n f)\| = 0 \tag{4.21}$$

holds.

Combining this with (4.20) we obtain that

$$\lim_{n\to\infty} \|P^n f - 1_G.P^n f\| = 0\ . \tag{4.22}$$

Let $f \in D$ and $\delta > 0$ be given and $m \in N$ is such that

$$\|P^m f - 1_G.P^m f\| < \delta/2M, \quad \|Af - A(1_G P^m f)\| \le \delta/2 \tag{4.23}$$

holds. Combining this with the obvious inequality

$$\|A_n f - A_n P^m f\| \le m.M/n \quad \text{for } n \in N \tag{4.24}$$

we get

$$\|A_n f - A_n(1_G.P^m f)\| \le \delta/2 + m.M/n \tag{4.25}$$

and

$$\limsup_n \|A_n f - A(1_G P^m f)\| \le \delta/2\ . \tag{4.26}$$

Finally, the second part of (4.23) yields

$$\limsup_n \|A_n f - Af\| \le \delta\ . \tag{4.27}$$

Since $\delta > 0$ was arbitrary, the sequence $\{A_n f\}$ converges strongly to Af. Applying Proposition 1.4 (ii) we obtain that P is weakly almost periodic.

Combining Propositions 4.1, 2.2, 1.5 and 1.6 we conclude that Theorem 2 holds for Markov operators. □

At the end of this chapter we present an example of operator that was obtained as a result of an unsuccessful attempt to find a weaker version of condition (1.15) for weak almost periodicity of Markov operators.

Example 4.1. Let $X = Z_+$, $\mathcal{B} = 2^X$ and $\mu(\{n\}) = 1$ for $n \in X$. Put

$$e_k(n) = \begin{cases} 1 & \text{for } k = n \\ 0 & \text{for } k \ne n \end{cases} \quad \text{for } k, n \in Z_+ \tag{4.28}$$

and

$$P(e_0) = e_0,\ P(e_k) = 2^{-k}.e_0 + (1 - 2^{-k}).e_{k+1} \quad \text{for } k \ge 1\ . \tag{4.29}$$

For any fixed $k \in N$, $f = e_k$ and arbitrary $n \in N$ we have

$$P^n f(n+k) = \prod_{i=k}^{n+k-1} (1 - 2^{-i})$$

thus

$$(1-2^{-k}) \geq P^n f(n+k) \geq (1-2^{-k}).\left(1 - \sum_{i=k+1}^{\infty} 2^{-i}\right) = (1-2^{-k})^2 \geq 1-2^{-k+1}.$$

Moreover $P^n f(m) = 0$ for $0 \neq m \neq n+k$. Hence $2^{-k} \leq P^n f(0) \leq 2^{-k+1}$.

The sequences $\{P^n f\}$ and $\{A_n f\}$ converge in the second dual $Z = [L^\infty(\mu)]'$ to the element z that determines an additive measure μ_z on $\mathcal{B}$ satisfying

$$\mu_z(X) = 1, \quad 2^{-k} \leq \mu_z(\{0\}) \leq 2^{k+1}$$

and

$$\mu_z(\{m\}) = 0, \quad 1 - 2^{-k+1} \geq \mu_z(\{m, \infty\}) \geq 1 - 2^{-k}.$$

Hence the operator P does not satisfy the condition (1.15.a). The weaker condition

$$\lim_{m\to\infty} \left[\liminf_{n} \int_{X-B_m} P^n f \, d\mu\right] > 0 \quad \text{for } f \in D,\ \{B_m\} \to \emptyset \tag{4.29}$$

does not imply weak almost periodicity.

5. Asymptotic Periodicity of Power Bounded Operators

In this chapter we complete the proof of Theorems 1 and 2. Let P be a constrictive (or quasi-constrictive) operator. According to Proposition 1.4 and Theorem 5 there exist a P-invariant density g and a power bounded operator P_g on $L^1(\mu_g)$ such that P_g is unity preserving and constrictive (or quasi-constrictive). Moreover, if P_g is asymptotically periodic or asymptotically stable then the same is true about P. Therefore, it suffices to restrict again our attention to the case when P is a unity preserving constrictive (or quasi-constrictive) operator.

Applying Sucheston Decomposition (see e.g. [12, Th. 3.1]) we can split the space X into two disjoint subsets X_0 and Y such that

$$f \in L^1 \quad \text{and} \quad \text{supp} f \subset X_0 \Rightarrow \text{supp} Pf \subset X_0 \quad \text{and} \quad \lim_{n\to\infty} \|P^n f\| = 0 \tag{5.1}$$

and

$$f \in L^1_+ \quad \text{and} \quad \text{supp} f \subset Y, \quad \|f\| > 0 \Rightarrow \liminf \|P^n f\| > 0 \tag{5.2}$$

hold.

If $\mu(Y) = 0$ then P is trivial. Further we suppose that $\mu(Y) > 0$.

Consider the operator T defined on

$$L^1_Y = \{f \in L^1 : \text{supp} f \subset Y\}$$

by

$$Tf = 1_Y . Pf . \tag{5.3}$$

Put

$$L^1_{Y,+} = L^1_Y \cap L^1_+ .$$

From (5.1) we easily obtain that the operator T satisfies

$$T^n f = 1_Y . P^n f \quad \text{for } f \in L^1_Y \quad \text{and} \quad n \in N \ , \tag{5.4}$$

as well as

$$\lim_{n \to \infty} \|P^n f - P^{n-m} T^m f\| = 0 \quad \text{for } f \in L^1_Y \quad \text{and} \quad m \in N \ . \tag{5.5}$$

Therefore, for every $f \in L^1_{Y,+}$, $\|f\| > 0$, we have

$$\liminf \|T^n f\| \geq M^{-1} . \liminf \|P^n f\| > 0 \tag{5.6}$$

where M is a constant satisfying (1.5).

Proposition 5.1. *(i) There exists $h \in L^1_{Y,+}$ such that*

$$h \leq M, \quad \text{supp} h = Y \tag{5.7}$$

and the operator T can be uniquely extended to a Markov operator on $L^1(\mu_h)$.

(ii) If P is quasi-constrictive and $\varepsilon > 0$ is a constant satisfying (1.14.a) then

$$h \geq \varepsilon \ a.e. \tag{5.8}$$

This clearly yields that P is asymptotically periodic.

(iii) If P is constrictive then the same is true about the operator T on $L^1(\mu_h)$. The resulting asymptotic periodicity of T on $L^1(\mu_h)$ yields the same property of P on $L^1(\mu)$.

Proof. (i) Let $U = T'$ be the dual operator that is defined on $L^\infty_Y = L^1_Y \cap L^\infty$. Since T is a unity preserving operator on L^1_Y, the operator U preserves the L^1 norm of nonnegative elements of L^∞_Y, thus it can be extended to a Markov operator on L^1_Y. Moreover, (1.5) implies the inequality

$$U^n 1_Y \leq M . 1_Y \text{ a.e. for } n \in N \ . \tag{5.9}$$

Hence the sequences $U^n 1_Y$ and $A'_n 1_Y = n^{-1} . \sum_{i=0}^{n-1} U^i 1_Y$ are weakly $i = 0$ precompact in L^1_Y. According to Mean Ergodic Theorem [5] there exists a function $h \in L^1_{Y,+}$ such that

$$\lim_{n \to \infty} \|A'_n 1_Y - h\| = 0 \ . \tag{5.10}$$

The inequality (5.9) implies

$$h \leq M . 1_Y \text{ a.e.} \tag{5.11}$$

and thus also

$$L^1_Y \subset L^1(\mu_h) \ . \tag{5.12}$$

Let $f \in L^1_{Y,+}$, $\|f\| > 0$ be a given function. According to Proposition 1.4 (ii) the trajectory $T^n f$ is weakly precompact, thus the sequence

$$A_n f = n^{-1} . \sum_{i=0}^{n-1} T^i f$$

converges strongly to a function $Af \in L^1_{Y,+}$.

We have

$$0 < \liminf \|T^n f\| \le \lim_{n\to\infty} \|A_n f\| = \|Af\| \ . \tag{5.13}$$

On the other hand

$$\|Af\| = \lim_{n\to\infty} (A_n f, 1_Y) = \lim_{n\to\infty} (f, A'_n 1_Y) = (f, h) \tag{5.14}$$

which yields

$$\int f \, d\mu_h > 0 \quad \text{for } f \in L^1_Y, \ \|f\| > 0 \ . \tag{5.15}$$

Putting $f = 1_B$, $\mu(B) > 0$ we get $0 < \int_B h \, d\mu$. This evidently yields

$$\text{supp} h = Y \quad \text{and} \quad L^\infty_Y = L^\infty(\mu_h) \ . \tag{5.16}$$

This together with (5.11) implies that L^1_Y is a dense subspace of $L^1(\mu_h)$. For arbitrary $f \in L^1_{Y,+}$ we have

$$\int Tf \, d\mu_h = (Tf, h) = (f, Uh) = (f, h) = \int f \, d\mu_h \ .$$

Therefore, T can be uniquely extended to a Markov operator on $L^1(\mu_h)$.

(ii) Let P be a quasi-constrictive operator. The inequality (1.14.a) evidently implies that P is strictly nontrivial, hence $Y = X$ and $P = T$. From (5.10) and (1.14.a) we obtain

$$\int_B h \, d\mu \ge \liminf \int P^n 1_B \, d\mu > \varepsilon . \mu(B) \ , \tag{5.17}$$

which yields (5.8).

Furthermore, the inequality (5.8) implies that the operator P on $L^1(\mu_h)$ satisfies the inequality (1.14.a) with ε replaced by ε^2/M. Therefore, P is a quasi-constrictive Markov operator on $L^1(\mu_h)$. According to Corollary 3.1 P is asymptotically periodic on $L^1(\mu_h)$. This means that the convergence in (1.8) holds true in $L^1(\mu_h)$. Combining this with the inequality (5.7) we obtain that P is asymptotically periodic on $L^1(\mu)$.

(iii) Now we show that if P is a constrictive operator on L^1, then T is a constrictive Markov operator on $L^1(\mu_h)$. Let F be a weakly compact constrictor for P. The set

$$F_Y = \{1_Y . f : f \in F\}$$

is obviously a weakly compact constrictor for T on L^1_Y.

The inequality (5.7) implies that F_Y is a weakly precompact subset of $L^1(\mu_h)$ and the same clearly holds for the subset

$$F_1 = \{f \in L^1_{Y,+} : f \le g \in F_Y\} \ .$$

We show that F_1 is a constrictor for T on $L^1(\mu_h)$.

Let $f \in L^1_Y$ be a density in $L^1(\mu_h)$. From (5.2), (5.13) and (5.14) we obtain that

$$0 < \liminf \|T^n f\| \leq \|Af\| = \int f \, d\mu_h = 1 \ .$$

Let $\varepsilon > 0$ be arbitrary. There exists $m \in N$ such that

$$\|T^m f\| < 1 + \varepsilon / M^2 \ ,$$

where M is the constant satisfying (1.5).

Put $f_0 = T^m f / \|T^m f\|$. We have $\|[T^m f - f_0]^+\| \leq \varepsilon / M^2$ and thus also $\|[T^{n+m} f - T^n f_0]^+\| \leq \varepsilon / M$ and $\int [T^{n+m} f - T^n f_0]^+ d\mu_h < \varepsilon$ for $n \in N$.

There exists a sequence g_n in F_Y such that

$$\lim_{n\to\infty} \|T^n f_0 - g_n\| = 0$$

and thus also $\lim_{n\to\infty} \int [T^n f_0 - g_n]^+ d\mu_h = 0$ as well as

$$\limsup_{n\to\infty} \int [T^{n+m} f - g_n]^+ d\mu_h < \varepsilon \ .$$

Since $\varepsilon > 0$ was arbitrary we conclude that $\lim_{n\to\infty} d(T^n f, F_1) = 0$. Hence T is a constrictive unity preserving Markov operator on $L^1(\mu_h)$. Consider the pairwise orthogonal bounded densities $\bar{g}_1, \ldots, \bar{g}_a$ and positive linear functionals $\lambda_1, \ldots, \lambda_a$ given by (3.26) that satisfy (1.8) for the operator T on $L^1(\mu_h)$. We show that they satisfy (1.8) for T on L^1_Y as well. Let $f \in L^1_Y$ and $\bar{Q}f$ be given by (3.27). The sequence

$$T^n(f - \bar{Q}f) \tag{5.18}$$

converges to 0 in the norm of $L^1(\mu_h)$. We show that this convergence holds also in the norm of L^1_Y. According to (5.16) the sequence $B_m = \{x : h(x) < m^{-1}\}$ converges to $\emptyset$.

Let $\varepsilon > 0$ be given. The fact that the sequence (5.18) is weakly precompact in L^1 implies that there exists $m \in N$ such that

$$\sup_n \int_{B_m} |T^n(f - Qf)| \, d\mu < \varepsilon \ . \tag{5.19}$$

We have

$$\limsup_n \int_X |T^n(f - \bar{Q}f)| \, d\mu \leq \limsup_n \int_{B_m} |T^n(f - \bar{Q}f)| \, d\mu$$
$$+ m. \lim_{n\to\infty} \int_X |T^n(f - \bar{Q}f)| \, d\mu_h < \varepsilon \ .$$

Since $\varepsilon > 0$ was arbitrary, $T^n(\bar{f} - Qf)$ converges strongly to 0 in L^1. Since (1.8.a), (1.8.b), (1.8.c) and (1.5) obviously imply (1.8) the operator T is asymptotically periodic on L^1_Y with a period $q \leq a!$. Now we show that the same holds for the operator P on L^1.

For $i = 1, \ldots, a$ and $n \in N$ we have

$$P^q \bar{g}_i \geq T^q \bar{g}_i = \bar{g}_i \ .$$

Therefore, the sequences $\{P^{n.q}\bar{g}_i\}_{n=1}^{\infty}$, $i = 1, \ldots, a$, are nondecreasing and bounded. Hence they converge strongly to the limits

$$g_i = \lim_{n\to\infty} P^{n.q}\bar{g}_i \ . \tag{5.20}$$

We have

$$1_Y.g_i = \lim_{n\to\infty} 1_Y.P^{n.q}\bar{g}_i = \lim_{n\to\infty} T^{n.q}\bar{g}_i = \bar{g}_i, \quad 1 \le i \le a \tag{5.21}$$

Therefore,

$$Y \cap \mathrm{supp} g_i \cap \mathrm{supp} g_j = \mathrm{supp}\bar{g}_i \cap \mathrm{supp}\bar{g}_j = \emptyset \quad \text{for } i \neq j \ . \tag{5.22}$$

From (5.1), (5.5) and (5.20) we get that

$$\begin{aligned} P^q g_i &= \lim_{n\to\infty} P^{n.q}P^q\bar{g}_i = \lim_{n\to\infty} P^{n.q}T^q\bar{g}_i + \lim_{n\to\infty} P^{n.q}(1_{X_0}.P^q\bar{g}_i) \\ &= \lim_{n\to\infty} P^{n.q}\bar{g}_i = g_i \quad \text{for } i = 1, \ldots, a \ . \end{aligned}$$

For any $f \in L^1$ put

$$Qf = \sum_{i=1}^{a} \lambda_i(1_Y.f).g_i \ .$$

Obviously $P^q(Qf) = Qf$; $1_Y.Qf = \bar{Q}(1_Y.f)$.

From (5.1), (5.5) and (1.5) we get that

$$\begin{aligned} \lim_{n\to\infty} \|P^{n.q}f - Qf\| &= \lim_{n\to\infty} \|P^{n.q}(1_{X_0}.(f-Qf)\| + \lim_{n\to\infty} \|P^{n.q}(1_Y.f - 1_Y.Qf)\| \\ &\le M. \lim_{n\to\infty} \|T^{n.q}(1_Y.f - \bar{Q}(1_Y f))\| = 0 \ . \end{aligned}$$

Hence P satisfies (1.8.c), thus it is asymptotically periodic. □

Proof of Proposition 1.1. The parts (i), (ii) and (iii) clearly follow from (5.22). The part (iv) is contained in Corollary 3.1. □

Proof of Proposition 1.3. Let P be a strictly nontrivial operator with a lower set B. From (5.1), (5.3) and (5.7) we obtain that $Y = X$, $T = P$ and B is a lower set for the Markov operator P on $L^1(\mu_h)$. Hence P is weakly asymptotically stable on $L^1(\mu_h)$. Finally, (5.16) yields that the same is true for P on L^1. □

Corollary 5.1. *Suppose that the operator P has a nontrivial lower-bound function h. Then P is quasi-constrictive. (It satisfies (1.14.a) with $\varepsilon = \|h\|/2$). Hence P is asymptotically periodic. Moreover it has a lower set* supp(h)*, hence it is asymptotically stable.*

6. Asymptotic Periodicity of Operators on Signed Measures

In this chapter we extend Theorems 1, 2, 4 and 5 to the class of operators on bands of signed measures.The nonnegative elements of such a band L will be denoted by L_+, the normalized elements of L_+ will be denoted by D and called probabilities. We shall use the following simple construction that enables us to utilise the results of preceding chapters.

Proposition 6.1. Let $\mu \in L_+$. Let $\{c_i\}_{i=0}^{\infty}$ be a sequence of positive real numbers such that

$$\sum_{i=0}^{\infty} c_i = 1 .$$

Put

$$\bar{\mu} = \sum_{i=0}^{\infty} c_i P^i \mu .$$

Then

$$M_{\bar{\mu}} = \{\nu : \nu \ll \bar{\mu}\}$$

is the smallest P-invariant band containing μ.

Proof. $M_{\bar{\mu}}$ is isomorphic with the Banach lattice $L^1_{\bar{\mu}}$, hence it is a band.

We show that it is P-invariant.

We have

$$P\bar{\mu} = \sum_{i=0}^{\infty} c_i P^{i+1} \mu \ll \bar{\mu} .$$

Moreover $\nu \in M_{\bar{\mu}}$ is equivalent to

$$\nu = \sup_n \{\nu \wedge n.\bar{\mu}\} .$$

We show that this implies $P\nu \in M_{\bar{\mu}}$. P clearly satisfies (2.4), hence $P(\nu \wedge n.\bar{\mu}) \leq P\nu \wedge n.P\bar{\mu}$ for any n.

Therefore,

$$0 \leq P\nu - P\nu \wedge n.P\bar{\mu} \leq P\nu - P(\nu \wedge n.\bar{\mu}) = P(\nu - \nu \wedge n.\bar{\mu})$$

and

$$\|P\nu - P\nu \wedge n.P\bar{\mu}\| \leq \|P(\nu - \nu \wedge n.\bar{\mu})\| \leq \|\nu - \nu \wedge n.\bar{\mu}\| .$$

Using Lebesgue bounded convergence theorem we get

$$P\nu = \sup_n \{P\nu \wedge n.P\bar{\mu}\}$$

hence

$$P\nu \ll P\bar{\mu} \ll \bar{\mu} .$$ □

Corollary 6.1. *Let P be a nontrivial constrictive or quasi-constrictive operator on a band L of signed measures. Then for every $\mu \in L$ there exist numbers $a \in N$,*

$q \leq a!$ and $\lambda_1, \ldots, \lambda_a \in R_+$ and periodic measures $\nu_1, \ldots, \nu_a$ (with period q) such that the periodic function

$$Q(\mu) = \sum_{i=1}^{a} \lambda_i \nu_i = \lim_{n \to \infty} P^{n.q} \mu \tag{6.1}$$

satisfies

$$\lim_{n \to \infty} \| P^n(\mu - Q(\mu)) \| = 0 \ . \tag{6.2}$$

Note that Proposition 6.1 immediately yields the announced extensions of Theorems 4 and 5.

The following result from [5, Th. IV.9.2] enables us to elaborate a simple proof of the extension of Theorem 1.

Lemma 6.1. *Let L be a band of signed measures and F be a weakly precompact subset of L. Then there exists $\mu_0 \in L_+$ such that the subband $L_0 = \{\mu \in L : \mu \ll \mu_0\}$ contains F.*

Proof of the Extension of Theorem 1. Suppose that P is a constrictive operator on L and F is a weakly compact constrictor for P. Let $L_0 \subset L$ be a P-invariant band containing F and isomorphic to $L^1(\mu_0)$ for some $\mu_0 \in L$. The restriction P_0 of P to L_0 is a constrictive positive operator. Therefore P_0 is asymptotically periodic on L_0. Hence there exist linearly independent periodic measures $\nu_1, \ldots, \nu_a \in L_0$ with a common period $q \leq a!$ such that

$$F_0 = \{\mu \in L_0 : \|\mu\| \leq M, \ \mu = \sum_{i=1}^{a} \lambda_i \nu_i; \ \lambda_1, \ldots, \lambda_a \in R_+\} \tag{6.3}$$

is a finite dimensional constrictor for P_0. It is evident that F_0 is a constrictor for the entire operator P. The number q satisfies (6.1) for all $\mu \in L$. Hence the convergence in (6.1) determines a positive operator from L to the finite dimensional linear envelope of F_0. Therefore, the coordinate maps $\lambda_i : L \to R$ are positive continuous linear functionals. □

Definition 6.1.

(i) We say that two measures μ, ν are proportional if $\nu = \gamma.\mu$ for some $\gamma \in R$, $\gamma \neq 0$.

(ii) We say that a periodic measure $\mu \in L_+$, $\mu \neq 0$ is minimal if any periodic measure $\nu \ll \mu$ is proportional to μ.

(iii) We say that $\mu = P\mu \in L_+$ is a minimal P-invariant measure if any P-invariant measure $\nu \ll \mu$ is proportional to μ.

Proposition 6.2. *Let P be a strictly nontrivial operator.*

(i) Two minimal periodic measures are either proportional or orthogonal.

(ii) Two minimal invariant measures are either proportional or orthogonal.

Proof. (i) Let $\mu, \nu \in L_+$ be minimal periodic measures and let n be their common period. Suppose that $\|\nu \wedge \mu\| > 0$. We show that $\nu \wedge \mu$ is periodic with period n. We have

$$P^n(\mu \wedge \nu) \leq P^n\mu \wedge P^n\nu = \mu \wedge \nu ,$$

hence the sequence $\{P^{n.m}(\mu \wedge \nu)\}_{m \in N}$ is nonincreasing and dominated by $\mu \wedge \nu$. Therefore, it converges to a P-invariant measure $\mu_0 \ll \mu \wedge \nu$. The sequence $\{P^{n.m}(\mu \wedge \nu - \mu_0)\}_{m \in N}$ converges to 0, hence $\mu_0 = \mu \wedge \nu = P^n(\mu \wedge \nu)$. ($P$ is strictly nontrivial.)

(ii) It suffices to repeat the above proof for $n = 1$. □

Proposition 6.3. *Let P be a quasi-constrictive power bounded operator.*

(i) Every periodic measure $\mu \in L_+$ is a linear combination of finitely many nonnegative minimal periodic measures.

(ii) If $\nu \in L_+$ is a minimal periodic measure with the period n then all images $P^k\mu : 1 \leq k \leq n-1$ are minimal periodic measures and the measure

$$A_n(\mu) = n^{-1}.\sum_{i=0}^{n-1} P^i(\mu) \tag{6.4}$$

is a minimal invariant measure.

(iii) Every invariant measure is a finite linear combination of minimal invariant measures. For every minimal invariant measure ν there exists a minimal periodic measure μ such that $\nu = A_n\mu$ for some $n \in N$.

(iv) The family of minimal invariant probabilities is finite.

Proof. (i) Let $\mu_0 \in L_+$ be a periodic measure with period n. The measure $\bar{\mu} = A_n(\mu_0)$ is invariant and the subband

$$L_0 = \{\nu \ll \bar{\mu}\} \approx L^1(\bar{\mu})$$

is P-invariant. The restriction P_0 of P to L_0 is power bounded and quasi-constrictive, thus it is asymptotically periodic. Hence it has a finite dimensional constrictor F_0 of the form (6.3), where $\nu_1, \ldots, \nu_a$ are periodic measures satisfying (6.1) and (6.2) for all $\mu \in L_0$. The difference $\mu_0 - Q(\mu_0)$ is periodic and at the same time it converges strongly to 0. Hence

$$\mu_0 = Q(\mu_0) = \sum_{i=1}^{a} \lambda_i \nu_i .$$

According to Proposition 1.1 (iii) the measures ν_i, $i = 1, \ldots, a$ have disjoint supports. We show that they are minimal periodic measures. Let $j \in \{1, \ldots, a\}$ be given and $\nu \ll \nu_j$ be a periodic measure. We have

$$\nu = Q(\nu) = \sum_{i=1}^{a} \lambda_i \nu_i \tag{6.5}$$

(since F_0 given by (6.3) is a constrictor for P_0). However, $\nu \ll \nu_j$, hence $\lambda_i = 0$ for $i \neq j$ and $\nu = \lambda_j.\nu_j$.

(ii) Let μ be a minimal periodic measure with period n. For any given $k \leq n-1$ and $\nu = P^m \nu \ll P^k \mu_0$ we have

$$P^{n.m-k}\nu \ll P^{m.n}\mu_0 = \mu_0 \ ,$$

thus $P^{n.m-k}\nu = \gamma.\mu_0$ for some $\gamma \neq 0$. Hence $\nu = \gamma.P^k\mu_0$, which means that $P^k\mu_0$ is a minimal periodic measure.

The measure $\bar{\mu} = A_n(\mu_0)$ given by (6.4) is obviously invariant and satisfies

$$A_n(P^k\mu_0) = \bar{\mu} \quad \text{for } 0 \leq k \leq n-1 \ . \tag{6.6}$$

The restriction P_0 of P to the subband $L_0 \approx L^1(\bar{\mu})$ is power bounded and quasi-constrictive, hence it is asymptotically periodic. Let $\nu_1, \ldots, \nu_a \in L_0$ be the minimal periodic measures satisfying (6.1) and (6.2) for P_0 on L_0. We have

$$\operatorname{supp}(\nu_i) \subset \operatorname{supp}(\bar{\mu}) = \bigcup_{k=0}^{n-1} \operatorname{supp}(P^k\mu_0) \quad \text{for } i = 1, \ldots, a \ .$$

According to Proposition 6.2 (i) each of the measures ν_i must be proportional to one of the measures $P^k\mu_0$, say

$$\nu_i = \gamma_i.P^{k(i)}\mu_0 \quad \text{for } i = 1, \ldots, a \ . \tag{6.7}$$

Making use of (6.6) we get

$$A_n(\nu_i) = \gamma_i.\bar{\mu} \quad \text{for } i = 1, \ldots, a \ . \tag{6.8}$$

If $\nu \ll \bar{\mu}$ is a P-invariant measure then it satisfies (6.5), hence

$$\nu = A_n(\nu) = \sum_{i=1}^{a} \lambda_i.A_n(\nu_i) = \left(\sum_{i=1}^{a} \lambda_i\gamma_i\right).\bar{\mu} \ .$$

Therefore the measure $\bar{\mu}$ is a minimal P-invariant measure.

(iii) If μ is a P-invariant measure then it satisfies (6.5) for some $a \in N$ and minimal periodic measures $\nu_1, \ldots, \nu_a$ with common period $q \leq a!$. Obviously,

$$\mu = A_q(\mu) = \sum_{i=1}^{a} \lambda_i A_q(\nu_i) \ .$$

(iv) If the number of minimal invariant probabilities is not finite, then we can inductively construct a countable sequence of pairwise orthogonal minimal invariant probabilities $\{\nu_n\}_{n=1}^{\infty}$. Put $B_m = \bigcup_{k=m}^{\infty} \operatorname{supp}(\nu_k)$ for $m \in N$. Obviously $B_m \to \emptyset$. However, for $k \geq m$ and $n \in N$ we have

$$P^n\nu_k(B_m) = \nu_k(X) = 1 \ .$$

Therefore, $\lim_{m\to\infty}[\inf_{\nu\in D}\{\lim_n \inf P^n\nu(X - B_m)\}] = 0$. This contradicts the assumption that P is quasi-constrictive. □

Combining Propositions 6.3 and 6.2 we directly obtain the following results.

Corollary 6.2. *If P is a quasi-constrictive operator then the system*

$$S = \{\nu : \exists n \in N,\ A_n\nu \quad \text{is a minimal invariant probability}\}$$

is finite.

Proof of the Extension of Theorem 2. Let P be a quasi-constrictive operator. Let $S = \{\nu_1, \ldots, \nu_a\}$ be a system of minimal periodic measures defined in Corollary 6.2. Combining Corollary 6.1, Proposition 6.2 and 6.3 we conclude that the set F_0 given by (6.3) is a constrictor for P. Therefore, P is a constrictive operator, hence it is asymptotically periodic. □

Corollary 6.3. *Suppose that an operator P has a nontrivial lower-bound measure $\nu \in L_+$. Then P is asymptotically stable.*

Proof. Arguing in the same way as in Corollary 5.1 we obtain that P is quasi-constrictive. Its asymptotic behavior is the same as the one of the restriction $P_0 = P_{|L_0}$, where $L_0 \approx L^1(\mu_0)$ is a subband containing a constrictor F. Since P_0 has a lower-bound function it is asymptotically stable and the same is true about P. □

Example 6.1. (Communicated by A. Lasota.) Let X be a compact metric space and $\mathcal{B}$ be the σ-algebra of Borel sets. Let $r \in N$ and T_i, $i = 1, \ldots, r$, be continuous transformations of X such that $T_1(x) \equiv x_0$ for $x \in X$. Let $p_1, \ldots, p_r \in R_+$, $p_1 > 0$ and $\sum_{i=1}^{r} p_i = 1$.

Let $C(x)$ be the Banach space of real continuous functions on X with the norm of uniform convergence. Its dual $\mathcal{M}$ consists of all regular Borel measures on $\mathcal{B}$. Consider the linear homomorphism

$$T : C(X) \to C(X), \quad Tf(x) = \sum_{i=1}^{r} p_i . f(T_i x)$$

and its dual given by

$$(P\mu)f = \mu(Tf) .$$

Obviously, P is a Markov operator on $\mathcal{M}$ having a lower-bound measure μ_0 given by

$$\mu_0(A) = \begin{cases} p_1 & \text{if } x_0 \in A \\ 0 & \text{otherwise} . \end{cases}$$

Hence P is asymptotically stable.

References

[1] W. Bartoszek, Asymptotic Periodicity of the Iterates of Positive Contractions on Banach Lattices, Studia Mathematica, **XCI** (1988), 179–188
[2] N. Bourbaki, Intégration, Hermann, Paris 1969
[3] A. Boyarski, R. Levesque, Spectral Decomposition for Combinations of Markov Operators, J. Math. Anal. Appl. **132** (1988), No. 1, 251–263
[4] J. L. Doob, Stochastic Processes, Wiley, New York 1953
[5] N. Dunford, J. T. Schwartz, Linear Operators, Part 1, 4th printing (1967) Interscience Publishers, New York
[6] S. R. Foguel, The Ergodic Theory of Markov Processes, Van Nostrand, New York 1969
[7] J. Komorník, Asymptotic Periodicity of the Iterates of Weakly Constrictive Markov Operators, Tôhoku Math. Journ. **38** (1986), 15–27
[8] J. Komorník, Asymptotic Decomposition of Positive Operators, [in press]
[9] J. Komorník, Asymptotic Decomposition of Smoothing Positive Operators, Acta Univ. Carolinae, Math. at Phys. **30** (1989), No. 2, 77–81
[10] J. Komorník, A. Lasota, Asymptotic Decomposition of Markov Operators, Bull. Pol. Ac. Math. **35** (1987), 321–327
[11] J. Komorník, E. G. F. Thomas, Asymptotic Periodicity of Markov Operators on Signed Measures, Časop. Pèst. Math. [in press]
[12] U. Krengel, Ergodic Theorems, deGruyter, 1985
[13] A. Lasota, Asymptotic Behaviour of Solutions: Statistical Stability and Chaos, Proc. ICM '82, PNW / North Holland, 1984
[14] A. Lasota, T. Y. Li, J. A. Yorke, Asymptotic Periodicity of the Iterates of Markov Operators, Trans. Amer. Math. Soc. **286** (1984), 751–764
[15] A. Lasota, M. C. Mackey, Globally Asymptotic Properties of Proliferating Cell Populations, J. Math. Biology **19** (1984), 43–62
[16] A. Lasota, M. C. Mackey, Probabilistic Properties of Deterministic Systems, Cambridge Univ. Press, 1985
[17] A. Lasota, M. C. Mackey, Noise and Statistical Periodicity, Physica **28D** (1987), 143–154
[18] A. Lasota, J. Socala, Asymptotic Properties of Constrictive Markov Operators, Bull. Pol. Ac. Math. **35** (1987), 71–76
[19] A. Lasota, J. A. Yorke, Statistical Periodicity of Deterministic Systems, Časop. Pèst. Mat. **111** (1986), 1–13
[20] M. Miklavčič, Asymptotic Periodicity of the Iterates of Positivity Preserving Operators, Trans. Amer. Math. Soc. **307** (1988), 469–480
[21] R. Rudnicky, Stability of the Iterates of Markov Operators, An. Polon. Math. **XLVIII** (1988), 95–104
[22] J. Socala, On Existence of Invariant Measures for Markov Operators, An. Polon. Math. **XLVIII** (1988), 51–56
[23] H. H. Schaefer, On Positive Contractions in L^p Spaces, Trans. Amer. Math. Soc. **257** (1980), 261–268
[24] Shu-Teh Chen Moy, Period of an Irreducible Positive Operator. Illinois J. Math. **11** (1967), 24–39
[25] R. Sine, Constricted Systems, [preprint]
[26] E. Straube, On the Existence of Invariant Absolutely Continuous Measures, Comm. Math. Phys. **81** (1981), 27–30

[27] B. Sz. Nagy, C. Foias, Analysa Harmonique des Operateuers de l'espace de Hilbert, Masson & Cie, Paris 1967

[28] A. E. Taylor, Introduction to Functional Analysis, John Wiley, New York 1967

[29] K. Yosida, E. Hewit, Finitely Additive Measures, Trans. Amer. Math. Soc. **72** (1952), 42–46

[30] A. Zalewska, A Generalization of the Lower Bound Theorem for Markov Operators. Univ. Jagelonicae Acta Math. [In press]

A Nekhoroshev-Like Theory of Classical Particle Channeling in Perfect Crystals

H. Scott Dumas

1. Introduction

Some twenty years ago, N.N. Nekhoroshev announced a theorem in Hamiltonian perturbation theory which may be paraphrased as follows.

Suppose the Hamiltonian

$$H(I, \theta) = h(I) + f(I, \theta) \tag{1.1}$$

is analytic on F (a certain complex neighborhood of the real domain $K \times \mathbf{T}^n$, $K \subset \mathbf{R}^n$, $\mathbf{T}^n = \mathbf{R}^n/\mathbf{Z}^n$). Suppose further that the perturbation f is of order ε ($\sup_F |f| = \varepsilon$), and that the integrable part h satisfies certain "steepness conditions" (for example h is convex). Then there exist positive numbers a, b, C, and ε_0, depending only on F and h, such that if $\varepsilon < \varepsilon_0$, then any solution $\big(I(t), \theta(t)\big)$ of Hamilton's equations

$$\dot{I} = -\partial_\theta H, \quad \dot{\theta} = \partial_I H \tag{1.2}$$

with arbitrary real initial conditions $\big(I(0), \theta(0)\big) \in K \times \mathbf{T}^n$ satisfies

$$\|I(t) - I(0)\| < \varepsilon^b \tag{1.3}$$

for all t in the exponentially long time interval $[0, T]$, $T = \exp(C\varepsilon^{-a})$.

Nekhoroshev's theorem has been described as an upper bound on the average speed of Arnol'd diffusion, as a physicist's KAM theorem, and even as the crowning achievement in classical perturbation theory for nearly integrable Hamiltonian systems. In spite of this, for reasons that will be discussed below, the theorem's importance remains somewhat unappreciated, especially when compared with attention focused on the closely related KAM theorem.

It is the aim of this article to present the complete proof of a Nekhoroshev-like result in the context of a compelling physical application. Mathematically, the classical particle channeling problem treated here provides a good introduction to Nekhoroshev theory in the convex case, as it is one of the simplest problems which retains a full nontrivial resonance structure. The techniques of proof are drawn mainly from the series of vivid articles by G. Benettin, L. Galgani, G. Gallovotti and A. Giorgilli, who – with others – have developed and applied Nekhoroshev-like results in the convex case to a wide range of physical problems.

Physically, this work sets up a mathematically rigorous framework for the motion of highly energetic charged particles in crystals, and is an outgrowth of the author's research with J.A. Ellison. The reader familiar with particle channeling will recognize that the treatment here begins with the three dimensional perfect crystal model and proceeds to a simultaneous derivation of generalized versions of J. Lindhard's continuum models for axial and planar channeling. Using a separate result in geometric number theory (the proof of which appears elsewhere), it is natural to add a "spatial" continuum model to the generalized axial and planar continuum models.

The remainder of this article is organized as follows. Section 2 provides brief introductions to both Hamiltonian perturbation theory and to the physics of particle channeling in crystals, as well as a preview, expressed mainly in physical terms, of the main results to be presented in Sect. 5. Section 3 formulates the mathematical model to be treated in the sequel, while the normal forms on which the continuum models are based are constructed in Sect. 4. In Sect. 5, once notation and elementary results concerning the geometry of the problem are established, the main results are presented as a series of theorems. Theorems 5.3 through 5.5 make up the spatial continuum model; Theorem 5.7 deals with both axial and planar channeling. Brief remarks on the channeling theorem conclude the article in Sect. 6.

2. Background and Outline of Main Results

2.1 Sketch of the Development of Nekhoroshev Theory

Nekhoroshev's result is the modern culmination of a long history of investigation into the behavior of solutions of "nearly integrable" Hamiltonian systems. The idea of integrability goes back to K.G.J. Jacobi and W.R. Hamilton at the beginning of the nineteenth century and, for a time, it was thought that all or nearly all Hamiltonian systems could be brought into integrable form. In other words, it was thought that for a system with bounded motion governed by an analytic n degree of freedom Hamiltonian of the form $H = H(p,q)$ $((p,q) \in \mathbf{R}^n \times \mathbf{R}^n)$, there is an open subset D of phase space $\mathbf{R}^n \times \mathbf{R}^n$ and an analytic canonical transformation $(p,q) \mapsto (I,\theta) \in \mathbf{R}^n \times \mathbf{T}^n$ on D under which H takes the integrable form $H' = H'(I)$. That is, H' is independent of θ, so that phase space is foliated into invariant n-tori, smoothly parametrized by the actions I, with each torus supporting quasiperiodic flow with frequency vector $\partial H'_I(I)$.

These hopes vanished at the turn of the century with Poincaré's seminal work [45], in which he shows that "most" (we would now say generically) nondegenerate Hamiltonian systems are not integrable. His results fueled hopes for the "ergodic hypothesis" of then-nascent statistical mechanics (a conjecture invalidated for generic Hamiltonians [38]), and they raised questions about the stability of motion for conservative systems and about the convergence of perturbation series used to approximate the motion of nearly integrable systems, especially in celestial mechanics.

Real progress in resolving these questions did not come for more than a half-century, when A.N. Kolmogorov announced [27] – and sketched the proof of – the celebrated KAM (Kolmogorov-Arnol'd-Moser) theorem. The content of this surprising result is that, when the integrability of a nondegenerate integrable Hamiltonian is broken in the classical sense by means of a sufficiently small perturbation, a large relative measure of the trajectories of the system retain their integrable behavior. More precisely, tori of the original integrable system supporting quasiperiodic flow with highly nonresonant frequency vectors $\partial_I H(I)$ suffer only small deformations under sufficiently small perturbations of the Hamiltonian, and continue to support quasiperiodic flow.

The activity spawned by Kolmogorov's announcement has continued up to the present day. Proof of a slightly more general result in the analytic case was given by V.I. Arnol'd [1] who went on to discuss some of its ramifications [2]; at nearly the same time, J. Moser [40] proved a closely related invariant curve theorem for perturbations of finitely differentiable area-preserving twist maps of the plane. The result for planar twist mappings is especially important for Hamiltonian systems with $n = 2$ (two degrees of freedom), as it ensures the existence of nested, two-dimensional invariant tori that partition each three-dimensional energy surface into narrow toroidal compartments to which nonintegrable trajectories are confined. In this way the global (or topological) stability of nearly integrable systems with two degrees of freedom is assured under sufficiently small perturbations away from integrability.

There is however a fundamental aspect of the case $n > 2$ (more than two degrees of freedom) which the elegant results on invariant tori fail to address. For $n > 2$, the n-dimensional KAM tori no longer partition the $2n - 1$-dimensional energy surfaces, so that trajectories not initially residing on invariant tori may "leak out" through the connected complement of the tori and suffer large deviations in action over finite times. That this leakage actually occurs was first demonstrated for a specific system with "$2\frac{1}{2}$" degrees of freedom by Arnol'd [3] shortly after his proof of the KAM theorem, and the phenomenon has since been known as "Arnol'd diffusion." Despite examples for systems with any finite number of degrees of freedom [26], detailed and insightful discussions of the mechanisms generating it [14], [15], [16], [33], and discussions of its genericity [15], [16], Arnol'd diffusion remains poorly understood. It is not entirely clear why this type of instability behaves like diffusion; its existence has not been rigorously demonstrated in many important examples where it is suspected; and above all, its "speed" has not been gauged accurately.

Nonetheless, for analytic nearly integrable systems, an upper bound on the average rate of diffusion was suspected early on. In the mid 1950's, Moser [39] and Littlewood [31], [32] conjectured that generically in such systems, the timescale required to observe an $O(1)$-displacement in the action variables was beyond all orders in inverse powers of the small parameter; Arnol'd [4] later conjectured a similar result. Nekhoroshev's theorem affirms this conjecture, and shows further that the timescale required for $O(1)$-displacement is exponentially long in an inverse power of the small parameter.

Nekhoroshev's theorem and the KAM theorem are complementary results; each represents a slightly different response to the question: In what sense is integrable behavior continuous with respect to perturbation? KAM theory assures the persistence of a Cantor set of invariant tori and the continuity of their measure under perturbation; Nekhoroshev's theorem takes the more prosaic but physically practical view of integrability as stability of the action variables, showing strong continuity with respect to perturbation – for *all* initial conditions – of the *timescale* of stability.

Though it relies partly on technology from Arnol'd's proof of the KAM theorem, Nekhoroshev's proof of his own result is more classical, more geometric, and more complex. Proof of the C^∞-genericity of "steepness" in the class of analytic functions [42] came two years after the theorem's announcement [41]; the theorem's proof, excepting certain technical results, appeared three years later [43], while the final technical estimates were published two years after that [44].

No doubt the complexity and length of the proof contributed to relative ignorance of the result among both physicists and mathematicians. The complexity was however more than justified by Nekhoroshev's achievement of the optimal geometric hypothesis (steepness – a generalization of convexity) on the unperturbed part of the Hamiltonian; the subsequent focus on the convex case in applications to particular problems (as in this article) confronts a much less daunting task geometrically.

In the earliest announcement of his result [41], Nekhoroshev formulated the steepness property for functions on a domain $K \subset \mathbf{R}^n$ as follows. Assume that the function f has nonvanishing gradient on K, and denote by $\{\lambda^r(x)\}$ the set of hyperplanes of dimension r, passing through $x \in K$ and perpendicular to $\mathrm{grad} f(x)$. The function f is steep on K if for each $r = 1, \dots, n-1$, there are positive constants C_r, δ_r (called steepness coefficients), and $\alpha_r \geq 1$ (the steepness indices) such that for all $x \in K$, all $\lambda^r(x) \in \{\lambda^r(x)\}$ and all $\xi \in (0, \delta_r]$,

$$\max_{0 \leq \eta \leq \xi} \min_{\substack{y \in \lambda^r(x) \cap K \\ \|x-y\|=\eta}} \|\mathrm{grad}\big(f|_{\lambda^r(x)}\big)(y)\| > C_r \xi^{\alpha_r} \ . \tag{2.1}$$

Conditions similar to steepness may be traced back to J. Glimm's work [24] on the formal stability of Hamiltonian systems; but it was in [42] that Nekhoroshev first showed steepness to be a strongly generic property of analytic functions, in the sense that the coefficients of the Taylor series of a nonsteep function satisfy an infinite number of algebraic conditions. For detailed discussions of steepness, its generalizations, and its role in the stability of Hamiltonian systems (including its relation to convexity and various nondegeneracy conditions), the reader is referred to [34], [35], [36], [42] and [43]. Here we limit ourselves to the following remarks.

In the proof of Nekhoroshev's theorem, steepness assures that the contact between resonant surfaces and the so-called planes of fast drift (the hyperplanes in action space where motion is unconstrained by a given resonant normal form; cf. Sect. 5.1 below) is weakly transverse; this allows for bounds on the size of the intersection of zones surrounding the resonant surfaces and planes of fast drift.

In this way steepness controls the passage of trajectories through resonance; without it, trajectories may quickly suffer large deviations in action as they move unimpeded along resonances. On the other hand, the complicated passage through resonance phenomena may be largely eliminated by strengthening the hypothesis on the unperturbed part of the Hamiltonian from steepness to convexity (convex functions have unit steepness indices and are consequently the "steepest"). In that case it may be shown that trajectories remain trapped, to a certain level of approximation, in a single resonance for exponentially long times. This stronger result is essential in the problem treated below, as the channeling phenomenon is itself interpreted as motion at resonance in the model system.

As his remarks indicate [43], Nekhoroshev was aware of the possibility of a simplified proof when h in (1.1) is convex or quasiconvex (i.e., has convex level sets). In the 1980's a group of Italian mathematicians and physicists began to exploit the simplifications afforded by convexity (or by cases where the unperturbed part consists of uncoupled harmonic oscillators, which is simpler still; cf. [11]), and they proved estimates for a number of physical models. Their results are extremely interesting, as they address delicate stability questions in physics that had not previously received rigorous treatment. These include estimates of the size of the region surrounding the Lagrange points of the Sun-Jupiter system which enjoy stability times on the order of the age of the universe [13], [23], investigations into the problem of holonomic constraint in mechanical systems [10], and a vindication and extension of some of the ideas of L. Boltzmann and J. Jeans concerning very slow relaxation times for high frequency degrees of freedom in statistical systems (as related, for example, to the ultraviolet catastrophe in classical physics and the problems concerning equipartition of energy and the behavior of classical polyatomic gases with various internal degrees of freedom) [9].

In spite of the novelty of these results, and the digestability of the simpler proofs (only somewhat more complex than proofs of KAM), there remained the problem of physically unrealistic values for the order constants of the theorems. For example, in early versions of Nekhoroshev-like theorems the "threshold of validity" ε_0 (from Sect. 1) is typically more than a dozen orders of magnitude smaller than what might be expected physically, and the stability exponents a and b are dismayingly far from optimal. (The dependence of a on the number of degrees of freedom n is especially critical in applications to statistical mechanics.)

As work on the long-time stability of conservative systems moves forward, much effort has gone toward making the estimates realistic for specific model systems [13], [23], [37], [46]. A significant improvement in the general convex case may be found in the recent work of P. Lochak [34], [35], who has produced nearly optimal stability exponents (Lochak believes the optimal exponents are very likely $a = 1/2n, b = 1/2$) using an entirely different method of proof. Lochak replaces the usual approach of Fourier series and estimates of small divisors with a study of the behavior of trajectories near periodic orbits of the unperturbed problem. More recently, J. Pöschel produced the exponents Lochak believes to be optimal by reformulating mainly the geometric part of Nekhoroshev's proof, using a new efficient method of partitioning action space into resonant blocks [46]. Pöschel's approach also appears to generalize to the steep case.

Finally, exponential estimates have recently been obtained for symplectic maps [7]; these were spurred by the need for a theory to support the numerical technique of symplectic integration, and by the need for estimates of the stability times of particle beams orbiting in accelerator storage rings. The application presented in this article treats the motion of charged particles as they *emerge* from accelerator storage rings or linear accelerators and impinge upon crystalline targets; the physics of the problem is sketched in the next subsection.

2.2 Brief Description of the Physics of Particle Channeling in Crystals

When a beam of energetic positive particles (e.g., MeV protons or GeV positrons) is directed at a crystalline target in a random direction, the beam and crystal interact strongly: beam particles are backscattered, matter is ejected from the crystal, nuclear reactions may even take place. Radiative and collisional energy loss eventually bring many particles to rest, so that beam matter is implanted in the crystal. But if the crystal is now repositioned so that the beam is incident in a "non-random" direction – in the direction of a low-order crystal axis or plane – the results observed are very different. The average depth of penetration into the crystal is greatly increased, and the rate of particle backscattering may decrease by as much as two orders of magnitude. The interaction of the beam with the crystal in this way is called *channeling*, and for beams with positive charge, it is not entirely naïve to imagine particles streaming through the channels between planes or rows of crystal nuclei, with soft collisions with these planes or rows guiding particles away from nuclei.

Channeling has proved to be a useful tool for understanding the properties of solids, and has had numerous technological applications. It has been used as a material analysis tool to study crystal defects, surfaces and interfaces, and to determine the location of crystal impurities. It has been used to measure nuclear lifetimes, to study the strain in "strained-layer superlattices," and to deflect high energy particle beams. These and many other applications of channeling – including more speculative possibilities such as monoenergetic gamma ray sources and cosmic-ray telescopes – are discussed in a large body of literature, much of which is cited in the bibliographies of [12] and [21], and in the excellent if by now somewhat early review article [22].

Since the early 1960's, theoretical investigations of channeling have relied on the so-called *continuum model*, in which channeling is described as the motion of particles moving in a *continuum potential* obtained by averaging the crystal potential over the axis or plane with which the particles' incident direction is most nearly parallel. This model was introduced independently by a number of theoreticians, but the most convincing arguments for it were given by J. Lindhard. His and his coworkers' papers [28], [29], [30] include detailed physical arguments; by translating the simplest classical cases considered in these articles into the language of Hamiltonian systems, we may paraphrase that part of the theory which ignores electron multiple scattering as follows. We begin by considering the perfect crystal model, in which particle motion is governed by the three degree of freedom Hamiltonian

$$\mathcal{H} = \frac{1}{2m} p^2 + V(q), \quad \text{where} \quad p, q \in \mathbf{R}^3, \quad \text{and} \quad p^2 = \sum p_i^2 \ . \tag{2.2}$$

Here V is the periodic crystal potential, usually expressed as a sum of thermally averaged screened Coulomb atomic potentials at the lattice sites. (The relevance of this model in channeling physics and some of the effects it ignores are discussed briefly in the next section.) Lindhard argued that for incident particle directions nearly parallel to low-index axes, the motions of $\mathcal{H}$ are well described by the so-called impulse-momentum approximation, wherein, for grazing angles, the influence of a string of atoms on a particle may be approximated by a sequence of soft collisions with individual atoms in the string. He went on to show that, because particles experience a large number of soft collisions with atoms in a given string, another approximation is justified in which each discrete string of atoms is replaced by a "continuum string." In terms of the Hamiltonian (2.2), this amounts to replacing the crystal potential V by the axial continuum potential V_A obtained from V by averaging along the direction of the strings considered. This greatly simplifies (2.2); the position q_3' along the axial direction is now a cyclic coordinate, so the momentum p_3' in this direction is constant, and the motion in the plane transverse to the axial direction is governed by the two degree of freedom axial continuum Hamiltonian

$$\mathcal{H}_A = \frac{1}{2m} \left({p_1'}^2 + {p_2'}^2 \right) + V_A(q_1', q_2') \ . \tag{2.3}$$

Though Lindhard did not express his result in this way originally, and though he did not transform system (2.2) into system (2.3) nor give rigorous estimates of how solutions of (2.3) approximate those of (2.2), it is clear that he did establish a strong link between the two systems, and that this link is one of the cornerstone achievements in channeling theory. It is not clear however, that a similarly convincing connection has ever been established between (2.2) and the one degree of freedom planar continuum Hamiltonian

$$\mathcal{H}_P = \frac{1}{2m} {p_1''}^2 + V_P(q_1'') \ , \tag{2.4}$$

in which the conjugate coordinates p_1'' and q_1'' are momentum and displacement in the direction transverse to the planes in question, and the planar continuum potential V_P is the average of V parallel to the planes. Although this planar continuum model shares the intuitive appeal of the axial continuum model and has gained the same basic currency among channeling theorists, the arguments used to derive it from the perfect crystal model have not been wholly convincing. The problem is that the impulse-momentum approximation breaks down, because, while it is clear how particles can undergo successive grazing collisions with a segment of collinear atoms, it is not clear how they can do so with a region of coplanar atoms. This criticism does not detract from the planar continuum model, the usefulness of which has been borne out through repeated experiment, but it does point out a basic theoretical difficulty in the way it is descended from the perfect crystal model.

In the 1970's, J.A. Ellison noticed that the equations of motion for the perfect crystal model could be scaled so as to appear in standard form for the method of averaging for ODEs; this led to new mathematical results in channeling theory, including general derivations of axial continuum models from perfect crystal models [20]. Inasmuch as Nekhoroshev-type perturbation theory may be viewed as a specialized averaging method for Hamiltonian systems, the results given here extend that work to a derivation of axial, planar, and "spatial" continuum models (the methods here are however not suited to practical calculation as are the averaging methods; see [20] for a brief discussion).

2.3 Outline of Main Results

Equipped with the rudiments of Nekhoroshev theory and of the physics of particle channeling, we may anticipate the main results to be presented in detail in Sect. 5.

The channeling theory is founded on the natural duality between the symmetry of the crystal (manifested in its planes and axes) and the resonances that arise in attempts to transform the perfect crystal potential to its average. This duality becomes apparent once the perfect crystal model is scaled so that the repulsive effects of crystal nuclei are viewed as perturbations to rectilinear motion of highly energetic particles. The key to this scaling is the so-called channeling criterion, which singles out channeling trajectories as those which avoid close encounters with crystal nuclei. Once the Hamiltonian has been scaled, Nekhoroshev-like normal forms may be constructed, each of them adapted to a particular resonance (i.e., to a particular axis or plane of symmetry). Geometry enters the picture by way of the domains on which each of the normal forms is valid (and where estimates are possible). It should be stressed that, because Nekhoroshev's Theorem is a global result, these domains (called blocks) are ordinarily required to cover all of action space; but this is not the case for the results presented here. Indeed, Nekhoroshev's Theorem bounds the drift in action for all initial conditions in phase space, while the focus here is on a special class of nearly resonant motions, only one feature of which is near-invariance of the initial actions. Thus channeling motions are governed by resonant normal forms (generalized planar and axial continuum models), and are seen to occur only for a small subset of initial conditions. The behavior of trajectories for the majority of remaining initial conditions is governed by a nonresonant normal form (the spatial continuum model). However, some initial conditions remain outside the union of the domains of validity of all the continuum models; this point is discussed further in Sect. 6.

In physical terms, this article is concerned with the behavior of highly energetic charged particles incident upon a perfect crystal in various directions. The results obtained may be summarized as follows.

(i) For particles incident upon a perfect crystal in nonchanneling directions, i.e. in directions sufficiently far from low-order crystal axes or planes, particle motions are nearly rectilinear until they encounter a crystal nucleus. These encounters occur quickly, within a time that may be estimated in terms of the way directions "far from low-order axes or planes" are defined. This result, called the spatial continuum model, is embodied in Theorems 5.3 through 5.5.

(ii) For positive particles incident in directions closely aligned with a low-order crystal plane, in appropriate initial position, and sufficiently far from low-order axes, subsequent motions remain between adjacent planes and avoid close encounters with crystal nuclei on a time interval which is exponentially long in the particle energy. On the same time interval, the transverse energy and longitudinal momentum of particles are approximately conserved, and trajectories are governed by a Hamiltonian which coincides, to leading order, with the appropriate planar continuum model Hamiltonian.

(iii) For positive particles incident in directions closely aligned with a low-order crystal axis, and in appropriate initial position, subsequent motions remain away from axes and so avoid close encounters with crystal nuclei on a time interval which is exponentially long in the particle energy. On the same time interval, the transverse energy and longitudinal momentum of particles are approximately conserved, and trajectories are governed by a Hamiltonian which coincides, to leading order, with the appropriate axial continuum model Hamiltonian.

Results (ii) and (iii) characterize channeling trajectories, and are expressed mathematically by Theorem 5.7. The remainder of this paper is devoted to establishing results (i), (ii) and (iii) as mathematical theorems bearing on the perfect crystal model described in the next section.

3. Formulation of the Channeling Problem

This section formulates the basic mathematical problem considered in the sequel. The perfect crystal model is introduced, and a criterion is given for identifying channeling motions in the model. This criterion leads naturally to a nearly integrable form of the perfect crystal model, and it is argued that this system is a good model of particle motion whenever the channeling criterion is satisfied.

3.1 The Perfect Crystal Model

Our approach to the problem of classical, positively charged particle motion in perfect cubic crystals begins with a classical Hamiltonian system with three degrees of freedom and with a periodic potential having finitely many (usually only a few) smooth repulsive sites in each lattice cell. In fact, throughout this work the crystal potential is assumed to be analytic, which makes possible the necessary technical estimates. The requirement of analyticity is not as severe as it might seem, as some of the best available perfect crystal potentials are analytic, essentially because of thermal averaging (cf. [30]).

In physical variables, the perfect crystal Hamiltonian represents the total energy of a point particle with mass m moving under the influence of the potential V, and may be written

$$\mathcal{H}(p,q) = \frac{1}{2m}p^2 + V(q) \tag{3.1}$$

where $p, q \in \mathbf{R}^3$ are the momenta and position of the particle under consideration. Here V has period d in each component of q, along with a small number of repulsive sites in each periodic cell. More precisely, V is constructed by summing screened, thermally averaged Coulombic atomic potentials centered on each lattice site (see Appendix 2 of [17] for a more complete discussion of the potential V).

Reducing the channeling problem to the study of (3.1) neglects a number of physical effects, among which are the following:

(i) Quantum and relativistic effects
(ii) Dynamic effects of thermal crystal lattice vibrations (Thermal effects are included in the potential in an average way only.)
(iii) Effects of close encounters with nuclei (The hard Coulombic singularities have been smoothed out of the potential, so close encounter processes, e.g. Rutherford scattering, are not represented.)
(iv) Effects of individual electrons (Their influence is instead incorporated into V by means of a screening function, which essentially smears each electron over its spatial probability distribution.)
(v) Energy loss effects, which arise, for example, through radiation emission and electron multiple scattering

The neglect of effects (i) is known to be approximately valid over a certain range of particle compositions and particle energies (see, e.g., [22], [28], [29]). On the other hand, it is hard to imagine circumstances in which a theory of particle channeling neglecting effects (ii) through (v) could be called complete. The reader may therefore be hesitant about the viability of (3.1) as a model for positive particle channeling. But the arguments for its relevance are strong. In fact, under conditions where effects (i) may be neglected, theories accounting for the remaining effects, however detailed, may be broadly described as perturbations to the perfect crystal model – it is the foundation for theories of particle channeling. In spite of this, no completely satisfactory account of the model has yet been given. This paper is meant to remedy this situation by providing a single, coherent mathematical framework in which to view the basic channeling phenomena. In retrospect, the fidelity with which the solutions of system (3.1) reproduce channeling effects is surprising, and is perhaps the most pleasing aspect of this work.

Nevertheless, the neglect of effect (iii) requires special consideration. Physicists are apt to say that the potential V used here is a good classical approximation to the true potential away from lattice sites, and a poor approximation in their close vicinity (cf. [30]). In this work, the preceding statement will be formalized by singling out neighborhoods of lattice sites inside of which the potential is "untrustworthy." The precise criterion is given in the next subsection.

3.2 The Channeling Criterion

We now introduce a simple criterion, key to much of what follows, for distinguishing channeling from nonchanneling trajectories of (3.1). We first assume that the potential V has been adjusted so that its minimum value is zero and its maximum value over $\mathbf{R}^3$ is $\mathcal{E}_M$, so that for energies $\mathcal{E}_\perp$ $(0 \leq \mathcal{E}_\perp \leq \mathcal{E}_M)$, we may consider the subsets of configuration space

$$\mathcal{B}(\mathcal{E}_\perp) = \{q \in \mathbf{R}^3 \mid V(q) \geq \mathcal{E}_\perp\} . \tag{3.2}$$

If, as is assumed here, the potential governs the motion of positively charged particles, then clearly for sufficiently large $\mathcal{E}_\perp < \mathcal{E}_M$, the set $\mathcal{B}(\mathcal{E}_\perp)$ is the disjoint union of (slightly deformed) balls centered on the lattice sites. By choosing a physically suitable value for $\mathcal{E}_\perp$, we may distinguish particle trajectories which come too close to nuclei to be governed by the thermally averaged potential as those which enter $\mathcal{B}(\mathcal{E}_\perp)$. More precisely, fix $\mathcal{E}_\perp$, and consider a solution $(p(\tau), q(\tau))$ of the equations of motion corresponding to (3.1). Such a solution is a *channeling solution on the time interval* $\mathcal{I}$ provided

$$q(\tau) \notin \mathcal{B}(\mathcal{E}_\perp), \quad \text{or equivalently} \quad V(q(\tau)) < \mathcal{E}_\perp \quad \forall \tau \in \mathcal{I} . \tag{3.3}$$

This is the *channeling criterion*, and it is assumed that the perfect crystal model is a good approximation for particle trajectories that satisfy it. A trajectory which first fails to satisfy this criterion at time t_1 is assumed to suffer a "close encounter" with a nucleus, and is not viewed as a good approximation to an actual particle trajectory for subsequent times $t > t_1$. While the channeling criterion is independent of the incident energy $\mathcal{E}$, channeling motions are such that $\mathcal{E} \gg \mathcal{E}_\perp$, and this defines the fundamental perturbation parameter in the analysis below.

3.3 Transformation to Nearly Integrable Form

The Hamiltonian (3.1) may be transformed to nondimensional, nearly integrable form as follows. Restricting attention to particles of fixed energy $\mathcal{E}$, i.e., to trajectories $\big(p(\tau), q(\tau)\big)$ satisfying $\mathcal{H}\big(p(\tau), q(\tau)\big) = \mathcal{E}$, we define the scaled momentum (actions) $I \in \mathbf{R}^3$, the scaled position (angles) $\theta \in \mathbf{T}^3$, the scaled potential W and the scaled time t by

$$I = (m\mathcal{E})^{-1/2}p, \quad \theta = \frac{1}{d}q, \quad W(\theta) = \frac{1}{\mathcal{E}_\perp}V(\theta d), \quad t = \frac{1}{\tau_0}\tau . \tag{3.4}$$

Here $\tau_0 = d\sqrt{m/\mathcal{E}}$ is the time required for a particle to travel a distance $\sqrt{2}d$ in the potential-free case. The choice of scaling is motivated by the assumption $\mathcal{E} \gg \mathcal{E}_\perp$, so that trajectories satisfying the channeling criterion maintain a kinetic energy of approximately $\mathcal{E}$. The transformed Hamiltonian H now reads $H(I, \theta) = \frac{1}{2}I^2 + (\mathcal{E}_\perp/\mathcal{E})W(\theta)$, or, writing $\varepsilon = \mathcal{E}_\perp/\mathcal{E} \ll 1$, this becomes

$$H(I, \theta) = \frac{1}{2}I^2 + \varepsilon W(\theta), \quad \text{where} \quad W \in C^\omega(\mathbf{T}^3) , \tag{3.5}$$

$$\text{and} \quad 0 \leq W(\theta) \leq \mathcal{E}_M/\mathcal{E}_\perp \equiv E_o > 1 \ , \tag{3.6}$$

and we are interested in the behavior of solutions on the surface $H = 1$ for small ε and long times.

It is no surprise that the scaled Hamiltonian appears in action-angle form, since this scaling views the potential as a perturbation of rectilinear motion in the lattice $\mathbf{T}^3$. System (3.5) is called the "nearly integrable form of the perfect crystal model" or the "scaled perfect crystal model."

The region of close encounter $\mathcal{B}(\mathcal{E}_\perp)$ described in (3.2) is immediately reformulated in terms of the scaled variables as

$$\mathcal{C}(1) = \{\theta \in \mathbf{T}^3 | W(\theta) \geq 1\} \ , \tag{3.7}$$

and so the criterion (3.3) for a solution $(I(t), \theta(t))$ of Hamilton's equations corresponding to (3.5) to be a channeling trajectory on the time interval $\mathcal{I}$ becomes

$$\theta(t) \notin \mathcal{C}(1), \quad \text{or equivalently} \quad W(\theta(t)) < 1 \quad \forall t \in \mathcal{I} \ . \tag{3.8}$$

Finally, we note that for fixed ε (i.e., for fixed values of $\mathcal{E}_\perp/\mathcal{E}$), it follows from energy conservation that for all t, real-valued motions $(I(t), \theta(t))$ of (3.5) are confined to a spherical shell in action space of thickness $O(\varepsilon)$ given by

$$\mathcal{S}^\varepsilon_{\sqrt{2}} = \{I \in \mathbf{R}^3 | 2 - 2\varepsilon E_o \leq \|I(t)\|^2 \leq 2\} \ , \tag{3.9}$$

where $\|I\|^2 = I_1^2 + I_2^2 + I_3^2$ is the Euclidean norm, and $E_o \equiv \mathcal{E}_M/\mathcal{E}_\perp > 1$. We will always assume that ε is small enough to ensure that $I \in \mathcal{S}^\varepsilon_{\sqrt{2}} \Rightarrow \|I\| \geq \sqrt{2} - \varepsilon E_o$; this fact will be useful in proving the stability of channeling motions in Sect. 5.

4. Construction of the Normal Forms

In this section we construct the normal forms on which the generalized continuum models are based. This construction corresponds to what is called the analytic lemma in the proof of Nekhoroshev's theorem, and the methods used here follow those used in proving the analytic lemma in [8]. The result obtained is however closer in spirit to the analytic lemma in [11]. The use of KAM-type resonant zones here is apparently unique in a Nekhoroshev-like result; use of these zones is not warranted in Nekhoroshev's theorem itself, as the aim is to obtain uniform estimates of the deviation of actions from their initial values. The goal here is instead to obtain separate descriptions of particle motions in regions of phase space with distinct resonance properties, and the KAM zones adhere to the observed decrease in size of planar and axial channeling zones with increasing order.

This rather lengthy and detailed section begins with a description of the methods to be used, and is followed by subsections devoted to notation, and to the Lie method for canonical transformations. The main result – the analytic lemma – is stated in §4.4; its proof requires the iterative lemma, whose statement and proof occupy §4.5 (with a few technical results relegated to §4.6). The analytic lemma is proved in §4.7.

4.1 Description of Methods

As discussed in Sect. 1, it is generally impossible to transform the nearly integrable Hamiltonian

$$H(I,\theta,\varepsilon) = h(I) + \varepsilon f(I,\theta,\varepsilon) \tag{4.1}$$

to completely integrable form $H'(I',\varepsilon)$ on an open subset of phase space. Various alternate approaches are possible; the approach used here, which yields exponentially long stability times, is classical, in that it requires transformations to be defined on subsets of phase space with nonempty interior. In this way (4.1) is brought into a resonant normal form (in this case Gustavson normal form)

$$H'(J,\phi) = h(J) + G(J,\phi^*) + R(J,\phi) \tag{4.2}$$

consisting of an effective Hamiltonian $h+G$ together with a small remainder R of order $O(\varepsilon^L)$. The dependence of the normal form on ε is intentionally suppressed because ε must be fixed before the transformation takes place; the transformation is analytic in (I,θ) and (J,ϕ), but is not even continuous in ε. This is because, in order to limit the encroachment of small divisors into the domain of the transformation, we truncate the Fourier series of f at some finite order and push the tail series into the remainder R before effecting the transformation; the truncation order must depend on ε to ensure that $R = O(\varepsilon^L)$, which introduces a discontinuous dependence of the transformation's domain on ε. (This procedure, sometimes called the "ultraviolet cutoff technique," was used by Arnol'd in his proof of the KAM theorem [1].) The remainder R is minimized by letting the exponent L depend on ε, and by balancing the need to make L as large as possible against the need to keep the small divisors from intruding too far into the domain. This balance is achieved for L equal to an inverse power of ε, so that R is indeed exponentially small.

The ϕ^* appearing in (4.2) represents the so-called "resonant variables"; these span a subspace of lower dimension than the original configuration variables θ, or in other words, the effective Hamiltonian is cyclic in one or more linear combinations of the transformed configuration variables ϕ. The precise composition of the resonant variables ϕ^* depends on a subset (called a resonant block; cf. Sect. 5.1) of the region of action space where the transformation is defined. Roughly speaking, these blocks are constructed by excluding from action space those points for which the gradient Dh is nearly orthogonal to the $k \in \mathbf{Z}^3$ (with norm less than the cutoff) which do not belong to a particular submodule $\mathcal{M}$ of $\mathbf{Z}^3$ (the modular structure of $\mathbf{Z}^3$ is recalled in the next subsection). These points must be excluded to control the (finitely many) small divisors arising in the homological equation defining the transformation. The resonant variables ϕ^* then belong to span$\mathcal{M}$. Given a nontrivial q-dimensional submodule $\mathcal{M}$, we thus have a normal form consisting of a q degree of freedom effective (or resonant) Hamiltonian and a small remainder. If $\mathcal{M}$ is the trivial submodule $\{0\}$, then the resonant variables are absent and the effective (nonresonant) Hamiltonian is completely integrable. The normal forms corresponding to nontrivial submodules of dimension 1 then govern motion inside a portion of the zones excluded from the domain of definition of the nonresonant Hamiltonian; 2 degree of freedom normal forms (i.e., with 2 degree of freedom effective Hamiltonians) govern motion inside zones excluded from 1 degree of freedom normal forms, and so on.

4.2 Notation

In this section we will always use the max norm $\|x\|_\infty = \max_i |x_i|$ for vectors x in $\mathbf{R}^3$ or in $\mathbf{C}^3$, while for integer vectors $k \in \mathbf{Z}^3$ we use the taxicab norm $|k| = |k_1| + |k_2| + |k_3|$. In succeeding sections, the Euclidean norm $\|x\| = (\sum_i x_i^2)^{1/2}$ will sometimes be used, both for real and for integer vectors. The action-angle variables (I, θ) on which (3.5) depends initially belong to $K \times \mathbf{T}^3$, where K is a compact subset of $\mathbf{R}^3$, and where the notation $\theta \in \mathbf{T}^3$ is used to indicate that any function depending on θ has unit periodicity in the real part of each of the components $(\theta_1, \theta_2, \theta_3)$.

We will take advantage of the fact that the Hamiltonian (3.5) is analytic by extending its domain of definition $K \times \mathbf{T}^3$ to complex phase space, which in turn allows us to use standard Cauchy estimates. Following [8] and [36], given $K \subset \mathbf{R}^3$ we define

$$F(K, \rho, \sigma) =$$
$$\{(I, \theta) \in \mathbf{C}^3 \times \mathbf{C}^3 \mid \mathrm{Re} I \in K,\ \|\mathrm{Im} I\|_\infty \leq \rho,\ \mathrm{Re}\theta \in \mathbf{T}^3,\ \|\mathrm{Im}\theta\|_\infty \leq \sigma\}\ . \tag{4.3}$$

Following Nekhoroshev, for $K \subset \mathbf{R}^3$ and $\delta \geq 0$ we next define $K - \delta$ as the set of centers of closed, max norm δ-neighborhoods contained in K, and $K + \delta$ as the union of δ-neighborhoods with centers in K. The following simple properties of these operations are immediately verified:

$$(K + \delta_1) + \delta_2 = K + (\delta_1 + \delta_2), \qquad (K - \delta_1) - \delta_2 = K - (\delta_1 + \delta_2) \tag{4.4a}$$

$$(K - \delta_1) + \delta_2 \subset K - (\delta_1 - \delta_2), \quad K + (\delta_1 - \delta_2) \subset (K + \delta_1) - \delta_2, \delta_1 \geq \delta_2 \tag{4.4b}$$

$$(K - \delta) \cap (L - \delta) = (K \cap L) - \delta, \qquad (K \cap L) + \delta \subset (K + \delta) \cap (L + \delta) \tag{4.4c}$$

$$K + \delta \subset L \iff K \subset L - \delta\ . \tag{4.4d}$$

For complex domains $D = F(K, \rho, \sigma)$ and $0 < \delta < \rho$, $0 < \sigma < \xi$ we set $D - (\delta, \xi) = F(K - \delta, \rho - \delta, \sigma - \xi)$.

For f analytic on D, we use $\|f\|_D$ to denote $\sup_D |f(I, \theta)|$ if f is scalar-valued, and $\max_i \sup_D |f_i(I, \theta)|$ if f is vector-valued. For nonnegative integers α, β, $\|\partial^{\alpha+\beta} f / \partial I^\alpha \partial \theta^\beta\|_D$ will denote $\max_{|r|=\alpha, |s|=\beta} \|\partial^{|r|+|s|} f / \partial I^r \partial \theta^s\|_D$, where $r = (r_1, r_2, r_3)$ and $s = (s_1, s_2, s_3)$ are multiindices with nonnegative integer entries, and where $I^r = I_1^{r_1} I_2^{r_2} I_3^{r_3}$ and $\theta^s = \theta_1^{s_1} \theta_2^{s_2} \theta_3^{s_3}$.

In order to discuss resonances, we recall that a finite subset $\{k^{(i)}\}_{i=1}^m$ of $\mathbf{Z}^3$ generates a submodule of $\mathbf{Z}^3$ defined as $\{z \in \mathbf{Z}^3 \mid z = \sum n_i k^{(i)}, n_i \in \mathbf{Z}\}$. Each submodule of $\mathbf{Z}^3$ has dimension 0, 1, 2, or 3, in the sense that the smallest vector subspace of $\mathbf{R}^3$ containing the submodule has that dimension. If $\{k^{(i)}\}_{i=1}^m$ generates a submodule of dimension n, then the maximal submodule generated by $\{k^{(i)}\}_{i=1}^m$ is the largest submodule of dimension n containing $\{k^{(i)}\}_{i=1}^m$. In the remainder of this article, the symbol $\mathcal{M}$ will refer to a 0-, 1- or 2-dimensional maximal submodule of $\mathbf{Z}^3$. Not surprisingly, every $\mathcal{M}$ is generated respectively by integer combinations of a 1- or 2-element basis in $\mathbf{Z}^3$ (cf. Appendix 1 of [17]); the *order* $|\mathcal{M}|$ of a maximal submodule is the smallest nonnegative integer r such that $\mathcal{M}$ admits a basis of vectors with norm less than or equal to r.

Finally, for functions f on D with Fourier coefficients $f_k(I)$, we introduce the notation

$$\Pi_{\mathcal{M}} f = \sum_{k \in \mathcal{M}} f_k(I) e^{2\pi i k \cdot \theta} \tag{4.5}$$

for the resonant subseries of f corresponding to the maximal submodule $\mathcal{M}$. $\Pi_{\mathcal{M}}$ is a projection operator, as the notation suggests, and, most important for the channeling problem, the subseries $\Pi_{\mathcal{M}} W$ is precisely the continuum potential $\overline{W}$ obtained by averaging W over its period(s) in the orthogonal complement (in $\mathbf{R}^3$) of $\mathcal{M}$. Thus when $\dim\mathcal{M} = 1$ or 2, $\Pi_{\mathcal{M}} W$ is a planar or axial continuum potential, respectively. The proofs of these facts are nearly obvious; full details may be found in Appendix 2 of [17].

4.3 Near Identity Canonical Transformations via the Lie Method

The following lemma expresses facts known about canonical transformations since the time of Sophus Lie, and it (or some variant) is proved in [8] and in [36]. Because of its importance, we also give a proof which references some standard results but is more detailed than the proofs just cited.

Lemma 4.1 (Lie method for canonical transformations). *Given a compact set $K \subset \mathbf{R}^3$ and positive numbers ρ and σ, consider the domain $D = F(K, \rho, \sigma)$ and an analytic function $\chi : D \to \mathbf{C}$ which is real for real arguments; i.e., for $(I, \theta) \in K \times \mathbf{T}^3$. Let $(I, \theta)(\tau) = (T_I^\tau, T_\theta^\tau) = T^\tau(J, \phi)$ denote the flow which takes the initial condition $(J, \phi) \in D$ to the solution (I, θ) at time τ of the Hamiltonian initial value problem*

$$\frac{dI}{d\tau} = -\frac{\partial \chi}{\partial \theta}, \quad \frac{d\theta}{d\tau} = \frac{\partial \chi}{\partial I} \tag{4.6a}$$

$$(I, \theta)(0) = (J, \phi) \ . \tag{4.6b}$$

Define $T = T^1$ to be the time-one map corresponding to the flow T^τ. If A, B, δ, ξ are positive numbers satisfying

$$A < \delta < \rho \quad \textit{and} \quad B < \xi < \sigma, \quad \textit{and if} \tag{4.7}$$

$$\left\| \frac{\partial \chi}{\partial I} \right\|_{D-(\delta/2, \xi/2)} \le \frac{B}{2}, \quad \left\| \frac{\partial \chi}{\partial \theta} \right\|_{D-(\delta/2, \xi/2)} \le \frac{A}{2} \ , \tag{4.8}$$

then for $0 \le \tau \le 1$, the flow T^τ is a well-defined analytic function of (J, ϕ) and τ on a suitable (I, θ)-domain $D' = F(K', \rho - \delta, \sigma - \xi)$ satisfying

$$D - (\delta, \xi) \subset D' \subset D - (\delta/2, \xi/2) \ . \tag{4.9}$$

Moreover for fixed $\tau \in [0, 1]$, the map $T^\tau : D' \to T(D')$ is a bijective canonical transformation taking real sets onto real sets, with a canonical analytic inverse

$(T^\tau)^{-1} : T(D') \to D'$ which coincides with $T^{-\tau}$. Each of these maps is close to the identity in the sense that

$$K - \delta \subset K' \subset K - \delta/2 \ , \tag{4.10a}$$

$$K - \delta \subset T_I^\tau(K') \subset K - \delta/2, \quad \textit{and} \tag{4.10b}$$

$$\|I - J\|_{D'} \le \frac{A}{2}, \quad \|\theta - \phi\|_{D'} \le \frac{B}{2} \ . \tag{4.11}$$

Finally, given any function f analytic on D, the analytic function $f' = f \circ T : D' \to \mathbf{C}$ satisfies

$$\|f'\|_{D'} \le \|f\|_D \quad \textit{and} \tag{4.12}$$

$$\|f' - f - \{f, \chi\}\|_{D'} \le \frac{1}{2}\|\{\{f, \chi\}, \chi\}\|_{D-(\delta/2,\xi/2)} \ , \tag{4.13}$$

where $\{,\}$ is the Poisson bracket

$$\{f, \chi\} = \sum_i \left(\frac{\partial f}{\partial \theta_i}\frac{\partial \chi}{\partial I_i} - \frac{\partial f}{\partial I_i}\frac{\partial \chi}{\partial \theta_i} \right) \ . \tag{4.14}$$

Proof. From the standard existence-uniqueness theory for ODE's (see, e.g., [5], §32.5) we know that, given $(J, \phi) \in Int(D)$, there is an open interval (a, b) containing 0 for which the flow $(I(\tau), \theta(\tau)) = T^\tau(J, \phi)$ is a one-to-one analytic function of τ and (J, ϕ) for all $\tau \in (a, b)$. Since χ is real for real arguments, if (J, ϕ) is real then $T^\tau(J, \phi)$ is real for all τ. By well-known continuation arguments (cf. [25], §1.2), the existence interval extends to a maximal existence interval (a', b') such that $T^\tau(J, \phi) \to \partial D$ as $\tau \searrow a'$ or $\tau \nearrow b'$. If the partial derivatives of χ are uniformly bounded on D or some appropriate subset, it is a simple matter to estimate the maximal existence time $\min\{|a'|, |b'|\}$ in terms of the distance from (J, ϕ) to ∂D. This observation may be turned around to estimate the size of D' such that T^τ is defined and $T^\tau(D') \subset D - (\delta/2, \xi/2)$ for all $\tau \in [0, 1]$. More precisely, set $D'' = D - (\delta/2, \xi/2)$ and $D' = \{(J, \phi) \in D'' | T^\tau(J, \phi) \in D''$ for $0 \le \tau \le 1\}$. Then $D - (\delta, \xi) \subset D'$ is easily seen as follows. Let $(J, \phi) \in D - (\delta, \xi) \subset D''$ and suppose $(I, \theta) = T^\tau(J, \phi) \notin D''$ for some $\tau \in [0, 1]$. Set $\tau_1 = \inf_{0 \le \tau' \le 1}\{T^{\tau'}(J, \phi) \in \partial D''\}$. ($\tau_1$ exists in $[0, 1]$ by continuity of the flow in D''.) By definition $\|T_I^{\tau_1}(J, \phi) - J\|_\infty \ge \delta/2$; on the other hand, by (4.6a), $T_I^{\tau_1}(J, \phi) - J = -\int_0^{\tau_1}(\partial\chi/\partial\theta)d\tau'$, and so using (4.7) and (4.8), we have $\|T_I^{\tau_1}(J, \phi) - J\|_\infty < \delta/2$. The contradiction shows that $T^\tau(J, \phi) \in D''$, which proves (4.9) and (4.10a) for $0 \le \tau \le 1$. A very similar argument also shows that $T^\tau(J, \phi) \in D''$ for $-1 \le \tau \le 0$; we thus have

$$T^\tau(D - (\delta, \xi)) \subset D - (\delta/2, \xi/2) \quad \text{for all } \tau \in [-1, 1] \ . \tag{4.15}$$

To prove (4.10b), let $\tau \in [0, 1]$. By definition $T^\tau(D') \subset D - (\delta/2, \xi/2)$; now let $(I, \theta) \in D - (\delta, \xi) \subset D'$. Then (4.15) $\Rightarrow$

$$T^{-\tau}(I, \theta) \in D'' \ . \tag{4.16}$$

For every $\eta \in [0,1]$, $|\eta - \tau| \le 1 \Rightarrow$

$$T^{\eta}T^{-\tau}(I,\theta) = T^{\eta-\tau}(I,\theta) \in D'' \ . \tag{4.17}$$

From (4.16), (4.17) and the definition of D', it follows that $T^{-\tau}(I,\theta) \in D'$, so $T^{-\tau}(D - (\delta,\xi)) \subset D'$, or $(D - (\delta,\xi)) \subset T^{\tau}(D')$, from which (4.10b) follows.

To establish the conclusions of the lemma concerning the inverse transformation, note that for any $\tau \in [0,1]$ we now have a domain $D' = F(K', \rho - \delta, \sigma - \xi)$ for T^{τ} where K' satisfies (3.7a). Since $T^{\tau} : D' \to T^{\tau}(D')$ is bijective, it has an inverse $(T^{\tau})^{-1}$. On the other hand, given $(I,\theta) \in T^{\tau}(D')$, suppose $(T^{\tau})^{-1}(I,\theta) = (J,\phi)$. Because the system (3.3) is autonomous we also have $T^{-\tau}(I,\theta) = (J,\phi)$. Therefore $(T^{\tau})^{-1} = T^{-\tau}$, which is analytic.

It has long been known that a well-defined flow such as T^{τ} given by an autonomous Hamiltonian system (4.6) provides a canonical transformation of its domain onto its range for fixed τ, and that its inverse does the same. By canonical, we mean that the symplectic form $dI \wedge d\theta$ is preserved ($dI \wedge d\theta = dJ \wedge d\phi$), and that any set of Hamilton's equations written in terms of the phase variables (I,θ) and the Hamiltonian H retain their form in terms of the variables (J,ϕ) with new Hamiltonian $H \circ T^{\tau}$. Proofs of this fact abound; the reader may consult [6], §38A and §44E for a modern treatment; detailed proofs in canonical coordinates may be found in older texts.

To complete the proof, note that the bounds (4.11) now clearly follow from (4.6), (4.7) and (4.8), while (4.12) is immediate, since $f' = f \circ T$ and $T(D') \subset D$. As for (4.13), using the familiar relation $\frac{d}{d\tau}(f \circ T^{\tau}) = \{f,\chi\} \circ T^{\tau}$ (see, e.g., [5], §40A), we have by Taylor's formula at second order: $f' = f \circ T^0 + \{f,\chi\} \circ T^0 + \frac{1}{2}\{\{f,\chi\},\chi\} \circ T^{\tau}$ for some $\tau \in (0,1) \Rightarrow \|f' - f - \{f,\chi\}\|_{D'} = \frac{1}{2}\|\{\{f,\chi\},\chi\} \circ T^{\tau}\|_{D'}$. But $(J,\phi) \in D'$ and $\tau \in (0,1) \Rightarrow T^{\tau}(J,\phi) \in D - (\delta/2,\xi/2)$, so this last quantity is less than or equal to $\frac{1}{2}\|\{\{f,\chi\},\chi\}\|_{D-(\delta/2,\xi/2)}$. □

4.4 Statement of the Analytic Lemma

The original version of the following result was called the "analytic lemma" by Nekhoroshev, because it is simply a statement about the transformation of a Hamiltonian on certain parts of phase space; no mention is made of the corresponding solutions to Hamilton's equations. Showing that solutions remain in regions where the transformation to normal form is valid is a separate issue to be discussed in the next section.

Using the notation of §4.2, we now state

Lemma 4.2 (Analytic lemma). *Let α, τ, d, $c > 1/2$, $p > 4$, and $a \equiv \alpha + (p+1)\tau$ be positive numbers satisfying the consistency relations*

$$1 - 2a - (p+5)\tau/2 - d > 0 \ , \tag{4.18a}$$

$$1 - a - (p+5)\tau/2 - c > 0 \ , \tag{4.18b}$$

$$1 - 4a - (p+6)\tau > 0 \ . \tag{4.18c}$$

(An example of such a set of numbers is $\alpha = 1/8$, $\tau = 1/72$, $d = 1/2$, $c = 5/8$, $p = 5$, $a = 5/24$.)
Let $0 < \sigma, A \le 1$, E be positive numbers and let $\varepsilon > 0$ be a small parameter satisfying the following restrictions (no attempt is made to combine them):

$$\varepsilon \le e^{-1/\tau} \,, \tag{4.19a}$$

$$\varepsilon \le \left(\frac{A^2 \sigma^{p+4}}{2^{2p+13} C_1 E} \right)^{\frac{1}{1-2a-(p+5)\tau/2-d}} \,, \tag{4.19b}$$

$$\varepsilon \le \left(\frac{A^2 \sigma^{p+4}}{2^{2p+12} C_1 E} \right)^{\frac{1}{1-a-(p+5)\tau/2-c}} \,, \tag{4.19c}$$

$$\varepsilon \le \left(\frac{A^2 \sigma^{2p+8}}{C_2 e 2^{4p+22}} \right)^{\frac{1}{1-4a-(p+6)\tau}} \,, \tag{4.19d}$$

$$\varepsilon^{-\tau/2} \ge \frac{8}{\pi\sigma} \left(1 - \log\left(\frac{(\pi\sigma)^3}{2^{13}[2(p+2)]^{p+2}} \right) - \frac{5\tau}{2} \log\varepsilon \right) \,, \tag{4.19e}$$

where

$$C_1 = \frac{2^7 [2(p+4)]^{p+4}}{\pi^{p+4} e^p} \,, \quad \textit{and} \tag{4.20a}$$

$$C_2 = 4 C_1 \left(\frac{E}{A} \right) \left[9A + 88 C_1 \left(\frac{E}{A} \right) \right] \,. \tag{4.20b}$$

Set $N = \varepsilon^{-\tau}$ and let $\mathcal{M} \ne \mathbf{Z}^3$ be a maximal submodule of $\mathbf{Z}^3$. Define $h(I) = \frac{1}{2} I^2$, and suppose $H(I,\theta) = h(I) + f(I,\theta)$ is analytic on $D = F(K, A\varepsilon^a, \sigma)$ where $K \subset \mathbf{R}^3$ is a compact set for which

$$I \in K \Rightarrow |k \cdot I| \ge \frac{3}{2} A \varepsilon^\alpha |k|^{-p} \quad \textit{for all } k \notin \mathcal{M} \quad \textit{with } |k| \le N \,. \tag{4.21}$$

Suppose that

$$\|H\|_D \le E \quad \textit{and} \tag{4.22a}$$

$$\|f\|_D \le \varepsilon E \tag{4.22b}$$

(in general, $E \ge E_o$, where E_o is defined in (3.6)). Then there exists a bijective transformation $T : D_\infty \to T(D_\infty)$ taking real sets onto real sets, with both T and T^{-1} canonical and analytic, and with

$$D - \left(\frac{A}{2} \varepsilon^a, \frac{\sigma}{2} \right) \subset D_\infty \subset D - \left(\frac{A}{8} \varepsilon^a, \frac{\sigma}{8} \right) \quad \textit{and} \tag{4.23a}$$

$$D - \left(\frac{A}{2} \varepsilon^a, \frac{\sigma}{2} \right) \subset T(D_\infty) \subset D - \left(\frac{A}{8} \varepsilon^a, \frac{\sigma}{8} \right) \,. \tag{4.23b}$$

On D_∞, T transforms H to the form $H' = H \circ T$, which may be written

$$H'(J,\phi) = h(J) + G(J,\phi) + R(J,\phi), \quad \textit{where} \tag{4.24}$$

$$G(J,\phi) = \sum_{k\in\mathcal{M}} G_k(J)e^{2\pi i k\cdot\phi}, \quad \textit{and} \tag{4.25a}$$

$$R(J,\phi) = \sum_{k\notin\mathcal{M}} R_k(J)e^{2\pi i k\cdot\phi} \ . \tag{4.25b}$$

Furthermore,

$$\|R\|_{D_\infty} \le 2\varepsilon E \exp(-\varepsilon^{-\tau/4}), \quad \textit{and} \tag{4.26}$$

$$\|G - \Pi_{\mathcal{M}} f\|_{D_\infty} \le E\varepsilon^{1+\tau} \ . \tag{4.27}$$

Finally, given $(J,\phi) \in D_\infty$ *and setting* $(I,\theta) = T(J,\phi)$*, we have*

$$\|I - J\|_\infty \le \frac{A}{4}\varepsilon^c, \quad \textit{and} \tag{4.28a}$$

$$\|\theta - \phi\|_\infty \le \frac{\sigma}{4}\varepsilon^d \ . \tag{4.28b}$$

Proof of the analytic lemma is deferred to the end of the section; it is first necessary to set up the "iterative lemma," which allows for a single step in the formation of the composite transformation T appearing above.

4.5 The Iterative Lemma

Lemma 4.3 (Iterative lemma). *Assume the hypotheses of Lemma 4.2, with the exception of (4.22b). Write*

$$H(I,\theta) = h(I) + G(I,\theta) + R(I,\theta) \ , \tag{4.29}$$

where $G = \Pi_{\mathcal{M}} f$ *and* $R = (1 - \Pi_{\mathcal{M}})f$*. Assume that*

$$\|G + R\|_D \le \gamma E \le 2\varepsilon E \quad \textit{and} \tag{4.30}$$

$$\|R\|_D \le \eta E \le 2\varepsilon E \ , \tag{4.31}$$

where γ*,* η *satisfy*

$$0 < \eta, \gamma \le 2\varepsilon \quad \textit{and} \quad \eta \le 2\gamma \ . \tag{4.32}$$

Then given A' *and* B' *satisfying* $\frac{1}{4}A\varepsilon^{\tau/2} \le A' < A$ *and* $\frac{1}{4}\sigma\varepsilon^{\tau/2} \le B' < \sigma$*, there is a bijective transformation* $T' : D' \to T'(D')$ *with both* T' *and* $(T')^{-1}$ *canonical and analytic and with*

$$D - (A'\varepsilon^a, B') \subset D' \subset D - \left(\frac{A'\varepsilon^a}{2}, \frac{B'}{2}\right) , \tag{4.33}$$

$$D - (A'\varepsilon^a, B') \subset T'(D') \subset D - \left(\frac{A'\varepsilon^a}{2}, \frac{B'}{2}\right) , \tag{4.34}$$

such that if $T'(J,\phi) = (I,\theta)$ *then*

$$\|I - J\|_{D'} \le \frac{A'}{2}\varepsilon^c, \quad \textit{and} \tag{4.35}$$

$$\|\theta - \phi\|_{D'} \le \frac{B'}{2}\varepsilon^d , \tag{4.36}$$

and such that the transformed Hamiltonian $H' = H \circ T'$ *takes the normal form*

$$H'(J,\phi) = h(J) + G'(J,\phi) + R'(J,\phi) \tag{4.37}$$

on D', *where*

$$\Pi_{\mathcal{M}} G' = G', \quad \Pi_{\mathcal{M}} R' = 0 , \tag{4.38}$$

$$\|H'\|_{D'} \le E , \tag{4.39}$$

$$\|G' + R'\|_{D'} \le \gamma' E , \tag{4.40}$$

$$\|R'\|_{D'} \le \eta' E = (\eta_1' + \eta_2')E, \quad \textit{and where} \tag{4.41}$$

$$\gamma' = \gamma + \frac{1}{2}\eta' = \gamma + \frac{1}{2}(\eta_1' + \eta_2') , \tag{4.42}$$

$$\eta_1' = \frac{2^6[2(p+2)]^{p+2}\eta e^{-N\pi B'/2}}{(\pi B')^3} \tag{4.43}$$

$$\eta_2' = C_2 \frac{\gamma\eta}{(A')^2(B')^{2p+8}\varepsilon^{4a}} \tag{4.44}$$

$$C_1 = \frac{2^7[2(p+4)]^{p+4}}{\pi^{p+4}e^p} \tag{4.45}$$

$$C_2 = 4C_1\left(\frac{E}{A}\right)\left[9A + 88C_1\left(\frac{E}{A}\right)\right] . \tag{4.46}$$

Proof. Proceeding according to Lemma 4.1, we construct the desired transformation by means of a Lie generating function $\chi : D \to \mathbf{C}$. Assume tentatively that χ is analytic and define T' as in Lemma 4.1. Setting $H' = H \circ T'$ and using (4.29), the identity $H' = H' + H - H + \{H,\chi\} - \{H,\chi\}$ may be written

$$\begin{aligned} H' &= h + G + R + \{h,\chi\} + \{G+R,\chi\} + H' - H - \{H,\chi\} \\ &= h + G + R^{\le} + \{h,\chi\} + S , \end{aligned} \tag{4.47}$$

where we have introduced the cut off part of R (recall $N = \varepsilon^{-\tau}$):

$$R^{\le}(J,\phi) = \sum_{\substack{k\notin\mathcal{M} \\ |k|\le N}} R_k(J)e^{2\pi i k\cdot\phi} , \tag{4.48}$$

and the N-tail of R:

$$R^{>} = R - R^{\le}, \quad \text{and where} \tag{4.49}$$

$$S = R^{>} + \{G+R,\chi\} + H' - H - \{H,\chi\} . \tag{4.50}$$

This suggests choosing χ so as to eliminate $R^{\le}$. We should then be able to estimate S by means of (4.13), (4.30) and the fact that $R^{>}$ is small. Proceeding this way, we are led to consider the homological equation $R^{\le} + \{h,\chi\} = 0$, or explicitly

$$\sum_{\substack{k \notin \mathcal{M} \\ |k| \leq N}} R_k(J) e^{2\pi i k \cdot \phi} - J \cdot \frac{\partial \chi}{\partial \phi} = 0 \,. \tag{4.51}$$

Expanding χ in its Fourier series $\chi = \sum_k \chi_k(J) e^{2\pi i k \cdot \phi}$, we find that (4.51) will be satisfied if

$$\chi_k(J) = -\frac{i R_k(J)}{2\pi (k \cdot J)} \qquad \text{for} \quad k \notin \mathcal{M}, |k| \leq N \,, \tag{4.52}$$

$$\chi_k(J) = 0 \qquad \text{otherwise} \,.$$

χ defined in this way is analytic on D, since it is a finite trigonometric sum and its coefficients χ_k are well-behaved analytic functions of J on D, by (4.21) and Proposition 4.5. (Note: Propositions 4.5 through 4.7 are in §4.6.) With a view to using Lemma 4.1, let us estimate the partial derivatives of χ. First, using the defining relation (4.52) together with the nonresonance condition (4.21) and Proposition 4.5, we find (recall $\alpha < a$)

$$\|\chi_k\|_D \leq \frac{|k|^p}{A\pi\varepsilon^\alpha} \|R_k\|_D \leq \frac{|k|^p}{A\pi\varepsilon^a} \|R_k\|_D \,. \tag{4.53}$$

Because R is analytic on D, it is a simple matter to show (see, e.g., Appendix 2.3 of [17]) that its Fourier coefficients decrease exponentially as

$$\|R_k\|_D \leq e^{-2\pi\sigma|k|} \|R\|_D \,, \tag{4.54}$$

and so combining the last two inequalities gives

$$\|\chi_k\|_D \leq \frac{|k|^p}{A\pi\varepsilon^a} e^{-2\pi\sigma|k|} \|R\|_D \,. \tag{4.55}$$

Proposition 4.6 and (4.31), (4.45), and (4.55) then give

$$\begin{aligned} \|\chi\|_{D-(0,\frac{B'}{2})} &\leq \left(\frac{2^5 [2(p+2)]^{p+2}}{\pi^{p+4} e^{p+2-\pi B'/2}} \right) \frac{\eta E}{A(B')^{p+3}\varepsilon^a} \\ &\leq C_1 \frac{\eta E}{A(B')^{p+3}\varepsilon^a} \,. \end{aligned} \tag{4.56}$$

A standard Cauchy estimate gives

$$\|\frac{\partial \chi}{\partial J}\|_{D-(\frac{A'}{2}\varepsilon^a, \frac{B'}{2})} \leq \frac{2}{A'\varepsilon^a} \|\chi\|_{D-(0,\frac{B'}{2})} \leq 2C_1 \frac{\eta E}{AA'(B')^{p+3}\varepsilon^{2a}} \,, \tag{4.57}$$

which, since $\eta \leq 2\varepsilon$ and $A' \geq \frac{1}{4} A\varepsilon^{\tau/2}, B' \geq \frac{1}{4}\sigma\varepsilon^{\tau/2}$, is bounded by

$$\frac{B'}{2} \left(\frac{2^{2p+13} C_1 E}{A^2 \sigma^{p+4}} \right) \varepsilon^{1-2a-(p+5)\tau/2} \leq \frac{B'}{2} \varepsilon^d \,, \tag{4.58}$$

where the last inequality follows from the restriction (4.19b).

The other partial derivative may be estimated with the help of (4.45), (4.55) and Proposition 4.7:

$$\left\| \frac{\partial \chi}{\partial \phi} \right\|_{D-(0,\frac{B'}{2})} \leq \frac{2^6 \|R\|_D [2(p+3)]^{p+3}}{A\varepsilon^a (\pi B')^{p+4}} e^{\pi B'/2-p-3} \leq$$

$$\left(\frac{2^6 [2(p+3)]^{p+3}}{\pi^{p+4} e^p} \right) \frac{\eta E}{A(B')^{p+4}\varepsilon^a} \leq C_1 \frac{\eta E}{A(B')^{p+4}\varepsilon^a} \, . \tag{4.59}$$

Again using $A' \geq \frac{1}{4} A\varepsilon^{\tau/2}$, $B' \geq \frac{1}{4}\sigma\varepsilon^{\tau/2}$, and $D - (\frac{A'}{2}\varepsilon^a, \frac{B'}{2}) \subset D - (0, \frac{B'}{2})$, we find that

$$\left\| \frac{\partial \chi}{\partial \phi} \right\|_{D-(\frac{A'}{2}\varepsilon^a, \frac{B'}{2})} \leq \left\| \frac{\partial \chi}{\partial \phi} \right\|_{D-(0,\frac{B'}{2})} \leq$$

$$\frac{A'}{2} \left(\frac{2^{2p+12} C_1 E}{A^2 \sigma^{p+4}} \right) \varepsilon^{1-a-(p+5)\tau/2} \leq \frac{A'}{2} \varepsilon^c \, , \tag{4.60}$$

where restriction (4.19c) was used in the last inequality.

Lemma 4.1 on canonical transformations now ensures that the transformation T' defined as the time-one map of the Hamiltonian flow generated by χ given in (4.52) is a well-defined, analytic canonical transformation, with a canonical analytic inverse, on a domain D' satisfying (4.33) and (4.34). The lemma also assures that T' satisfies (4.35) and (4.36).

On D', we have by construction (see 4.47)

$$H' = H \circ T' = h + G + S \, , \tag{4.61}$$

where S is given by (4.50). Let us estimate the terms comprising S. First, since $D' \subset D - (0, \frac{B'}{2})$,

$$\|R^{>}\|_{D'} \leq \|R^{>}\|_{D-(0,\frac{B'}{2})} \leq$$

$$\frac{2^5 [2(p+2)]^{p+2} \|R\|_D}{(\pi B')^3} e^{-N\pi B'/2} \leq \frac{2^5 [2(p+2)]^{p+2} \eta e^{-N\pi B'/2}}{(\pi B')^3} E = \frac{1}{2}\eta_1' E \, . \tag{4.62}$$

The second inequality derives from (4.54) and Proposition 4.6; the third inequality derives from (4.31). (Proposition 4.6 applies to $R^{>}$ as well as to $R^{\geq}$.)

To estimate the next term of S, we abbreviate $D_2 = D - (\frac{A'}{2}\varepsilon^a, \frac{B'}{2})$ and write

$$\|\{G+R, \chi\}\|_{D'} \leq \|\{G+R, \chi\}\|_{D_2} \leq$$

$$\sum_{j=1}^{3} \left(\left\| \frac{\partial(G+R)}{\partial \phi_j} \right\|_{D_2} \left\| \frac{\partial \chi}{\partial J_j} \right\|_{D_2} + \left\| \frac{\partial(G+R)}{\partial J_j} \right\|_{D_2} \left\| \frac{\partial \chi}{\partial \phi_j} \right\|_{D_2} \right) \leq$$

$$3 \left(\frac{2}{B'} \|G+R\|_D \left\| \frac{\partial \chi}{\partial J} \right\|_{D-(0,\frac{B'}{2})} + \frac{2}{A'\varepsilon^a} \|G+R\|_D \left\| \frac{\partial \chi}{\partial \phi} \right\|_{D-(0,\frac{B'}{2})} \right) \, , \tag{4.63}$$

where the last inequality was obtained with Cauchy estimates. By (4.30), (4.57) and (4.59), we then have

$$\|\{G+R, \chi\}\|_{D'} \leq 18 C_1 \left(\frac{E}{A} \right) \frac{E\gamma\eta}{A'(B')^{p+4}\varepsilon^{2a}} \, . \tag{4.64}$$

We now proceed to the last term of S, which is estimated here by brute force. Using (4.13) with $\delta = A\varepsilon^a$, $\xi = \sigma$, we have

$$\|H' - H - \{H,\chi\}\|_{D'} \leq \frac{1}{2}\|\{\{H,\chi\},\chi\}\|_{D_2} \; . \tag{4.65}$$

Expanding the double Poisson bracket and collecting terms, this is

$$\leq \frac{9}{2}\left(\left\|\frac{\partial\chi}{\partial J}\right\|_{D_2}\left\|\frac{\partial H}{\partial\phi}\right\|_{D_2}\left\|\frac{\partial^2\chi}{\partial J\partial\phi}\right\|_{D_2} + \left\|\frac{\partial\chi}{\partial J}\right\|_{D_2}\left\|\frac{\partial^2\chi}{\partial\phi^2}\right\|_{D_2}\left\|\frac{\partial H}{\partial J}\right\|_{D_2} + \right.$$

$$\left\|\frac{\partial\chi}{\partial J}\right\|_{D_2}^2\left\|\frac{\partial^2 H}{\partial\phi^2}\right\|_{D_2} + 2\left\|\frac{\partial\chi}{\partial J}\right\|_{D_2}\left\|\frac{\partial^2 H}{\partial J\partial\phi}\right\|_{D_2}\left\|\frac{\partial\chi}{\partial\phi}\right\|_{D_2} + \left\|\frac{\partial\chi}{\partial\phi}\right\|_{D_2}^2\left\|\frac{\partial^2 H}{\partial J^2}\right\|_{D_2}$$

$$\left. + \left\|\frac{\partial\chi}{\partial\phi}\right\|_{D_2}\left\|\frac{\partial H}{\partial J}\right\|_{D_2}\left\|\frac{\partial^2\chi}{\partial J\partial\phi}\right\|_{D_2} + \left\|\frac{\partial\chi}{\partial\phi}\right\|_{D_2}\left\|\frac{\partial^2\chi}{\partial J^2}\right\|_{D_2}\left\|\frac{\partial H}{\partial\phi}\right\|_{D_2}\right) . \tag{4.66}$$

Most of the work needed to gauge the factors in the preceding expression has already been done. $\|\frac{\partial\chi}{\partial J}\|_{D_2}$ and $\|\frac{\partial\chi}{\partial\phi}\|_{D_2}$ are estimated via (4.57) and (4.59); these inequalities together with Cauchy estimates then give

$$\left\|\frac{\partial^2\chi}{\partial J\partial\phi}\right\|_{D_2} \leq \frac{2C_1\eta E}{AA'(B')^{p+4}\varepsilon^{2a}} \tag{4.67}$$

$$\left\|\frac{\partial^2\chi}{\partial J^2}\right\|_{D_2} \leq \frac{8C_1\eta E}{A(A')^2(B')^{p+3}\varepsilon^{3a}} \; , \tag{4.68}$$

while

$$\left\|\frac{\partial^2\chi}{\partial\phi^2}\right\|_{D_2} \leq \left(\frac{2^7[2(p+4)]^{p+4}}{\pi^{p+4}e^{p+2}}\right)\frac{\eta E}{A(B')^{p+5}\varepsilon^a} \leq C_1\frac{\eta E}{A(B')^{p+5}\varepsilon^a} \tag{4.69}$$

follows from Proposition 4.7 and (4.31) and (4.55). Finally, the partial derivatives of H are all bounded using Cauchy estimates:

$$\left\|\frac{\partial H}{\partial J}\right\|_{D_2} \leq \frac{2E}{A'\varepsilon^a}, \qquad \left\|\frac{\partial H}{\partial\phi}\right\|_{D_2} \leq \frac{2E}{B'} \tag{4.70}$$

$$\left\|\frac{\partial^2 H}{\partial J^2}\right\|_{D_2} \leq \frac{8E}{(A')^2\varepsilon^{2a}}, \qquad \left\|\frac{\partial^2 H}{\partial\phi^2}\right\|_{D_2} \leq \frac{8E}{(B')^2} \tag{4.71}$$

$$\left\|\frac{\partial^2 H}{\partial J\partial\phi}\right\|_{D_2} \leq \frac{4E}{A'B'\varepsilon^a} \tag{4.72}$$

Inserting these expressions into (4.66) and simplifying yields the required estimate of the last term in S:

$$\|H' - H - \{H,\chi\}\|_{D'} \leq 88\left(\frac{C_1E}{A}\right)^2\frac{\eta^2E}{(A')^2(B')^{2p+8}\varepsilon^{4a}} \; . \tag{4.73}$$

Now combining (4.64) and (4.73) results in

$$\|\{G+R,\chi\}\|_{D'} + \|H' - H - \{H,\chi\}\|_{D'} \leq$$

$$2C_1\left(\frac{E}{A}\right)\left[9A'(B')^{p+4}\varepsilon^{2a} + 44C_1\left(\frac{E}{A}\right)\frac{\eta}{\gamma}\right]\frac{\gamma\eta E}{(A')^2(B')^{2p+8}\varepsilon^{4a}} \ . \tag{4.74}$$

Using $A' < A$, $B' < 1$, $\varepsilon \leq 1$ and $\eta \leq 2\gamma$, the last expression is

$$\leq 2C_1\left(\frac{E}{A}\right)\left[9A + 88C_1\left(\frac{E}{A}\right)\right]\frac{\gamma\eta E}{(A')^2(B')^{2p+8}\varepsilon^{4a}} =$$

$$\frac{1}{2}C_2\frac{\gamma\eta E}{(A')^2(B')^{2p+8}\varepsilon^{4a}} = \frac{1}{2}\eta_2' E \ . \tag{4.75}$$

Putting together (4.62) and (4.75) gives the final bound on S:

$$\|S\|_{D'} \leq \|R^{>}\|_{D'} + \|\{G+R,\chi\}\|_{D'} + \|H' - H - \{H,\chi\}\|_{D'} \leq$$

$$\frac{1}{2}\eta_1' E + \frac{1}{2}\eta_2' E = \frac{1}{2}\eta' E \ . \tag{4.76}$$

The iterative lemma may now be concluded by identifying the correct expressions for G' and R' and by verifying the final estimates. We take

$$G' = G + \Pi_{\mathcal{M}} S \quad \text{and} \quad R' = (1 - \Pi_{\mathcal{M}})S \ , \tag{4.77}$$

so that (4.37) and (4.38) are clearly satisfied, and

$$\|G' + R'\|_{D'} = \|G + S\|_{D'} \leq$$

$$\|G\|_{D'} + \|S\|_{D'} \leq \gamma E + \frac{1}{2}\eta' E = \gamma' E \ , \tag{4.78}$$

which proves (4.40). On the other hand,

$$\|R'\|_{D'} \leq \|S\|_{D'} + \|\Pi_{\mathcal{M}} S\|_{D'} \leq 2\|S\|_{D'} \leq \eta' E \tag{4.79}$$

proves (4.41). Finally,

$$\|H'\|_{D'} = \|H \circ T'\|_{D'} \leq \|H\|_D \leq E \ , \tag{4.80}$$

since $T'(D') \subset D$, by (4.34). The iterative lemma is proved. □

4.6 Technical Estimates

This subsection establishes some of the inequalities used in the proof of the iterative lemma. We begin by showing that a real subset of action space which is nonresonant (to a certain order and with respect to a particular submodule) remains nonresonant when complexified.

Proposition 4.5. *Given a maximal submodule $\mathcal{M}$ of $\mathbf{Z}^3$ and an ultraviolet cutoff $N = \varepsilon^{-\tau}$, suppose $K \subset \mathbf{R}^3$ is such that $I \in K \Rightarrow |k \cdot I| \geq \frac{3}{2}A\varepsilon^{\alpha}|k|^{-p}$ for some*

$\alpha, p > 0$ and for all $k \notin \mathcal{M}$ with $|k| \leq N$. If $a = \alpha + (p+1)\tau$, $\sigma \geq 0$, and $D = F(K, A\varepsilon^a, \sigma)$, then given $(I, \theta) \in D$, the complex action variable I satisfies

$$|k \cdot I| \geq \frac{A}{2}\varepsilon^\alpha |k|^{-p} \quad \mathit{for} k \notin \mathcal{M}, |k| \leq N \; . \tag{4.81}$$

Proof. $|k \cdot I| \geq \big||k \cdot \mathrm{Re}I| - |k \cdot \mathrm{Im}I|\big| \geq \left|\frac{3}{2}A\varepsilon^\alpha |k|^{-p} - |k| A\varepsilon^a\right| \geq \left|\frac{A}{2}\varepsilon^\alpha |k|^{-p} + A\varepsilon^{\alpha+p\tau} - A\varepsilon^{a-\tau}\right| = \frac{A}{2}\varepsilon^\alpha |k|^{-p}$. □

The following all-purpose proposition is adapted from [36], and estimates the tail of a Fourier series whose coefficients satisfy a generalized exponential decay law.

Proposition 4.6. *Given numbers $N \geq 0$, $p \geq 0$, and $0 < \beta < \sigma \leq 1$, let $f(I, \theta) = \sum_k f_k(I) e^{2\pi i k \cdot \theta}$ be analytic on the domain $D = F(K, \rho, \sigma)$ defined in (3.1), and suppose that $\|f_k\|_D \leq C|k|^p e^{-2\pi\sigma|k|}$. If the N-tail of f is given by $f^{\geq}(I, \theta) = \sum_{|k| \geq N} f_k(I) e^{2\pi i k \cdot \theta}$, then*

$$\|f^{\geq}\|_{D-(0,\beta)} \leq \frac{16C(p+2)^{p+2} e^{\pi\beta - p - 2}}{(\pi\beta)^{p+3}} e^{-\pi\beta N} \; . \tag{4.82a}$$

In particular, for $N = 0$ we recover $f^{\geq} = f$, and

$$\|f\|_{D-(0,\beta)} \leq \frac{16C(p+2)^{p+2}}{(\pi\beta)^{p+3}} e^{\pi\beta - p - 2} \; . \tag{4.82b}$$

Proof. We have

$$\|f^{\geq}\|_{D-(0,\beta)} \leq \sum_{|k| \geq N} \|f_k\|_{D-(0,\beta)} \|e^{2\pi i k \cdot \theta}\|_{D-(0,\beta)} \; . \tag{4.83}$$

Now since $D - (0, \beta) \subset D$ and since $(I, \theta) \in D - (0, \beta) \Rightarrow \|\mathrm{Im}\theta\|_\infty \leq \sigma - \beta$, the expression above is less than or equal to

$$\sum_{|k| \geq N} C|k|^p e^{-2\pi\sigma|k|} e^{2\pi(\sigma-\beta)|k|} = C \sum_{|k| \geq N} |k|^p e^{-2\pi\beta|k|} \; . \tag{4.84}$$

Using an elementary counting procedure, it is easy to see that the number of elements k in $\mathbf{Z}^3$ with index norm $|k| = j$ is precisely $4j^2 + 2$ for $j \geq 1$. For any nonnegative integer j, we may thus overestimate the number of elements in $\mathbf{Z}^3$ with norm j by $4(j+1)^2$. The desired sum is therefore bounded by

$$C \sum_{j=N}^{\infty} 4(j+1)^2 j^p e^{-2\pi\beta j} \leq 4C \sum_{j=N}^{\infty} (j+1)^{2+p} e^{-2\pi\beta j} \; . \tag{4.85}$$

Using $xe \leq e^x$ to write $[\pi\beta(j+1)/p+2]e \leq \exp[\pi\beta(j+1)/p+2] \Rightarrow (j+1)^{p+2} \leq \left(\frac{p+2}{\pi\beta e}\right)^{p+2} e^{\pi\beta(j+1)}$, we see that the last sum is less than or equal to

$$4C\left(\frac{p+2}{\pi\beta e}\right)^{p+2} e^{\pi\beta}\sum_{j=N}^{\infty} e^{-\pi\beta j} = 4C\left(\frac{p+2}{\pi\beta e}\right)^{p+2} e^{\pi\beta}\left(\frac{e^{-\pi\beta N}}{1-e^{-\pi\beta}}\right) . \tag{4.86}$$

Finally, using $\pi\beta/4 \leq 1 - e^{-\pi\beta}$ for $0 \leq \beta < 1$, we arrive to the desired estimate (4.82a), from which (4.82b) follows in turn. □

The preceding proposition is now used to prove

Proposition 4.7. *Given $0 < \beta < \sigma \leq 1$, let $f(I,\theta) = \sum_k f_k(I)e^{2\pi i k\cdot\theta}$ be analytic on the domain $D = F(K,\rho,\sigma)$ defined in (3.1), and suppose that $\|f_k\|_D \leq C|k|^p e^{-2\pi\sigma|k|}$. Then for any positive integer s,*

$$\left\|\frac{\partial^s f}{\partial\theta^s}\right\|_{D-(0,\beta)} \leq 16(2\pi)^s C(s+p+2)^{(s+p+2)}(\pi\beta)^{-s-p-3}e^{\pi\beta-s-p-2} . \tag{4.87}$$

Proof. Given any multiindex r, the Fourier coefficient of the partial derivative $\frac{\partial^{|r|} f}{\partial\theta^r}$ is

$$\left(\frac{\partial^{|r|} f}{\partial\theta^r}\right)_{k(I)} = (2\pi i)^{|r|}k^r f_k(I) , \tag{4.88}$$

$$\text{where} \quad k^r = k_1^{r_1}k_2^{r_2}k_3^{r_3} .$$

Thus $\|(\frac{\partial^{|r|} f}{\partial\theta^r})_k\|_D \leq (2\pi)^{|r|}|k|^{|r|}\|f_k\|_D \leq (2\pi)^{|r|}C|k|^{|r|+p}e^{-2\pi\sigma|k|}$. Now by Proposition 4.6,

$$\left\|\frac{\partial^{|r|} f}{\partial\theta^r}\right\|_{D-(0,\beta)} \leq 16(2\pi)^{|r|}C(|r|+p+2)^{(|r|+p+2)}(\pi\beta)^{-|r|-p-3}e^{\pi\beta-|r|-p-2} , \tag{4.89}$$

and the result follows. □

4.7 Proof of the Analytic Lemma

Proof of the analytic lemma is of course accomplished through repeated application of the iterative lemma. We will show under the present hypotheses that the iterative lemma may be applied $L = [\varepsilon^{-\tau/4}]$ times, where $[\varepsilon^{-\tau/4}]$ denotes the greatest integer in $\varepsilon^{-\tau/4}$.

We first show that the iterative lemma may be applied once. In accordance with (4.29), we prepare the original Hamiltonian $H = h + f$ for the first application of the iterative lemma by writing $H = H^{(0)} = h + G^{(0)} + R^{(0)}$, where $G^{(0)} = \Pi_{\mathcal{M}} f$ and $R^{(0)} = (1-\Pi_{\mathcal{M}})f$. We then have $\|G^{(0)}+R^{(0)}\|_D = \|f\|_D \leq \varepsilon E$ and $\|R^{(0)}\|_D \leq \|f\|_D + \|\Pi_{\mathcal{M}} f\|_D \leq 2\|f\|_D \leq 2\varepsilon E$, so that (4.30), (4.31) and (4.32) are satisfied with $\gamma_0 = \varepsilon$ and $\eta_0 = 2\varepsilon$. We then take $A' = A_1 = A/4$ and $B' = B_1 = \sigma/4$. Under the additional hypotheses of the analytic lemma, the iterative lemma may now be applied to give a canonical transformation $T^{(1)} : D^{(1)} \to T^{(1)}(D^{(1)})$ and a corresponding transformed Hamiltonian $H^{(1)} =$

$h + G^{(1)} + R^{(1)}$ satisfying (4.33) through (4.34), where each of the primed parameters now carries a subscript of 1.

Now define $A_j = A/(4j^2)$ and $B_j = \sigma/(4j^2)$, (note that $\frac{1}{4}A\varepsilon^{\tau/2} \le A_j < A$ and $\frac{1}{4}\sigma\varepsilon^{\tau/2} \le B_j < \sigma$ for $j \le L$). Assume that the iterative lemma has been successfully applied $l-1$ times ($l \le L$), where at the jth application we have used $A' = A_j$ and $B' = B_j$. We will show that it can be applied again. After the $l - 1^{st}$ transformation, the Hamiltonian $H^{(l-1)} = H^{(0)} \circ T^{(1)} \circ \cdots \circ T^{(l-1)}$ has the form $H^{(l-1)} = h + G^{(l-1)} + R^{(l-1)}$ and is analytic on the domain $D^{(l-1)}$ satisfying

$$D - \left(\sum_{j=1}^{l-1} A_j\varepsilon^a, \sum_{j=1}^{l-1} B_j\right) \subset D^{(l-1)} \subset D - \left(\frac{1}{2}\sum_{j=1}^{l-1} A_j\varepsilon^a, \frac{1}{2}\sum_{j=1}^{l-1} B_j\right) . \tag{4.90}$$

From (4.42) we also have

$$\gamma^{(l-1)} = \gamma^{(0)} + \frac{1}{2}\sum_{j=1}^{l-1}\eta^{(j)} , \tag{4.91}$$

and the remaining conclusions of the iterative lemma hold, where the primed parameters now carry the subscript $(l-1)$. To apply the iterative lemma once more, the only hypothesis which is not immediately verified is (4.32). We therefore assume provisionally that

$$\eta^{(j)} \le 2\varepsilon^{1+\tau}e^{-j} \quad \text{for } 1 \le j \le l-1 ; \tag{4.92}$$

this assumption will also be justified as part of the finite induction. With this assumption, we find that $\eta^{(l-1)} \le 2\varepsilon^{1+\tau}e^{l-1} < 2\varepsilon = 2\gamma^{(0)} < 2\gamma^{(l-1)}$, while (4.91) with $\gamma^{(0)} = \varepsilon$ ensures that

$$\gamma^{(l-1)} \le \gamma^{(0)} + \varepsilon^{1+\tau}\sum_{j=1}^{l-1} e^{-j} < 2\varepsilon . \tag{4.93}$$

The remaining hypothesis (4.32) is thus verified, and the iterative lemma may be applied again.

Let us now prove (4.92) for $j = l$ (we skip the proof for the starting value $j = 1$, since it is nearly identical, assuming $\gamma^{(0)} = \varepsilon$ and $\eta^{(0)} = 2\varepsilon$). After the lth application of the iterative lemma, we have $\eta^{(l)} = \eta_1^{(l)} + \eta_2^{(l)}$, where, by (4.43),

$$\eta_1^{(l)} \le \frac{2^6[2(p+2)]^{p+2}\eta^{(l-1)}}{(\pi B_l)^3} e^{-N\pi B_l/2} . \tag{4.94}$$

Using $N = \varepsilon^{-\tau}$, $B_l \ge \frac{1}{4}\sigma\varepsilon^{\tau/2}$, and $\eta^{(l-1)} \le 2\varepsilon e^{l-1}$, this expression is bounded by

$$\left(\frac{2^{13}[2(p+2)]^{p+2}}{(\pi\sigma)^3}\varepsilon^{-5\tau/2}e^{1-\frac{1}{8}\pi\sigma\varepsilon^{-\tau/2}}\right)\varepsilon^{1+\tau}e^{-l} \le \varepsilon^{1+\tau}e^{-l} , \tag{4.95}$$

the last inequality being a consequence of restriction (4.19e). As for $\eta_2^{(l)}$, by (4.44)

$$\eta_2^{(l)} = C_2 \frac{\gamma^{(l-1)}\eta^{(l-1)}}{(A_l)^2(B_l)^{2p+8}\varepsilon^{4a}} \; . \tag{4.96}$$

Now $N = \varepsilon^{-\tau}$, $A_l \geq \frac{1}{4}A\varepsilon^{\tau/2}$, $B_l \geq \frac{1}{4}\sigma\varepsilon^{\tau/2}$, $\gamma^{(l-1)} \leq 2\varepsilon$, and $\eta^{(l-1)} \leq 2\varepsilon e^{l-1}$ imply that this is less than or equal to

$$\frac{4C_2 e^{1-l}\varepsilon^{2-4a-(p+5)\tau}}{(A/4)^2(\sigma/4)^{2p+8}} = \left(\frac{C_2 e\, 2^{4p+22}\varepsilon^{1-4a-(p+6)\tau}}{A^2\sigma^{2p+8}}\right)\varepsilon^{1+\tau}e^{-l} \leq \varepsilon^{1+\tau}e^{-l} \; , \tag{4.97}$$

where the final inequality follows from (4.19d).

We have succeeded in showing that the iterative lemma may be applied l times in succession, for any $l \leq L$, and for such l we have $\eta^{(l)} = \eta_1^{(l)} + \eta_2^{(l)} \leq 2\varepsilon^{1+\tau}e^{-l}$, by (4.95) and (4.97). We construct the transformation $T : D_\infty \to T(D_\infty)$ by setting $T = T^{(1)} \circ T^{(2)} \circ \cdots \circ T^{(L)}$ and $D_\infty = D^{(L)}$. Since $\sum_{j=1}^{L} A_j = (A/4)\sum_{j=1}^{L} j^{-2} < (A/4)(\pi^2/6) < A/2$ and $\sum_{j=1}^{L} A_j > A/4$ (Similarly $\sigma/4 < \sum_{j=1}^{L} B_j < \sigma/2$), we see that (4.23a) and (4.23b) follow from (4.90). On the other hand, setting $T(J,\phi) = (I,\theta)$ for $(J,\phi) \in D_\infty$ and applying (4.35) and (4.36) recursively to the definition of T gives $\|I - J\|_\infty \leq \frac{\varepsilon^c}{2}\sum_j A_j < \frac{A}{4}\varepsilon^c$ and $\|\theta - \phi\|_\infty \leq \frac{\varepsilon^d}{2}\sum_j B_j < \frac{\sigma}{4}\varepsilon^d$, which establishes (4.28a) and (4.28b).

Next, we take $H' = H \circ T$ and $G = G^{(L)}$, $R = R^{(L)}$, so the transformed Hamiltonian is properly separated into its resonant and nonresonant parts as prescribed by (4.24), (4.25a) and (4.25b). The norm of R is exponentially small as required, since

$$\begin{aligned} \|R\|_{D_\infty} &= \|R^{(L)}\|_{D^{(L)}} \leq E\eta^{(L)} \leq 2E\varepsilon^{1+\tau}e^{-L} = \\ &2E\varepsilon^{1+\tau}e^{-[\varepsilon^{-\tau/4}]} \leq 2E\varepsilon^{1+\tau}e^{1-\varepsilon^{-\tau/4}} \leq 2E\varepsilon e^{-\varepsilon^{-\tau/4}} \; , \end{aligned} \tag{4.98}$$

the last inequality following from (4.19a). This proves (4.26).

For the remaining inequality (4.27), note that by construction,

$$G - \Pi_{\mathcal{M}} f = \sum_{j=1}^{L} \Pi_{\mathcal{M}} S_j \; , \tag{4.99}$$

where S_j is given by (4.50) for the jth application of the iterative lemma. Thus

$$\begin{aligned} \|G - \Pi_{\mathcal{M}} f\|_{D_\infty} &= \|\sum_{j=1}^{L} \Pi_{\mathcal{M}} S_j\|_{D_\infty} \leq \sum_{j=1}^{L} \|\Pi_{\mathcal{M}} S_j\|_{D_\infty} \leq \\ \sum_{j=1}^{L} \|S_j\|_{D_\infty} &\leq \frac{E}{2}\sum_{j=1}^{L} \eta^{(j)} \leq E\varepsilon^{1+\tau}\sum_{j=1}^{L} e^{-j} < E\varepsilon^{1+\tau} \; , \end{aligned} \tag{4.100}$$

where the third to last inequality follows from (4.76). This establishes (4.27), and the analytic lemma is proved. □

5. The Generalized Continuum Models

The article culminates in this section with the description of channeling (and certain nonchanneling) motions of particles in classical perfect crystals. This is accomplished by dividing phase space into three kinds of nonoverlapping regions, or "blocks," where trajectories are governed by the normal forms constructed in Sect. 4. For initial conditions in the degenerate (or nonresonant) block, an auxiliary result in geometric number theory shows that trajectories are initially nearly rectilinear but quickly fail the channeling criterion. For initial conditions in the two other kinds of resonant blocks, energy conservation and convexity of the unperturbed Hamiltonian are combined with analytic results to show that trajectories remain trapped at resonance for exponentially long times, during which they execute channeling motions as conceived in Lindhard's continuum models.

We begin with terminology concerning the geometry of the resonances.

5.1 Resonant Zones and Resonant Blocks

To each maximal submodule $\mathcal{M}$ of $\mathbf{Z}^3$ corresponds a resonance in action space, namely the subspace of all J orthogonal to each $k \in \mathcal{M}$. Actions in the neighborhood of a low-order resonance (and sufficiently far from other low-order resonances) are channeling directions; this is formalized by first defining the *resonant zone corresponding to* $\mathcal{M}$ by

$$\mathcal{Z}_{\mathcal{M}}(C,\alpha) = \{J \in \mathbf{R}^3 \mid |k \cdot J| \leq C\varepsilon^{\alpha}|k|^{-p} \forall k \in \mathcal{M}, 0 < |k| \leq |\mathcal{M}|\} \qquad (5.1)$$

for appropriate positive values of C, α and p (recall that $|\mathcal{M}|$ is the smallest nonnegative integer r such that $\mathcal{M}$ admits a basis of vectors with norm less than or equal to r). In the case where $\mathcal{M}$ is one-dimensional, the zone $\mathcal{Z}_{\mathcal{M}}(C,\alpha)$ will also be denoted $\mathcal{Z}_k(C,\alpha)$, where k is the unique generator (up to inversion) of $\mathcal{M}$. The case $\mathcal{M} = \{0\}$ also deserves comment; since $|\{0\}| = 0$, no $k \in \mathbf{Z}^3$ satisfy $0 < |k| \leq |\{0\}|$, and the "zone" $\mathcal{Z}_{\{0\}}(C,\alpha)$ degenerates to $\mathbf{R}^3$ for any C, α.

Next, given $\mathcal{M}$, its associated *resonant block of order* $N \geq |\mathcal{M}|$ is defined as the subset of action space

$$\widehat{\mathcal{Z}}_{\mathcal{M}}^{C,\alpha}(N, C_1, \alpha_1) = \mathcal{Z}_{\mathcal{M}}(C_1,\alpha_1) \backslash \bigcup_{\substack{k \notin \mathcal{M} \\ |k| \leq N}} \operatorname{Int} \mathcal{Z}_k(C,\alpha) \ , \qquad (5.2)$$

where *Int* denotes interior. These actions correspond to channeling directions; physically, it is interesting to note that when $\dim\mathcal{M} = 1$ (corresponding to a planar direction), low-order "axial" zones are removed, as they should be. In the degenerate case $\mathcal{M} = \{0\}$, the corresponding block is called a "*nonresonant* block," and since the underlying zone is all of $\mathbf{R}^3$, it is denoted $\widehat{\mathcal{Z}}_0^{C,\alpha}(N)$ without reference to C_1 or α_1. Explicitly,

$$\widehat{\mathcal{Z}}_0^{C,\alpha}(N) = \mathbf{R}^3 \backslash \bigcup_{0<|k|\leq N} \operatorname{Int} \mathcal{Z}_k(C,\alpha) \ . \qquad (5.3)$$

Of course, the resonant blocks may be empty in certain parameter regimes; however, because they are made up of intersections of elementary geometric objects (planar slabs) in $\mathbf{R}^3$, it is not difficult to show that the blocks are nonempty as $\varepsilon \to 0^+$. In fact, we may estimate the volume occupied by the resonant blocks inside the spherical shell $\mathcal{S}^\varepsilon_{\sqrt{2}} = \{I \in \mathbf{R}^3 | 2 - 2\varepsilon E_o \leq \|I(t)\|^2 \leq 2\}$ of admissible actions for the Hamiltonian (3.5); for $p > 4$ and for positive C, α, C_1, α_1, it is easy to show that

$$\mathrm{Vol}\big(\hat{\mathcal{Z}}^{C,\alpha}_{\mathcal{M}}(N, C_1, \alpha_1) \cap \mathcal{S}^\varepsilon_{\sqrt{2}}\big) \to \mathrm{Vol}\big(\mathcal{Z}_{\mathcal{M}}(C_1, \alpha_1) \cap \mathcal{S}^\varepsilon_{\sqrt{2}}\big) \tag{5.4}$$

as $\varepsilon \to 0^+$, independently of N. In other words, for the parameter values of interest, as $\varepsilon \to 0^+$, the relative volume of a block approaches the relative volume of the zone on which it is based.

In Sect. A1.5 of Appendix 1 of [17], explicit (but crude) estimates are given for the threshold $\varepsilon^* > 0$ such that $0 < \varepsilon \leq \varepsilon^* \Rightarrow \mathrm{Vol}\big(\hat{\mathcal{Z}}^{C,\alpha}_{\mathcal{M}}(N, C_1, \alpha_1) \cap \mathcal{S}^\varepsilon_{\sqrt{2}}\big) > 0$. Separate thresholds are given for each of the cases $\dim\mathcal{M} = 0$, 1, or 2, and for $\dim\mathcal{M} = 2$, it is shown that, not only does (5.4) hold in the small ε limit, but there is in fact an ε-threshold below which $\hat{\mathcal{Z}}^{C,\alpha}_{\mathcal{M}}(N, C_1, \alpha_1) \cap \mathcal{S}^\varepsilon_{\sqrt{2}} = \mathcal{Z}_{\mathcal{M}}(C_1, \alpha_1) \cap \mathcal{S}^\varepsilon_{\sqrt{2}}$. In other words, an axial resonant block coincides with the zone on which it is based for small enough ε.

The explicit thresholds ε^* are not repeated here since they constitute much weaker restrictions on ε than the restrictions (4.19).

5.2 Geometric Considerations

Following [8], we now prove a result analogous to Nekhoroshev's "geometric lemma," which says that as long as trajectories remain inside the domain of a particular normal form, they very nearly move in an affine subspace in action space which is parallel to the submodule $\mathcal{M}$ defining the normal form.

First, some terminology concerning the transformed Hamiltonian (4.24). By removing the remainder R we obtain the so-called effective Hamiltonian

$$\widetilde{H}(J, \phi) = h(J) + G(J, \phi) = \frac{1}{2} J^2 + \sum_{k \in \mathcal{M}} G_k(J) e^{2\pi i k \cdot \phi} \,. \tag{5.5}$$

which is cyclic in certain linear combinations of ϕ; i.e., $\widetilde{H}$ is independent of any ϕ in the orthogonal complement of $\mathcal{M}$. In particular, if $\mathcal{M} = \{0\}$ is the trivial submodule, its orthogonal complement is all of $\mathbf{R}^3$ and (5.5) is completely cyclic, or independent of ϕ. In general, to emphasize the dependence of G on only those angle variables lying in $\mathrm{span}\mathcal{M}$, we will often write $G(J, \phi) = G(J, \phi^*)$, where ϕ^* denotes the "resonant part" of ϕ, defined as follows. We perform a linear coordinate transformation on $\mathbf{T}^3$, by means of a unimodular matrix, such that $r = \dim\mathcal{M}$ of the new angles have the form $\phi^* = k \cdot \phi$, with k a basis element of $\mathcal{M}$. Since $\mathcal{M}$ is maximal, such a unimodular matrix always exists, and the splitting of ϕ into a resonant part ϕ^* and a nonresonant part $\widehat{\phi} = \phi - \phi^*$ may be extended canonically to the action variables $I = I^* + \widehat{I}$. (In Appendix 2 of [17], this splitting is constructed explicitly and is used to show that $\Pi_{\mathcal{M}} W$ is the continuum potential $\overline{W}$ corresponding to $\mathcal{M}$.) To show that the cyclic property of (5.5) is nearly true of (4.24), we prove the following

Proposition 5.1 (Geometric proposition). *Assume the hypotheses of the analytic lemma (Lemma 4.2). Let $(J_0, \phi_0) \in K_\infty \times \mathbf{T}^3$ be any initial condition in the real part $K_\infty \times \mathbf{T}^3$ of the domain D_∞ of the transformed Hamiltonian (4.24) and let (J, ϕ) be the solution of Hamilton's equations corresponding to (4.24). Denote by T_0 the solution's (possibly infinite) first time of escape from $K_\infty \times \mathbf{T}^3$. If $\mathcal{M}$ is the maximal submodule to which (4.24) is adapted, denote by $J^* \in \operatorname{span}\mathcal{M}$ the resonant part of J described above and set $\widehat{J} = J - J^*$, so that we have the orthogonal splitting $J = J^* + \widehat{J}$. Then for $0 \le t \le T_0$,*

$$\|\widehat{J}(t) - \widehat{J}_0\|_\infty \le \frac{4E}{\sigma} t\varepsilon e^{-\varepsilon^{-\tau/4}} . \tag{5.6}$$

Proof. By the analytic lemma, for $0 \le t \le T_0$ we have

$$J(t) = \widetilde{J}(t) - \int_0^t \frac{\partial R}{\partial \phi}(J(t'), \phi(t'))dt' , \tag{5.7}$$

where

$$\widetilde{J}(t) \equiv J_0 - \int_0^t \frac{\partial G}{\partial \phi}(J(t'), \phi(t'))dt' =$$

$$J_0 - 2\pi i \int_0^t \sum_{k \in \mathcal{M}} k G_k(J(t')) e^{2\pi i k \cdot \phi(t')} dt' =$$

$$\widehat{J}_0 + J_0^* - 2\pi i \sum_{k \in \mathcal{M}} k \int_0^t G_k(J(t')) e^{2\pi i k \cdot \phi(t')} dt' . \tag{5.8}$$

The last term belongs to $\operatorname{span}\mathcal{M}$, and we see that the motion generated by (5.8) takes place entirely in the affine subspace parallel to $\mathcal{M}$ passing through J_0:

$$\widetilde{J}(t) = \widehat{J}_0 + (\widetilde{J}(t))^* . \tag{5.9}$$

From the equation

$$J(t) = \widehat{J}(t) + J^*(t) = \widehat{J}_0 + (\widetilde{J}(t))^* - \widehat{\left(\int_0^t \frac{\partial R}{\partial \phi}(t')dt'\right)} - \left(\int_0^t \frac{\partial R}{\partial \phi}(t')dt'\right)^* , \tag{5.10}$$

we identify

$$\widehat{J}(t) = \widehat{J}_0 - \widehat{\left(\int_0^t \frac{\partial R}{\partial \phi}(t')dt'\right)} \quad \text{and} \tag{5.11}$$

$$J^*(t) = (\widetilde{J}(t))^* - \left(\int_0^t \frac{\partial R}{\partial \phi}(t')dt'\right)^* . \tag{5.12}$$

Now because the initial condition (J_0, ϕ_0) is real, the solution (J, ϕ) is real for all t. Thus for $0 \le t \le T_0$,

$$\|\widehat{J}(t)-\widehat{J}_0\|_\infty \le \left\|\left(\int_0^t \frac{\partial R}{\partial \phi}(t')dt'\right)^{\frown}\right\|_{K_\infty\times\mathbf{T}^3} \le \left\|\int_0^t \frac{\partial R}{\partial \phi}(t')dt'\right\|_{K_\infty\times\mathbf{T}^3} \le$$

$$t\left\|\frac{\partial R}{\partial \phi}\right\|_{K_\infty\times\mathbf{T}^3} \le \frac{2t}{\sigma}\|R\|_{D_\infty} \le \frac{4tE}{\sigma}\varepsilon e^{-\varepsilon^{-\tau/4}}\,, \tag{5.13}$$

where the second to last inequality follows from a Cauchy's estimate and (4.23a), and the last inequality uses (4.26). □

Using this or a similar proposition, it is possible to prove simultaneously the confinement of motions of (4.24) within blocks corresponding to any submodule $\mathcal{M}$ (cf. [8]). Our purpose here however is to examine these motions separately, and for the sake of the channeling problem, the nonresonant case $\mathcal{M}=\{0\}$ is treated separately in the next subsection.

5.3 The Spatial Continuum Model

In the case $\mathcal{M}=\{0\}$, every vector $J\in\mathbf{R}^3$ is orthogonal to $\mathcal{M}$ and the splitting $J=J^*+\widehat{J}$ described in Proposition 5.1 degenerates to $J=\widehat{J}$. Therefore for any initial condition (J_0,ϕ_0) lying in the real part $K_\infty\times\mathbf{T}^3$ of the domain D_∞ of (4.24), by Proposition 5.1 the real solution (J,ϕ) of (4.24) satisfies

$$\|J(t)-J_0\|_\infty \le \frac{4tE}{\sigma}\varepsilon e^{-\varepsilon^{-\tau/4}} \tag{5.14}$$

for as long as it remains in $K_\infty\times\mathbf{T}^3$.

This near constancy of the transformed actions will be exploited by way of a "bootstrap argument" to show that solutions of (4.24) beginning with actions well inside a nonresonant block $\widehat{\mathcal{Z}}_0$ remain inside the block for exponentially long times. It is first necessary to show that points beginning inside a block remain nonresonant when allowed to wander slightly outside the block in $\mathbf{R}^3$. We use the abbreviated neighborhood notation introduced in §4.2, along with the manipulations listed in (4.4).

Proposition 5.2a. *Let $\mathcal{M}\neq\mathbf{Z}^3$ be a maximal submodule of $\mathbf{Z}^3$. Suppose $N=\varepsilon^{-\tau}$ and A, α, a, c and τ satisfy the hypotheses of the analytic lemma, and C_1, α_1 are any positive numbers (or are absent, if $\mathcal{M}=\{0\}$). Then*

$$\widehat{\mathcal{Z}}_{\mathcal{M}}^{\frac{5}{2}A,\alpha}(N,C_1,\alpha_1)+A\varepsilon^a \subset \mathbf{R}^3\setminus \bigcup_{\substack{k\notin\mathcal{M}\\ |k|\le N}} \operatorname{Int}\mathcal{Z}_k(\frac{3}{2}A,\alpha)\,. \tag{5.15}$$

In other words, if $I_0\in\widehat{\mathcal{Z}}_0^{\frac{5}{2}A,\alpha}(N)$ and if $\|I_0-J_0\|_\infty\le A\varepsilon^a$, then for every $k\notin\mathcal{M}$ with $|k|\le N$, $|k\cdot J_0|\ge\frac{3}{2}A\varepsilon^\alpha$.

Proof. Let $I_0\in\widehat{\mathcal{Z}}_0^{\frac{5}{2}A,\alpha}(N)$ and suppose $\|I_0-J_0\|_\infty\le A\varepsilon^a$. For any $k\notin\mathcal{M}$ with $|k|\le N$, we have

$$|k \cdot J_0| \geq |k \cdot I_0| - |k \cdot (I_0 - J_0)| \geq \frac{5}{2} A\varepsilon^\alpha |k|^{-p} - |k| \, \|I_0 - J_0\|_\infty \geq$$

$$\frac{5}{2} A\varepsilon^\alpha |k|^{-p} - |k|^{p+1}|k|^{-p} A\varepsilon^a \geq \frac{5}{2} A\varepsilon^\alpha |k|^{-p} - A\varepsilon^{a-(p+1)\tau}|k|^{-p} =$$

$$\frac{5}{2} A\varepsilon^\alpha |k|^{-p} - A\varepsilon^\alpha |k|^{-p} = \frac{3}{2} A\varepsilon^\alpha |k|^{-p} \, . \qquad \square$$

The following corollary simply restates Proposition 5.2a as it is used in the upcoming proofs; it follows from properties (4.4).

Corollary 5.2b. *Under the same hypotheses,*

$$\widehat{\mathcal{Z}}_0^{\frac{5}{2}A,\alpha}(N) \subset \widehat{\mathcal{Z}}_0^{\frac{3}{2}A,\alpha}(N) - A\varepsilon^a =$$

$$\left(\mathbf{R}^3 \backslash \bigcup_{\substack{k \neq 0 \\ |k| \leq N}} Int \mathcal{Z}_k(\frac{3}{2}A, \alpha) \right) - A\varepsilon^a \quad \textit{and} \tag{5.16}$$

$$\widehat{\mathcal{Z}}_{\mathcal{M}}^{\frac{5}{2}A,\alpha}(N, C_1, \alpha_1) \subset \left(\mathbf{R}^3 \backslash \bigcup_{\substack{k \notin \mathcal{M} \\ |k| \leq N}} Int \mathcal{Z}_k(\frac{3}{2}A, \alpha) \right) - A\varepsilon^a \, . \tag{5.17}$$

We may now prove

Theorem 5.3 (Rectilinear trajectories). *Assume the hypotheses of the analytic lemma with H given by (3.5), $\mathcal{M} = \{0\}$, and with*

$$K = \widehat{\mathcal{Z}}_0^{\frac{3}{2}A,\alpha}(N) \cap (\mathcal{S}_{\sqrt{2}}^\varepsilon + A\varepsilon^a) \ , \tag{5.18}$$

which is clearly compact and satisfies (4.21). Let

$$(I_0, \theta_0) \in \left(\widehat{\mathcal{Z}}_0^{\frac{5}{2}A,\alpha}(N) \cap \mathcal{S}_{\sqrt{2}}^\varepsilon \right) \times \mathbf{T}^3 \tag{5.19}$$

be an initial condition for (3.5) which is nonresonant to order $N = \varepsilon^{-\tau}$. Then for $0 \leq t \leq (A\sigma/16E)\varepsilon^{c-1} e^{\varepsilon^{-\tau/4}}$, the solution (I, θ) of (3.5) with initial condition (I_0, θ_0) satisfies

$$\|I(t) - I_0\|_\infty \leq \frac{3}{4} A\varepsilon^c \quad \text{and} \tag{5.20}$$

$$\|\theta(t) - \theta_0 - tI_0\|_\infty \leq \frac{3}{4} At\varepsilon^c \, . \tag{5.21}$$

Proof. Let $(J_0, \phi_0) = T^{-1}(I_0, \theta_0)$. To see that (J_0, ϕ_0) belongs to the real part $K_\infty \times \mathbf{T}^3$ of D_∞ (and so is an admissible initial condition for (4.24)), first note that by (5.16a) and (5.19), we have

$$I_0 \in \left(\widehat{\mathcal{Z}}_0^{\frac{3}{2}A,\alpha}(N) - A\varepsilon\right) \cap \mathcal{S}^\varepsilon_{\sqrt{2}} \subset \left(\widehat{\mathcal{Z}}_0^{\frac{3}{2}A,\alpha}(N) \cap \left(\mathcal{S}^\varepsilon_{\sqrt{2}} + A\varepsilon^a\right)\right) - A\varepsilon^a$$

$$= K - A\varepsilon^a \ . \tag{5.22}$$

Since $\theta_0 \in \mathbf{T}^3$, it follows from (4.23b) that $(I_0, \theta_0) \in T(D_\infty)$ (the domain of T^{-1}), and so (J_0, ϕ_0) is well-defined and real, since (I_0, θ_0) is real. Furthermore, by (4.28a), $\|I_0 - J_0\|_\infty \le \frac{A}{4}\varepsilon^c \le \frac{A}{4}\varepsilon^a$ ($c > a$ by (4.18b) and $c > 1/2$), and so $I_0 \in K - A\varepsilon^a \Rightarrow J_0 \in K - \frac{3}{4}A\varepsilon^a \subset K_\infty - \frac{A}{4}\varepsilon^a$ (by (4.23a)). Therefore $(J_0, \phi_0) \in D_\infty$ is in the domain of the transformed Hamiltonian H' given by (4.24).

Now suppose the solution (J, ϕ) of (4.24) with initial condition (J_0, ϕ_0) first leaves D_∞ at $T_0 < (A\sigma/16E)\varepsilon^{c-1}e^{\varepsilon^{-\tau/4}}$. Since (J, ϕ) is real for all t, ϕ cannot exit $\mathbf{T}^3$ and we must have $J(T_0) \in \partial K_\infty$. But by (5.14), $\|J(T_0) - J_0\|_\infty \le (4E/\sigma)T_0\varepsilon e^{-\varepsilon^{-\tau/4}} < \frac{A}{4}\varepsilon^c < \frac{A}{4}\varepsilon^a$, which contradicts $J(T_0) \in \partial K_\infty$, since then $J_0 \in K_\infty - \frac{A}{4}\varepsilon^a$ would imply $\|J(T_0) - J_0\|_\infty \ge \frac{A}{4}\varepsilon^a$. Thus $T_0 \ge (A\sigma/16E)\varepsilon^{c-1}e^{\varepsilon^{-\tau/4}}$, and so $(J(t), \phi(t)) \in K_\infty \times \mathbf{T}^3$ for $t \in [0, (A\sigma/16E)\varepsilon^{c-1}e^{\varepsilon^{-\tau/4}}]$; and the inequality (5.14) also holds on this time interval.

On this same time interval, we have the solution $(I(t), \theta(t)) = T(J(t), \phi(t))$ of (3.5) satisfying $(I(0), \theta(0)) = T(J_0, \phi_0) = TT^{-1}(I_0, \theta_0) = (I_0, \theta_0)$. Using (4.28a), (5.14) and the identity $I - I_0 = I - I_0 + J - J + J_0 - J_0$ we have, for any $t \in [0, (A\sigma/16E)\varepsilon^{c-1}e^{\varepsilon^{-\tau/4}}]$,

$$\|I(t) - I_0\|_\infty \le \|J_0 - I_0\|_\infty + \|J(t) - J_0\|_\infty + \|I(t) - J(t)\|_\infty \le$$

$$\frac{A}{4}\varepsilon^c + \frac{4E}{\sigma}t\varepsilon e^{-\varepsilon^{-\tau/4}} + \frac{A}{4}\varepsilon^c \le \frac{3A}{4}\varepsilon^c \ , \tag{5.23}$$

which proves (5.20).

On the other hand, from (3.5)

$$\theta(t) = \theta_0 + \int_0^t I(t')dt' = \theta_0 + \int_0^t I_0 dt' + \int_0^t (I(t') - I_0)dt' =$$

$$\theta_0 + tI_0 + \int_0^t (I(t') - I_0)dt' \quad \Rightarrow$$

$$\|\theta(t) - \theta_0 - tI_0\|_\infty \le \int_0^t \|I(t') - I_0\|_\infty dt' \le \frac{3}{4}At\varepsilon^c \tag{5.24}$$

for $t \in [0, (A\sigma/16E)\varepsilon^{c-1}e^{\varepsilon^{-\tau/4}}]$, as claimed in (5.21). □

Theorem 5.3 says that for initial conditions which are nonresonant to a certain order N, motions of (3.5) have nearly constant action $I \simeq I_0$ on an exponentially long time interval, while the configuration or angle variable θ is approximated by the rectilinear motion

$$\bar{\theta}(t) = \theta_0 + tI_0 \tag{5.25}$$

on any time interval $[0, C\varepsilon^{-b}]$ with $C > 0$, $0 < b < c$. This Nekhoroshev-like result is close to the KAM theorem, and complements it by showing that trajectories beginning *near* invariant tori remain near them and exhibit integrable-like behavior for very long times.

Mathematically, Theorem 5.3 characterizes the behavior of trajectories of (3.5) for small ε and for initial conditions in the nonresonant block. However, in terms of the channeling problem, more can be shown by recalling the postulate that solutions of (3.5) model particle motions in crystals only in so far as they satisfy the channeling criterion (3.8).

Now it is well known that linear flow $\bar{\theta}(t) = \theta_0 + tI_0$ with nonresonant direction vector I_0 densely fills $\mathbf{T}^3$, and it is reasonable to suspect that linear flow with "nearly" highly nonresonant direction vector $I_0 \in \widehat{\mathcal{Z}}_0^{C,\alpha}(N)$ will quickly fill "enough" of $\mathbf{T}^3$ to fail the channeling criterion. The following two theorems (the proofs of which appear elsewhere) confirm this suspicion.

Theorem 5.4 (Close encounters). *Assume the hypotheses of Theorem 5.3; i.e., assume the hypotheses of the analytic lemma with H given by (3.5), $\mathcal{M} = \{0\}$, and with $K = \widehat{\mathcal{Z}}_0^{\frac{3}{2}A,\alpha}(N) \cap (\mathcal{S}^{\varepsilon}_{\sqrt{2}} + A\varepsilon^a)$. Let $R > 0$ be the radius of the largest ball $\mathcal{B}_R$ contained in the region of close encounter $\mathcal{C}(1) \subset \mathbf{T}^3$ described in (3.7). Then for $\varepsilon > 0$ sufficiently small, the initial actions (or initial directions)*

$$I_0 \in \widehat{\mathcal{Z}}_0^{\frac{5}{2}A,\alpha}(N) \cap \mathcal{S}^{\varepsilon}_{\sqrt{2}} \tag{5.26}$$

are nonchanneling directions in the sense that given any initial condition

$$(I_0, \theta_0) \in \left(\widehat{\mathcal{Z}}_0^{\frac{5}{2}A,\alpha}(N) \cap \mathcal{S}^{\varepsilon}_{\sqrt{2}}\right) \times \mathbf{T}^3 , \tag{5.27}$$

the corresponding solution (I, θ) of (3.5) fails to satisfy the channeling criterion (3.8) at some t' in the (brief) time interval

$$[0, C_3\varepsilon^{-\alpha}], \qquad \textit{where} \tag{5.28a}$$

$$C_3 = \frac{2(2\sqrt{3})^p \|V_*\|_\triangle}{AR^{p+3/2}} , \tag{5.28b}$$

and the constant $\|V_*\|_\triangle$ is described following Theorem 5.5.

The proof of this theorem depends on the rate at which linear flow fills the torus when subject to truncated Diophantine conditions. In order to state a theorem which estimates this rate, we introduce the following terminology.

For vectors $\alpha \in \mathbf{R}^3$ with unit length (denoted $\alpha \in S^2$), we first consider the one-parameter family of translation maps $\alpha_t : \mathbf{T}^3 \to \mathbf{T}^3$ defined by $\alpha_t(\theta) = \theta + t\alpha$ (the vector arithmetic takes place on the torus, i.e., is first performed in the universal covering space $\mathbf{R}^3$, then projected onto $\mathbf{T}^3$). A rectilinear orbit of $\mathbf{T}^3$ with direction vector α and initial condition θ is then the image of θ under the flow defined by α_t over some closed interval in $\mathbf{R}$:

$$\bigcup_{a\leq t\leq b} \alpha_t(\theta) \ . \tag{5.29}$$

Next, given $R > 0$, the direction vector $\alpha \in S^2$ is said to *ergodize* $\mathbf{T}^3$ *to within* R *after time* T if no point in $\mathbf{T}^3$ is more than R-distant (Euclidean norm) from any rectilinear orbit $\cup_{0\leq t\leq T}\alpha_t(\theta)$.

Finally, for convenience we define the set $\mathcal{D}(p,C,N)$ of direction vectors satisfying a truncated Diophantine condition of order N as

$$\mathcal{D}(p,C,N) = \left\{\alpha \in S^2 \middle| \, |\alpha\cdot k| > \frac{C}{|k|^p} \forall k \in \mathbf{Z}^n, 0 < |k| \leq N\right\} . \tag{5.30}$$

It is intuitively clear that, for fixed $R > 0$, rectilinear orbits arising from resonant direction vectors will ergodize the torus, provided they are resonant at sufficiently high order. Such orbits are indeed degenerate (i.e., fill only lower dimensional tori), but at high enough order, they build a "web" with holes no larger than R. In fact, given R and given a Diophantine condition, it is possible to show that for $N > N^*$, where

$$N^* \equiv \kappa(p) R^{-\left(\frac{p+3/2}{p-3/2}\right)} + 3/2 \ , \tag{5.31}$$

when the Diophantine condition is truncated at order N, the ergodization time may be explicitly estimated. (Here $\kappa(p)$ is a constant depending on p; it is given explicitly in [18].) This result is stated as

Theorem 5.5. *Let* $0 < R \leq 1$, *and choose* $N > N^*$. *Then given* $\alpha \in \mathcal{D}(p,C,N)$, T^n *will be ergodized to within* R *by rectilinear orbits with direction vector* α *after time*

$$T = \frac{2(3^{p/2})\|V_*\|_\triangle}{C\pi R^{p+3/2}} \left[1 - \left(\frac{N^* - 3/2}{N-3/2}\right)^{p-\frac{3}{2}}\right]^{-1/2} . \tag{5.32}$$

Here the constant $\|V_*\|_\triangle$ is a Sobolev norm of a certain maximally smooth test function V_* on the unit ball in $\mathbf{R}^3$; it depends only on p and is discussed further in [18], where complete proofs of this and related results appear. The important feature of this theorem for our purposes is the appearance of C in the denominator of the estimate of the ergodization time T.

It is now straightforward to prove Theorem 5.4 from Theorem 5.5 by first normalizing the initial action vector I_0 and the time t, taking $C = (5/2\sqrt{2})A\varepsilon^\alpha$, $N = \varepsilon^{-\tau}$, and by using the rectilinear motion (5.25) to approximate the true motion on the time interval $[0, C_3\varepsilon^{-\alpha}]$. Because the ergodization time is less than the length of this interval, it is possible to produce a $t' \in [0, C_3\varepsilon^{-\alpha}]$ at which the angle θ belongs to $\mathcal{B}_R \subset \mathcal{C}(1)$. The solution (I,θ) therefore fails the channeling criterion at time t'. A proof along these lines (but with less wieldy constants) may be found in [19].

Theorems 5.3 and 5.4 taken together provide the picture of the "spatial continuum model" mentioned in the introduction. Namely, sufficiently energetic particles injected into a classical crystal in the nonresonant (or nonchanneling) directions $I_0 \in \widehat{\mathcal{Z}}_0^{\frac{5}{2}A,\alpha}(N) \cap \mathcal{S}^{\varepsilon}_{\sqrt{2}}$ will follow roughly straight paths to their first collisions with a region of close encounter $\mathcal{C}(1)$, and these collisions happen quickly, at times no later than $C_3\varepsilon^{-\alpha}$ from the time of injection. Moreover as mentioned in §5.1, with increasing particle energy, the volume in action space of the admissible nonresonant actions $\widehat{\mathcal{Z}}_0^{\frac{5}{2}A,\alpha}(N) \cap \mathcal{S}^{\varepsilon}_{\sqrt{2}}$ approaches the volume of the set of all admissible actions $\mathcal{S}^{\varepsilon}_{\sqrt{2}}$ for the perfect crystal Hamiltonian (3.5). Finally, it should be mentioned that although the emphasis in this article is on positively charged particles, Theorems 5.3 and 5.4 make no use of the nonnegativity of W; they hold irrespective of particle charge.

5.4 Channeling

With the addition of a few definitions and a physically motivated assumption about the potential W, we will be in position to prove the existence of solutions of the scaled perfect crystal Hamiltonian (3.5) which satisfy the channeling criterion (3.8) on exponentially long time intervals. These solutions have characteristics which physicists recognize as the traits of axial or planar channeling; namely, they have a low probability of encountering nuclei, their "transverse energy" and "longitudinal momenta" are nearly conserved, and they are approximated in some sense by solutions of an appropriate generalized (axial or planar) continuum Hamiltonian, which in turn is the ordinary (axial or planar) continuum Hamiltonian to leading order.

To begin with, we recall that given a 1- or 2-dimensional maximal submodule $\mathcal{M}$ of $\mathbf{Z}^3$ and given $\theta \in \mathbf{T}^3$, we write $\theta = \theta^* + \widehat{\theta}$, where $\theta^* \in \mathrm{span}\mathcal{M}$ and $\widehat{\theta} \in (\mathrm{span}\mathcal{M})^\perp$ are the resonant and nonresonant parts of θ described just before Proposition 5.1 above. We also recall that the resonant subseries $\Pi_{\mathcal{M}}W$ of the potential W depends only on θ^* and is equal to the function $\overline{W}$ obtained by averaging over $\widehat{\theta}$. This function $\Pi_{\mathcal{M}}W(\theta) = \Pi_{\mathcal{M}}W(\theta^*) = \overline{W}(\theta^*)$ is the (scaled) continuum potential corresponding to $\mathcal{M}$; if $\dim\mathcal{M} = 1$ it is a planar continuum potential, if $\dim\mathcal{M} = 2$ it is an axial continuum potential.

In order to state the needed assumption concerning W, we consider regions in configuration space bounded by equipotential surfaces of the continuum potentials. Given $\mathcal{M}$, $\dim\mathcal{M} = 1$ or 2, and any number $Q \geq 0$, define the closed subset $\mathcal{A}_{\mathcal{M}}(Q)$ as

$$\mathcal{A}_{\mathcal{M}}(Q) = \{\theta \in \mathbf{T}^3 \mid \Pi_{\mathcal{M}}W(\theta^*) \leq Q\} . \tag{5.33}$$

Our assumption about W is then the following:

Assumption 5.6. *There exists a critical order $M^* \geq 1$ such that for any 1- or 2-dimensional maximal submodule $\mathcal{M} \subset \mathbf{Z}^3$ with order $|\mathcal{M}| \leq M^*$, there exist numbers Q' and Q^*, $0 \leq Q' < Q^* < 1$ with the property that given any $Q \in [Q', Q^*]$, $\mathcal{A}_{\mathcal{M}}(Q) \neq \emptyset$ and $\mathcal{A}_{\mathcal{M}}(Q) \cap \mathcal{C}(1) = \emptyset$, where $\mathcal{C}(1) = \{\theta \in \mathbf{T}^3 \mid W(\theta) \geq 1\}$ is the region of close encounter described in (3.7).*

This physical assumption says that at sufficiently low order, there are equipotential surfaces of the continuum potential $\Pi_{\mathcal{M}}W$ which do not intersect the restricted region $\mathcal{C}(1)$. If $\dim\mathcal{M} = 1$, these surfaces are planes; if $\dim\mathcal{M} = 2$, the surfaces are cylindrical sheets or tubes. In other words, at sufficiently low order, there are clear planar or axial pathways through the crystal which do not meet the close encounter region $\mathcal{C}(1)$. We will not examine this assumption further, since a calculation of M^* requires specific knowledge about the location of lattice sites in the crystal which we do not assume here. We simply remark that the assumption is satisfied by any physical cubic crystal for reasonable values of $\mathcal{E}_\perp$, which ultimately defines the size of $\mathcal{C}(1)$.

We next define the transverse energy of a trajectory with respect to a particular continuum potential by

$$E_\perp(I,\theta) = \frac{1}{2}(I^*)^2 + \varepsilon \Pi_{\mathcal{M}}W(\theta^*) \ , \tag{5.34}$$

where again I^* and θ^* denote the projections of I and θ onto span$\mathcal{M}$. (It should be pointed out that if (I,θ) is a solution corresponding to (3.5) then (I^*,θ^*) is a pair of conjugate variables for (3.5), while if (J,ϕ) is a solution corresponding to (4.24), (J^*,ϕ^*) is *not* (quite) a conjugate pair; this will not be troublesome.)

It should also be pointed out that the set $\{(I,\theta) \mid E_\perp(I,\theta) \le Q\varepsilon\}$ is *not* a Cartesian product of a subset of $\mathbf{R}^3$ with a subset of $\mathbf{T}^3$, but it is a subset of the product $\mathcal{Z}_{\mathcal{M}}(|\mathcal{M}|^{p+1}(2Q/3)^{1/2}, 1/2) \times \mathcal{A}_{\mathcal{M}}(Q)$, where the zone $\mathcal{Z}_{\mathcal{M}}(C,\alpha)$ is defined in§5.1. In fact, every $\theta \in \mathcal{A}_{\mathcal{M}}(Q)$ is the second factor of the point $(0,\theta) \in \{(I,\theta) \mid E_\perp(I,\theta) \le Q\varepsilon\}$.

We may now state the theorem which gives this article its name,

Theorem 5.7 (Channeling). *Let $\mathcal{M}$ be a 1- or 2-dimensional submodule of $\mathbf{Z}^3$ with order $|\mathcal{M}| \le M^*$. Assume the hypothesis of the analytic lemma with*

$$K = (\mathcal{S}^\varepsilon_{\sqrt{2}} + A\varepsilon^a) \setminus \bigcup_{\substack{k\notin\mathcal{M} \\ |k|\le N}} \operatorname{Int} \mathcal{Z}_k(\frac{3}{2}A, \alpha) \ . \tag{5.35}$$

Fix the maximum initial (scaled) transverse energy $Q \ge Q'$ and the maximum change in (scaled) transverse energy $\delta > 0$ such that $Q + \delta \le Q^*$, where $Q' < Q^* < 1$ are defined in Assumption 5.6 ; set $C = |\mathcal{M}|^{p+1}(2Q/3)^{1/2}$. In addition to the restrictions (4.19) on ε, assume the (possibly weaker) further restrictions

$$\varepsilon \le \frac{\delta}{3}\left(\frac{3}{32}A^2 + \frac{3\sqrt{2}}{4}A + \frac{C_4\sigma}{4}\right)^{-1/b} \qquad (b = \min\{c - 1/2, d\}) \ , \tag{5.36a}$$

$$\varepsilon \le \min\left\{\left(\frac{1}{E}\right)^2, \left(\frac{2}{A}\right)^{1/c}, \left(\frac{\delta}{6}\right)^{1/\tau}, \left(\frac{A}{12E}\right)^{\frac{1}{8\tau}}\right\} , \tag{5.36b}$$

where C_4 is the Lipschitz constant for $\Pi_{\mathcal{M}}W$ with respect to the sup norm:

$$|\Pi_{\mathcal{M}}W(\theta^*) - \Pi_{\mathcal{M}}W(\phi^*)| \le C_4\|\theta^* - \phi^*\|_{\mathbf{T}^3} \le C_4\|\theta - \phi\|_{\mathbf{T}^3} \ . \tag{5.37}$$

Then any real initial condition (I_0, θ_0) for (3.5) with initial transverse energy

$$E_\perp(I_0, \theta_0) \le \varepsilon Q \tag{5.38}$$

and with suitable initial direction

$$I_0 \in \widehat{\mathcal{Z}}_{\mathcal{M}}^{\frac{5}{2}A,\alpha}(N, C, 1/2) \cap \mathcal{S}_{\sqrt{2}}^{\varepsilon} \tag{5.39}$$

gives rise to a solution (I, θ) of (3.5) which satisfies the channeling criterion (3.8) on the exponentially long time interval $[0, T_0]$, where

$$T_0 = \frac{\sigma}{24}\varepsilon^\tau e^{\varepsilon^{-\tau/4}} - 1 \ . \tag{5.40}$$

This solution is approximated by a "generalized continuum model" solution in the sense that $(I, \theta) = T(J, \phi)$, where T is the near-identity transformation defined in the analytic lemma and (J, ϕ) is the solution to (4.24) with initial condition $(J_0, \phi_0) = T^{-1}(I_0, \theta_0)$. Furthermore, on the interval $[0, T_0]$, the longitudinal momentum $\widehat{I}(t)$ is nearly constant:

$$\|\widehat{I}(t) - \widehat{I}_0\|_\infty \le \frac{3}{4}A\varepsilon^c \ , \tag{5.41}$$

as is the transverse energy:

$$|E_\perp(I(t), \theta(t)) - E_\perp(I_0, \theta_0)| \le \varepsilon\delta \ . \tag{5.42}$$

Proof. We first verify that $(I_0, \theta_0) \in T(D_\infty)$, so that (J_0, ϕ_0) may be defined as claimed. By (5.39) we have $I_0 \in \widehat{\mathcal{Z}}_{\mathcal{M}}^{\frac{5}{2}A,\alpha}(N, C, 1/2)$ which implies, using (5.17), that

$$I_0 \in \left[\left(\mathbf{R}^3 \setminus \bigcup_{\substack{k \notin \mathcal{M} \\ |k| \le N}} \operatorname{Int} \mathcal{Z}_k\left(\frac{3}{2}A, \alpha\right)\right) - A\varepsilon^a\right] \cap \mathcal{S}_{\sqrt{2}}^{\varepsilon} \ . \tag{5.43}$$

Using the properties (4.4), this is in turn contained in

$$I_0 \in \left[\left(\mathbf{R}^3 \setminus \bigcup_{\substack{k \notin \mathcal{M} \\ |k| \le N}} \operatorname{Int} \mathcal{Z}_k\left(\frac{3}{2}A, \alpha\right)\right) \cap (\mathcal{S}_{\sqrt{2}}^{\varepsilon} + A\varepsilon^a)\right] - A\varepsilon^a =$$

$$K - A\varepsilon^a = \left(K - \frac{A}{2}\varepsilon^a\right) - \frac{A}{2}\varepsilon^a \subset K_\infty - \frac{A}{2}\varepsilon^a \ , \tag{5.44}$$

where $K_\infty \times \mathbf{T}^3$ is the real part of the domain D_∞, and where $K - \frac{A}{2}\varepsilon^a \subset K_\infty$ by (4.23a). Since by (4.23b), $K - \frac{A}{2}\varepsilon^a \times \mathbf{T}^3 \subset T(D_\infty)$, we see that $(I_0, \theta_0) \in T(D_\infty)$, and we may apply T^{-1} to (I_0, θ_0) to obtain $(J_0, \phi_0) = T^{-1}(I_0, \theta_0)$. Since (I_0, θ_0) is real, (J_0, ϕ_0) is real, and we need only check that $J_0 \in K_\infty$ to see that (J_0, ϕ_0) belongs to the domain D_∞ of (4.24). But by (4.28a), $\|I_0 - J_0\|_\infty \le \frac{A}{4}\varepsilon^c \le \frac{A}{4}\varepsilon^a$

($c > a$ by (4.18b) and because $c > 1/2$), and so using (5.44), we have $J_0 \in K_\infty - \frac{A}{4}\varepsilon^a$, and $(J_0, \phi_0) \in (K_\infty - \frac{A}{4}\varepsilon^a) \times \mathbf{T}^3 \subset K_\infty \times \mathbf{T}^3$ as required.

If (J, ϕ) is the solution of (4.24) with initial condition (J_0, ϕ_0), let T^* be the time of first exit of (J, ϕ) from $K_\infty \times \mathbf{T}^3$. By Proposition 5.1,

$$\|\widehat{J}(t) - \widehat{J}_0\|_\infty \le \frac{4E}{\sigma} t\varepsilon e^{-\varepsilon^{-\tau/4}} \quad \text{for } 0 \le t \le T^* . \tag{5.45}$$

We proceed to estimate the change in the transformed transverse energy $E_\perp(J, \phi)$ by first noting that by (5.34), $\frac{1}{2}(I_0^*)^2 = E_\perp(I_0, \theta_0) - \varepsilon \varPi_{\mathcal{M}} W(\theta_0^*)$. But $\varPi_{\mathcal{M}} W$ is nonnegative, since it is the average of W over $\widehat{\phi} = \phi - \phi^*$, and W is nonnegative by construction. Therefore$\frac{1}{2}(I_0^*)^2$ is bounded by $E_\perp(I_0, \theta_0) \le \varepsilon Q < \varepsilon$, by (5.38) and $Q < 1$. Thus $\|I_0^*\|_\infty^2 \le \|I_0^*\|^2 = (I_0^*)^2 < 2\varepsilon \Rightarrow \|I_0^*\|_\infty < \sqrt{2}\varepsilon^{1/2}$, and so

$$\begin{aligned} \|J_0^*\|_\infty \le \|J_0^* - I_0^*\|_\infty + \|I_0^*\|_\infty = \|(J_0 - I_0)^*\|_\infty + \|I_0^*\|_\infty \le \\ \|J_0 - I_0\|_\infty + \|I_0^*\|_\infty \le \frac{A}{4}\varepsilon^c + \sqrt{2}\varepsilon^{1/2} , \end{aligned} \tag{5.46}$$

and therefore

$$\begin{aligned} |E_\perp(J_0, \phi_0) - E_\perp(I_0, \theta_0)| = \left| \frac{1}{2}\left((J_0^*)^2 - (I_0^*)^2\right) + \varepsilon\Big(\varPi_{\mathcal{M}} W(\phi_0^*) - \varPi_{\mathcal{M}} W(\theta_0^*)\Big) \right| \le \\ \frac{1}{2}\left((J_0^* - I_0^*) \cdot (J_0^* + I_0^*)\right) + \varepsilon \left| \varPi_{\mathcal{M}} W(\phi_0^*) - \varPi_{\mathcal{M}} W(\theta_0^*) \right| \le \\ \frac{3}{2}\|J_0^* - I_0^*\|_\infty \|J_0^* + I_0^*\|_\infty + C_4\varepsilon\|\phi_0^* - \theta_0^*\|_\infty \le \\ \frac{3}{2}\left(\frac{A}{4}\varepsilon^c\right)\left(\|J_0^*\|_\infty + \|I_0^*\|_\infty\right) + C_4\varepsilon\|\phi_0 - \theta_0\|_\infty \le \\ \frac{3A}{8}\varepsilon^c\left(\frac{A}{4}\varepsilon^c + 2\sqrt{2}\varepsilon^{1/2}\right) + C_4\varepsilon\frac{\sigma}{4}\varepsilon^d \le \\ \varepsilon^{1+b}\left(\frac{3}{32}A^2 + \frac{3\sqrt{2}}{4}A + \frac{C_4\sigma}{4}\right) \le \varepsilon\frac{\delta}{3} , \end{aligned} \tag{5.47}$$

where $b = \min\{c - 1/2, d\}$ and where the last inequality follows by restriction (5.36a). For $0 \le t \le \min\{T_0, T^*\}$, we now estimate

$$\begin{aligned} |E_\perp(J(t), \phi(t)) - E_\perp(J_0, \phi_0)| = \\ \left| \frac{1}{2}\left((J^*(t))^2 - (J_0^*)^2\right) + \varepsilon\left(\varPi_{\mathcal{M}} W(\phi^*(t)) - \varPi_{\mathcal{M}} W(\phi_0^*)\right) \right| . \end{aligned} \tag{5.48}$$

Using (4.24) and $J^2 = (J^*)^2 + \widehat{J}^2$, we see that this is equal to

$$\begin{aligned} \left| \frac{1}{2}\left(\widehat{J}_0^2 - (\widehat{J}(t))^2\right) + \Big(G(J_0, \phi_0^*) - \varepsilon\varPi_{\mathcal{M}} W(\phi_0^*)\Big) - \right. \\ \left. \Big(G(J(t), \phi^*(t)) - \varepsilon\varPi_{\mathcal{M}} W(\phi^*(t))\Big) + R(J_0, \phi_0) - R(J(t), \phi(t)) \right| \le \end{aligned}$$

$$\frac{1}{2}|\widehat{J}_0^2 - (\widehat{J}(t))^2| + 2\|G - \varepsilon \Pi_{\mathcal{M}} W\|_{D_\infty} + 2\|R\|_{D_\infty} \leq$$
$$\frac{3}{2}\|\widehat{J}_0 + \widehat{J}(t)\|_{D_\infty}\|\widehat{J}_0 - \widehat{J}(t)\|_{D_\infty} + 2\|G - \varepsilon \Pi_{\mathcal{M}} W\|_{D_\infty} + 2\|R\|_{D_\infty} \leq$$
$$\frac{3}{2}\left(2\|\widehat{J}_0\|_\infty + \|\widehat{J}(t) - \widehat{J}_0\|_{D_\infty}\right)\|\widehat{J}_0 - \widehat{J}(t)\|_{D_\infty} + 2\|G - \varepsilon \Pi_{\mathcal{M}} W\|_{D_\infty} + 2\|R\|_{D_\infty} . \tag{5.49}$$

We now use (4.26) to estimate $\|R\|_{D_\infty}$, (4.27) for $\|G - \varepsilon \Pi_{\mathcal{M}} W\|_{D_\infty}$, the geometric proposition (Proposition 5.1) for $\|\widehat{J}_0 - \widehat{J}(t)\|_{D_\infty}$, and finally the fact that $J_0 \in \mathcal{S}^{\varepsilon}_{\sqrt{2}} + \frac{A}{4}\varepsilon^c$ for $\|\widehat{J}_0\|_\infty$. (5.49) is therefore bounded by

$$\frac{3}{2}\left(2\left(\sqrt{2} + \frac{A}{4}\varepsilon^c\right) + \frac{4E}{\sigma}t\varepsilon e^{-\varepsilon^{-\tau/4}}\right)\frac{4E}{\sigma}t\varepsilon e^{-\varepsilon^{-\tau/4}} + 2E\varepsilon^{1+\tau} + 4E\varepsilon e^{-\varepsilon^{-\tau/4}} . \tag{5.50}$$

Since $0 \leq t \leq \min\{T_0, T^*\}$, where T_0 is given by (5.40), this expression is less than or equal to

$$\frac{3}{2}\left(2(\sqrt{2} + \frac{A}{4}\varepsilon^c) + \frac{E}{6}\varepsilon^{1+\tau}\right)\frac{E}{6}\varepsilon^{1+\tau} + 2E\varepsilon^{1+\tau} . \tag{5.51}$$

Using (5.36b), this simplifies to

$$\frac{3}{2}\left(2\sqrt{2} + 1 + \frac{1}{6}\right)\left(\frac{1}{6}\varepsilon^{1+\tau} - \frac{8E}{\sigma}\varepsilon e^{-\varepsilon^{-\tau/4}}\right) + 2E\varepsilon^{1+\tau} , \tag{5.52}$$

so finally, using (5.36b) ($\varepsilon \leq (\delta/6)^{1/\tau}$) and the fact that $2\sqrt{2} + 1 + 1/6 < 4$, we arrive to

$$|E_\perp(J(t), \phi(t)) - E_\perp(J_0, \phi_0)| \leq 3E\varepsilon^{1+\tau} \leq \varepsilon\frac{\delta}{3} . \tag{5.53}$$

Now combining (5.47) and (5.53), we get

$$E_\perp(J(t), \phi(t)) \leq E_\perp(I_0, \theta_0) + \left|E_\perp(J_0, \phi_0) - E_\perp(I_0, \theta_0)\right| +$$
$$\left|E_\perp(J(t), \phi(t)) - E_\perp(J_0, \phi_0)\right| \leq \varepsilon(Q + \frac{2}{3}\delta) \leq \varepsilon Q^* \leq \varepsilon \tag{5.54}$$

for $0 \leq t \leq \min\{T^*, T_0\}$, from which we find

$$\frac{1}{2}\|J^*(t)\|_\infty^2 \leq \frac{1}{2}(J^*(t))^2 \leq E_\perp(J(t), \phi(t)) \leq \varepsilon$$
$$\Rightarrow \|J^*(t)\|_\infty \leq \sqrt{2}\,\varepsilon^{1/2} \tag{5.55}$$

on the same interval.

We can now show that the time of escape $T^* \geq T_0$. By way of contradiction, suppose not; suppose $T^* < T_0$. Then $J(T^*) \in \partial K_\infty$ and $J_0 \in K_\infty - \frac{A}{4}\varepsilon^a$ (see above) imply that $\|J(T^*) - J_0\|_\infty \geq \frac{A}{4}\varepsilon^a$. But

$$\|J(T^*) - J_0\|_\infty \leq \|\widehat{J}(T^*) - \widehat{J}_0\|_\infty + \|J^*(T^*) - J_0^*\|_\infty \leq$$
$$\|\widehat{J}(T^*) - \widehat{J}_0\|_\infty + \|J^*(T^*)\|_\infty + \|J_0^*\|_\infty \leq$$

$$\frac{4E}{\sigma} T^* \varepsilon e^{-\varepsilon^{-\tau/4}} + 2\sqrt{2}\,\varepsilon^{1/2} \,, \tag{5.56}$$

by (5.13) and (5.55). Now from the assumption $T^* < T_0$ we find that this is

$$\leq \frac{E}{6}\varepsilon^{1+\tau} + 2\sqrt{2}\,\varepsilon^{1/2} < 3\varepsilon^{1/2} < \frac{A}{4}\varepsilon^a \frac{12}{A}\varepsilon^{8\tau} < \frac{A}{4}\varepsilon^a \,. \tag{5.57}$$

The second to last inequality derives from the fact that $a + 8\tau < 1/2$ (use (4.18c) and $p \geq 4$); the last inequality is a consequence of (5.36b) and $E \geq 1$. Thus $\|J(T^*) - J_0\|_\infty < \frac{A}{4}\varepsilon^a$ contradicts $J(T^*) \in \partial K_\infty$, so $T^* \geq T_0$, and we see that the solution (J, ϕ) of (4.24) with initial condition $(J_0, \phi_0) = T^{-1}(I_0, \theta_0)$ remains in $K_\infty \times \mathbf{T}^3 \subset D_\infty$ on the time interval $[0, T_0]$. We thus have a well-defined solution $(I, \theta) = T(J, \phi)$ of (3.5) on the same interval. Furthermore, for $0 \leq t \leq T_0$,

$$\begin{aligned}
\|\widehat{I}(t) - \widehat{I}_0\|_\infty &\leq \|\widehat{I}(t) - \widehat{J}(t)\|_\infty + \|\widehat{J}(t) - \widehat{J}_0\|_\infty + \|\widehat{J}_0 - \widehat{I}_0\|_\infty \leq \\
&\|I(t) - J(t)\|_\infty + \|\widehat{J}(t) - \widehat{J}_0\|_\infty + \|J_0 - I_0\|_\infty \leq \\
&\frac{A}{4}\varepsilon^c + \frac{4E}{\sigma} t\varepsilon e^{-\varepsilon^{-\tau/4}} + \frac{A}{4}\varepsilon^c \leq \\
&\frac{A}{2}\varepsilon^c + \frac{E}{6}\varepsilon^{1+\tau} < \frac{A}{2}\varepsilon^c + \frac{A}{2}\varepsilon^c\left(\frac{E}{3A}\varepsilon^{8\tau}\right) < \frac{3}{4}A\varepsilon^c \,.
\end{aligned} \tag{5.58}$$

The next to last inequality holds because $c + 8\tau < 1 + \tau$ by (4.18b); the last inequality follows from (5.36b) ($\varepsilon \leq (A/12E)^{\frac{1}{8\tau}} < (3A/E)^{\frac{1}{8\tau}}$); and (5.41) is proved.

The following estimate for $0 \leq t \leq T_0$, which parallels (5.47), is needed to conclude the proof:

$$\begin{aligned}
\left|E_\perp(I,\theta) - E_\perp(J,\phi)\right| &\leq \frac{3}{2}\|I^* - J^*\|_\infty\left(\|I^* - J^*\|_\infty + 2\|J^*\|_\infty\right) + C_4\varepsilon\|\theta^* - \phi^*\|_\infty \\
&\leq \frac{3}{2}\left(\frac{A}{4}\varepsilon^c\right)\left(\frac{A}{4}\varepsilon^c + 2\sqrt{2}\,\varepsilon^{1/2}\right) + C_4\frac{\sigma}{4}\varepsilon^{1+d} \\
&\leq \varepsilon^{1+b}\left(\frac{3}{32}A^2 + \frac{3\sqrt{2}}{4}A + \frac{C_4\sigma}{4}\right) \leq \varepsilon\frac{\delta}{3} \,,
\end{aligned} \tag{5.59}$$

as in (5.47). Combining (5.47), (5.53), and (5.59), we find

$$\left|E_\perp(I(t),\theta(t)) - E_\perp(I_0,\theta_0)\right| \leq \varepsilon\delta \,, \tag{5.60}$$

which proves (5.42). It follows that for $0 \leq t \leq T_0$,

$$\begin{gathered}
\varepsilon\Pi_{\mathcal{M}} W(\theta^*) \leq E_\perp(I(t),\theta(t)) \leq (Q+\delta)\varepsilon \\
\Rightarrow \theta \in \mathcal{A}(Q+\delta), \quad \text{and} \\
Q + \delta \leq Q^* \Rightarrow \mathcal{A}(Q+\delta) \cap \mathcal{C}(1) = \emptyset \,.
\end{gathered}$$

The solution (I, θ) thus satisfies the channeling criterion as claimed for $0 \leq t \leq T_0$, and the theorem is proved. □

6. Concluding Remarks

A few remarks should help clarify the meaning of the channeling theorem. First, the statement that the solutions (I, θ) discussed in the theorem "are approximated by solutions of a generalized continuum model" simply indicates that the difference between (I, θ) and (J, ϕ) is uniformly small on $[0, T_0]$, while (J, ϕ) is a solution of (4.24), which on the same interval may be written

$$H'(J, \phi) = \frac{1}{2}J^2 + \varepsilon \Pi_{\mathcal{M}} W(\phi^*) + O(\varepsilon^{1+\tau}) \tag{6.1}$$

by virtue of (4.26) and (4.27). Removing the $O(\varepsilon^{1+\tau})$ term from (6.1) and renaming its phase variables (p, q), we recover the ordinary continuum model

$$H_c(p, q) = \frac{1}{2}p^2 + \varepsilon \Pi_{\mathcal{M}} W(q^*) \tag{6.2}$$

which splits into the independent systems

$$H_\perp(p^*, q^*) = \frac{1}{2}(p^*)^2 + \varepsilon \Pi_{\mathcal{M}} W(q^*) \tag{6.3}$$

$$H_\parallel(\hat{p}, \hat{q}) = \frac{1}{2}\hat{p}^2 \tag{6.4}$$

((p^*, q^*) and $(\hat{p}, \hat{q})$ are the transverse and longitudinal phase variables, respectively; cf. §5.2.) If $\dim\mathcal{M} = 1$, $H_\perp$ is a 1 degree of freedom planar continuum Hamiltonian, the scaled version of (2.4); if $\dim\mathcal{M} = 2$ it is the scaled version of a 2 degree of freedom axial continuum Hamiltonian (2.3). This is the sense in which (4.24) is a generalized continuum model, and it is not surprising that the $O(\varepsilon^{1+\tau})$ remainder (comprising $[\varepsilon^{-\tau/4}]$ terms) is needed to obtain the uniform approximation on exponentially long time intervals. In general, the continuum solution (p, q) can uniformly approximate the exact solution (I, θ) to (3.5) on time intervals of length at most $O(\varepsilon^{-1/2})$ (see [20]), although it may be possible to construct a "phase-adjusted" solution based on the continuum solution which works on longer intervals, at least in the planar case.

A second point of interest is the set of admissible initial conditions

$$\left(\left(\widehat{\mathcal{Z}}_{\mathcal{M}}^{\frac{5}{2}A,\alpha}(N, C, 1/2) \cap \mathcal{S}^{\varepsilon}_{\sqrt{2}}\right) \times \mathbf{T}^3\right) \cap \mathcal{Q}\,, \tag{6.5}$$

$$\text{where} \qquad \mathcal{Q} = \{(I_0, \theta_0) \mid E_\perp(I_0, \theta_0) \le \varepsilon Q\}\,. \tag{6.6}$$

In physical terms, the set (6.5) represents particles with suitable initial transverse energy (which excludes particles initially directed at nuclei), and with suitable initial directions. It is interesting to note that the ε-dependence of the directions determined by (6.5) is in accordance with the incident energy dependence of channeling directions determined by Lindhard's critical angle [29] (of course, this is just a consequence of the choice of scaling; cf. §3.3).

There is also the related question of what happens to trajectories with initial directions not governed by either Theorem 5.3 or Theorem 5.7. Because of the requirement $c > \alpha$ in the analytic lemma (see (4.18b)), for sufficiently small ε there is a subset of action space with relative volume $O(\varepsilon^\alpha - \varepsilon^c)$ which is excluded from the initial conditions of both theorems. This in turn may be traced to the proof of the channeling theorem, and ultimately to the need to avoid close encounters as demanded by the channeling criterion. It must be stressed that this gap in the set of initial directions for which particle motions are described is almost certainly not an artifact of the rough estimates carried out here (although the estimates are certainly crude); instead, it appears to be an unavoidable consequence of the channeling criterion. This points out the essential way that the channeling theorem differs from other Nekhoroshev-like results, which are not called upon to provide information about trajectories in configuration space. It also leaves open the intriguing possibility that certain "dechanneling" phenomena may be represented rigorously in the perfect crystal model (some dechanneling is observed experimentally "on the shoulders" of axial directions; these shoulders appear to coincide roughly with the gaps in initial directions governed by the channeling theorem).

Much remains to be done in the mathematical theory of the motion of charged particles in crystals; some directions for further research are outlined in [19] and [20]. As it stands, the analysis of the perfect crystal model presented here establishes a general mathematical framework for particle channeling and should serve as the foundation for more comprehensive theories. Finally, it was a pleasure and a surprise for this author to uncover – with considerable assistance – the link between the vastly different but equally remarkable works of J. Lindhard and N. Nekhoroshev.

Acknowledgements. The results presented here are a condensation of the PhD thesis [17]. Accordingly, I am indebted to more individuals than I can name; many of them were acknowledged in [17]. I thank A. Giorgilli, P. Lochak, W.T. Kyner, E. Zehnder, and especially J.A. Ellison for their help; I also thank the American Fulbright Commission for the grant which allowed me to spend 1986–87 at the Ecole Normale Supérieure in Paris, where much of this work was carried out.

References

[1] V.I. Arnol'd, Proof of A.N. Kolmogorov's theorem on the preservation of quasi-periodic motions under small perturbations of the Hamiltonian, Russian Math. Surveys **18** (5) (1963), 9–36 (Russian original: Usp. Mat. Nauk. SSSR **18** (5) (1963), 13–40)

[2] V.I. Arnol'd, Small denominators and problems of stability of motion in classical and celestial mechanics, Russian Math. Surveys **18** (6) (1963), 85–191 (Russian original: Usp. Mat. Nauk. SSSR **18** (6) (1963), 91–192)

[3] V.I. Arnol'd, Instability of dynamical systems with many degrees of freedom, Soviet Math. Dokl. **5** (1964), 581–585 (Russian original: Dokl. Akad. Nauk. SSSR **156** (1964), 9–12)

[4] V.I. Arnol'd, A stability problem and ergodic properties of classical dynamical systems, Amer. Math. Soc. Trans. (2) **70** (1968), 5–11 (Russian original: in Proc. Inter. Congress Mathematicians pp. 387–392, Mir, Moscow, 1968)

[5] V.I. Arnol'd, Ordinary Differential Equations, MIT, Cambridge, Mass., 1973 (Russian original: Nauka, Moscow, 1973)

[6] V.I. Arnol'd, Mathematical Methods of Classical Mechanics, Springer, New York, 1978 (Russian original: Nauka, Moscow, 1974)

[7] A. Bazzani, S. Marmi and G. Turchetti, Nekhoroshev estimates for isochronous nonresonant symplectic maps, Celestial Mech. Dynam. Astronom. **47** (1989–90), 333–359

[8] G. Benettin, L. Galgani and A. Giorgilli, A proof of Nekhoroshev's theorem for the stability times in nearly integrable Hamiltonian systems, Celestial Mech. **37** (1985), 1–25

[9] G. Benettin, L. Galgani and A. Giorgilli, Exponential law for the equipartition times among translational and vibrational degrees of freedom, Phys. Lett. A **120** (1) (1987), 23–27

[10] G. Benettin, L. Galgani and A. Giorgilli, Realization of holonomic constraints and freezing of high frequency degrees of freedom in the light of classical perturbation theory, Part I, Comm. Math. Phys. **113** (1987), 87–103; ibid., Part II, Comm. Math. Phys. **121** (1989), 557–601

[11] G. Benettin and G. Gallavotti, Stability of motion near resonances in quasi-integrable Hamiltonian systems, J. Stat. Phys. **44** (1986), 293–338

[12] R.A. Carrigan and J.A. Ellison, Eds. Relativistic Channeling, Plenum, New York, 1987

[13] A. Celletti and A. Giorgilli, On the stability of the Lagrange points in the spatially restricted problem of three bodies, Celestial Mech. Dynam. Astronom. **50** (1991), 31–58

[14] B.V. Chirikov, A universal instability of many dimensional oscillator systems, Phys. Rep. **52** (1979), 265–379

[15] A. Delshams, Porque la difusion de Arnol'd aparece genericamente en los sistemas hamiltonianos con màs de dos grados de liberdad, Thesis, University of Barcelona, 1984

[16] R. Douady, Stabilité ou instabilité des points fixes elliptiques, Ann. scient. Éc. Norm. Sup., 4me série, tome 21 (1988), 1–46

[17] H.S. Dumas, A Mathematical Theory of Classical Particle Channeling in Perfect Crystals, PhD Thesis, University of New Mexico, 1988 (UMI, Ann Arbor, Michigan, 1988)

[18] H.S. Dumas, Ergodization rates for linear flow on the torus, J. Dyn. Diff. Equations **3** (1991), 593–610

[19] H.S. Dumas and J.A. Ellison, Nekhoroshev's theorem, ergodicity, and the motion of energetic charged particles in crystals, in Essays on Quantum and Classical Dynamics, J.A. Ellison and H. Überall, Eds., pp. 17–56, Gordon and Breach, Philadelphia, 1991

[20] H.S. Dumas, J.A. Ellison and A.W. Sáenz, Axial channeling in perfect crystals, the continuum model and the method of averaging, Annals of Physics (NY) **209** (1991), 97–123

[21] L.C. Feldman, J.W. Mayer and S.T. Picraux, Material Analysis by Ion Channeling: Submicron Crystallography, Academic Press, New York, 1982

[22] D.S. Gemmell, Channeling and related effects in the motion of charged particles through crystals, Rev. Mod. Phys. **46** (1974), 129–227

[23] A. Giorgilli, A. Delshams, E. Fontich, L. Galgani and C. Simó, Effective stability for a Hamiltonian system near an elliptic fixed point, with an application to the restricted three body problem, J. Diff. Eqs. **77** (1989), 167–198

[24] J. Glimm, Formal stability of Hamiltonian systems, Comm. Pure Appl. Math. **17** (1964), 509–526

[25] J.K. Hale, Ordinary Differential Equations, Wiley-Interscience, New York, 1969

[26] P.J. Holmes and J.E. Marsden, Melnikov's method and Arnol'd diffusion for perturbations of integrable Hamiltonian systems, J. Math. Phys. **23** (4) (1982), 669–675

[27] A.N. Kolmogorov, Preservation of conditionally periodic movements with small change in the Hamilton function, in Stochastic Behavior in Classical and Quantum Hamiltonian Systems, G. Casati and J. Ford, Eds., Lecture Notes in Physics **93** pp. 51–54, Springer, Berlin, 1979 (Russian original: Dokl. Akad. Nauk. SSSR **98** (4) (1954), 527–530)

[28] Ph. Lervig, J. Linhard and V. Nielsen, Quantal treatment of directional effects for energetic particles in crystal lattices, Nucl. Phys. **A 96** (1967), 481

[29] J. Lindhard, Influence of crystal lattice on motion of energetic charged particles, Mat. Fys. Medd. Dan. Vid. Selsk. **34**, no. 14 (1965)

[30] J. Lindhard, V. Nielsen and M. Scharff, Approximation method in classical scattering by screened Coulomb fields, Mat. Fys. Medd. Dan. Vid. Selsk. **36**, no. 10, 1968

[31] J.E. Littlewood, On the equilateral configuration in the restricted problem of three bodies, Proc. London Math. Soc. (3) **9** (1959), 342–372

[32] J.E. Littlewood, The Lagrange configuration in celestial mechanics, Proc. London Math. Soc. (3) **9** (1959), 525–543

[33] P. Lochak, Effective speed of Arnol'd's diffusion and small denominators, Phys. Lett. **A 143** (1) (1990), 39–42

[34] P. Lochak, Stabilité en temps exponentiels des systèmes hamiltoniens proches des systèmes integrables: résonances et orbites fermées (ENS preprint, 1990)

[35] P. Lochak, Canonical perturbation theory through simultaneous approximation (ENS preprint, 1991)

[36] P. Lochak and C. Meunier, Multiphase Averaging for Classical Systems With Applications to Adiabatic Theorems, Springer-Verlag, New York, 1988

[37] P. Lochak and A. Porzio, A realistic exponential estimate for a paradigm Hamiltonian, Annales IHP, Phys. théorique **51** (1989), 199–219

[38] L.M. Markus and K.R. Meyer, Generic Hamiltonian systems are neither integrable nor ergodic, Mem. Am. Math. Soc. **144** (1974)

[39] J.K. Moser, Stabilitätsverhalten kanonischer differentialgleichungssysteme, Nachr. Akad. Wiss. Göttingen, Math. Phys. Kl. II (1955), 87–120

[40] J.K. Moser, On invariant curves of area-preserving maps of an annulus, Nachr. Akad. Wiss. Göttingen, Math. Phys. Kl. II (1962), 1–20

[41] N.N. Nekhoroshev, Behavior of Hamiltonian systems close to integrable, Funct. Anal. **5** (1971), 338–339 (Russian original: Fun. Anal. Pril. **5** (4) (1971), 82–84)

[42] N.N. Nekhoroshev, Stable lower estimates for smooth mappings and for gradients of smooth functions, Math. USSR Sbornik **19** (3) (1973), 425–467 (Russian original: Mat. Sbornik **90** (132) (1973), 425–467)

[43] N.N. Nekhoroshev, An exponential estimate of the time of stability of nearly integrable Hamiltonian systems, Russian Math. Surveys **32** (6) (1977), 1–65 (Russian original: Usp. Mat. Nauk. SSSR **32** (6) (1977), 5–66)

[44] N.N. Nekhoroshev, An exponential estimate of the time of stability of nearly integrable Hamiltonian systems II, in Topics in Modern Mathematics, Petrovskii Seminar No. 5 (O.A. Oleinik, Ed.), Consultants Bureau, London, 1980 (Russian original: Tr. Sem. Petrows. **5** (1979), 5–62)

[45] J.H. Poincaré, Les méthodes nouvelles de la mécanique céleste, tome 1, Gauthier-Villars, Paris, 1892
[46] J. Pöschel, On Nekhoroshev's estimate for quasi-convex Hamiltonians, to appear in Math. Zeitschr.

The Adiabatic Invariant in Classical Mechanics

J. Henrard

Introduction

The adiabatic invariant theory in Classical Mechanics emerged from a very rich, but somewhat murky, sea of analogies with other types of problems in Theoretical Physics. We recall in the introduction of the first part how it derives from the "adiabatic principle" of Ehrenfest (1916) in the frame of the old quantum theory.

But this principle itself and the name attached to it has a much older story. The word "adiabatic" was used by the very first thermodynamicists to mean a thermic insulation. It comes from the greek $\alpha\delta\iota\alpha\beta\alpha\tau\iota\chi o\varsigma$, "which cannot be crossed" (specifically by heat in this context). From there it started to be used to characterize thermodynamical transformations where no heat is exchanged. A paradigm of such transformations is of course the rapid expansion or compression of a gas in the atmosphere or in a steam machine. Although these expansions or compressions are relatively rapid, they are still slow enough that the gas stays in the state of statistical equilibrium and the transformation is thus "isentropic" (meaning the entropy of the gas stays constant). The force of the paradigm has been such that the word "adiabatic" is now for many authors synonymous with "isentropic" although the two words refer to very different concepts.

The "isentropicity" of such thermodynamical transformations is explained in statistical mechanics by their "quasistatic" character. The entropy of the gas (the exponential of which is related to the measure of a volume in phase space) is conserved during the *slow variation* of the environment of the system.

From there we arrive by analogy to the "adiabatic principle" used in Quantum and then Classical Mechanics. It is based upon the fact that the harmonic oscillator (and other simple dynamical systems as it was found later) submitted to slow variations of its parameters modifies its energy but keeps its action (energy divided by frequency) constant. The action is also related to the volume in phase space.

As we can see, the path from the word "adiabatic" used in thermodynamics to the above "adiabatic principle" is tortuous and our greek colleagues are certainly puzzled by sentences such as "the changes in the adiabatic invariant due to separatrix crossing" which we shall use later. Keep in mind also that the word "adiabatic" suggests to a thermodynamicists or a meteorologist a fast expansion or compression of gas while on the contrary we are thinking of slow variations.

The adiabatic principle in Classical Mechanics applies essentially to non-autonomous Hamiltonian systems with one degree of freedom. However, it was realized (see for instance Kruskal, 1962) that it can be considered as a special case of the general perturbation theory for Hamiltonian systems. Indeed a non-autonomous Hamiltonian system with one degree of freedom can be considered as an autonomous system with two degrees of freedom. The name "adiabatic invariant" has then be used by some authors to designate the "quasi-integrals" (the adelphic integrals of Whittaker, 1916, 1927) found in this larger context.

We shall not follow this practice and we shall not consider the adiabatic invariant theory as just a special case of the general perturbation theory for Hamiltonian systems. We do not wish to deny by that the interest of the larger context (we shall use freely its results) but we think that the specific properties and the number of possible applications of the restricted frame (non autonomous Hamiltonian systems with one degree of freedom) are rich enough to warrant a special treatment.

Our description of the adiabatic invariant theory in Classical Mechanics is divided into four parts.

The first one describes the action-angle variables and the adiabatic invariance of the mean action. We assume that the reader is familiar with the classical tools of Hamiltonian mechanics: canonical transformations, contact transformations, Poisson's brackets, $\cdots$. We apply them and the basic ideas of general perturbation theory (which we summarize briefly) to arrive at the classical results of the adiabatic invariant theory. Most of the ideas developed in this first part are well-known and we discuss them only briefly. We emphasize the computational aspects of the theory which we feel have been neglected. Indeed we think that the action-angle variables and the classical adiabatic invariant theory should be used for other systems than the textbook harmonic oscillator or simple pendulum.

The second part extends the theory to the special but important case of "separatrix crossing". This special case is very delicate and has not been treated satisfactorily until recently (Cary et al., 1986). At the risk of being boring, we have tried there to be complete and to justify the various estimates as completely as possible. Several technical points are developed in the appendices.

The main thrust of the argumentation and the principal results are inspired from the work of Cary et al. (1986). Our presentation is nevertheless different. We have tried as far as possible to use the classical tools of perturbation theory: Birkhoff's normalization, Remainder functions, etc. This leads to some improvement in the estimates and hopefully it may be useful for further generalizations or significant improvements of the results. Indeed we believe that this part of the theory and the connections it presents with the field of chaotic behaviour in Hamiltonian systems could be improved and amplified leading to a better understanding of this fascinating field.

Following Cary et al. (1986) we show that seven quantities ω, $\partial J_i^\star/\partial\lambda$, $h_i^\star$, g_i $(1 \leq i \leq 2)$, to be evaluated along the separatrices, are instrumental in describing the behaviour of the system.

These quantities are evaluated in the third part for some "paradigms": these are model problems, well-known ones like the pendulum or less well-known like the "Colombo's top". They are simple enough that the evaluations can be carried out analytically and close enough to realistic problems that they can at least provide qualitative answers to interesting questions. The third part can be considered as "classroom" applications of the previously developed theories but also as a collection of formulae which will be used in part four.

The fourth part is devoted to the applications. Of course we cannot be exhaustive there. We made a selection of applications simple enough that they do not need long developments but hopefully varied enough that we can show through them the power of the method and that we can illustrate its fine points.

Our text is certainly not perfect but we hope it will be useful in drawing the attention to this powerful method and encourage further developments and applications. This text owes very much to Dr. Escande who all along its writing helped us with his comments, suggestions, references, ... We thank also Dr. Lemaître, Y. Elskens and an anonymous referee for their help in improving the final version.

Part I The Classical Adiabatic Invariant Theory

1. Introduction

The Classical Adiabatic Invariant theory has been known for a long time and is now well-established as an important tool in the study of the long term evolution of dynamical systems.

As far as we know, the first mention of it is to be found in the "adiabatic principle" of Ehrenfest (1916) derived in the frame of the "Old Quantum Theory". A new formulation in the frame of the "New Quantum Theory" was given by Born and Fock (1928). In the mean time, Kneser (1924) had given the first mathematically satisfactory proof of first order adiabatic invariance in classical mechanics.

It is only much later, in the problem of the acceleration of cosmic rays (Alfven, 1950; Helwig, 1955) that the importance of adiabatic invariant theory in Classical Mechanics was felt. In this instance the adiabatic invariant is the magnetic moment of a charged particle in a strong magnetic field.

At about the same time, it was recognized that the theory of virtually every device for the production of useful energy from controlled thermonuclear fusion leaned heavily on the constancy of this magnetic moment. It was seen that the requirement that particles remain confined for period of time encompassing many millions of gyrations could generally be met only if the magnetic moment were in fact constant to a much higher approximation than the first order (Kruskal, 1952).

Kulsrud (1957) considered the simpler problem of an harmonic oscillator and proved that its adiabatic invariant (ratio of energy over frequency) was constant to all orders. Kruskal (1957) proved the analogous result for the gyrating particle and Lenard (1959) and later Arnol'd (1964) showed the relationship between adiabatic invariant and classical perturbation theory.

Since then, the adiabatic invariant has proved itself to be a very versatile tool especially in the study of the motion of a charged particle in an electromagnetic field (Best, 1968; Aamodt, 1971, 1972; Tennyson 1979; Littlejohn, 1983; Cary et al., 1984; Menyuk, 1985) with applications to plasma physics and high energy accelerators.

Continuing interest in the theoretical aspect of the adiabatic invariant theory has also been shown (Littlewood, 1963; Leung and Meyer, 1975; Wasow, 1973, 1976; Stengle, 1977; Meyer, 1976, 1980) and has led to sharper (exponential) estimates of the invariance in special circumstances.

It took some time for the adiabatic invariant to migrate (slowly !) from Particle Physics to Celestial Mechanics. First order adiabatic invariant is mentioned by Goldreich and Peale (1966) and really used to its full power by Goldreich and Toomre (1969).

Since then it has shown itself to be an important tool in the study of the long term evolution of the Solar System (Peale, 1974, 1976; Ward et al., 1979; Yoder, 1979a; Henrard and Lemaître, 1983; Lemaître, 1985).

The basic idea of the adiabatic invariant theory is very simple.

Consider a simple dynamical system (such as the pendulum) and make it dependent upon a slowly varying parameter (say λ).

For a short period of time the motion will not depart significantly from a trajectory of the "frozen" system (the simple dynamical system where the parameter is constant: $\lambda = \lambda_0$). We shall call this trajectory a "guiding trajectory".

This will be true for any short period of time. But at different times the guiding trajectory will not be the same: it corresponds each time to a solution of the "frozen" system but frozen at different values of the parameter.

The adiabatic invariant theory will tell us implicitly how, on the long time scale, the guiding trajectory evolves by determining a function of the guiding trajectory and the parameter which stays approximately constant.

A very powerful feature of such a scheme is that we do not even have to know in detail the variations of the parameter. Knowing the guiding trajectory when $\lambda = \lambda_0$ (at time t_0), we can find the guiding trajectory at time t_1 (when $\lambda = \lambda_1$) without knowing the path between λ_0 and λ_1.

Notwithstanding the fact that the classical adiabatic invariant theory is well-established, we shall describe it in some details in this first part.

In Sect. 2, we introduce the classical angle-action variables which will be used throughout these notes. The concept is well-known but some fine points (as, for instance, the role of the curve of initial conditions) have not drawn much attention and may need at least a brief description. Also some formulae about the remainder function associated with the angle-action variables are new and will be useful later on.

In Sect. 3, we describe how classical perturbation theory leads to the asymptotic invariance of the averaged action-angle variable at any order.

In Sect. 4, we make a distinction between two possible definitions of the n-th order adiabatic invariant. The classical action-variable itself is an n-th order adiabatic invariant of some sort (we shall call it a "crude" adiabatic invariant of

n-th order) but is not an adiabatic invariant according to a sharper, more useful and more often quoted definition.

This point may clarify some confusion stemming from the much quoted paper of Lenard (1959). In this work an ad-hoc assumption blurs the distinction and seems to make the classical action-variable itself an n-th order adiabatic invariant of the second type.

The implicit character of the usual definition of the action-angle variables may lead to practical difficulties. First it is not always obvious how to compute the path integrals involved in the definition. See for instance, in Part III, how even for the pendulum problem, a well-chosen change of variables has to be made to transform the path integrals into quadratures.

A more important point is that, except in a few particular cases (actually we know of only three: the harmonic oscillator, the pendulum and the two body problem) the path integrals cannot be solved analytically. In more general circumstances, the action-angle variables seem thus to be reduced to a nice theoretical concept which cannot be used practically. But it does not need to be so.

In Sect. 5, we shall rephrase the definition of action-angle variables in such a way that it leads to explicit quadratures that can be easily implemented numerically. Furthermore the derivatives of functions of the action-angle variables can be expressed (and computed) in terms of solutions of the variational equations of the dynamical system.

In this way, the theoretical use of the action-angle variables is made easier because the definitions are explicit (and we shall make use of this in Part II) but also their practical use is now possible for a larger class of problems as we have shown elsewhere (Henrard and Lemaître, 1986b).

In Sect. 6, we come back briefly upon the averaging procedure which defines the n-th order adiabatic invariant to show how it has to be modified when the domain of interest approaches a critical curve (where the period goes to infinity). We shall conclude that classical adiabatic invariant theory is valid at least in a domain which excludes a neighborhood of order ε^q (where $q < 1$) of the critical curve.

2. Action-Angle Variables

Let the one-degree of freedom Hamiltonian function H the value of which is h,

$$H(q, p, \lambda) = h \tag{1}$$

be an analytical function of the variable q, its conjugate momentum p and a parameter λ, defined for (q, p) belonging to an open domain D of a two dimensional manifold and λ belonging to an interval $I \subset \mathbf{R}$. We assume that all the trajectories of the corresponding differential equations:

$$\dot{q} = \frac{\partial H}{\partial p}, \quad \dot{p} = -\frac{\partial H}{\partial q} \tag{2}$$

are periodic of period $T(h, \lambda)$.

The two-dimensional manifold may be simply $\mathbf{R}^2$ but it is not unusual that it is something else. This is indeed the case for the pendulum defined on the cylinder or for the "Colombo's top" problem (see Sect. III.4) defined on the sphere.

The canonical transformation from (q, p) to the action-angle variables (ψ, J) is usually defined by means of a generating function $S(q, J, \lambda)$

$$p = \frac{\partial S}{\partial q}, \quad \psi = \frac{\partial S}{\partial J} \tag{3}$$

where $S(q, J, \lambda)$ is a solution of the Hamilton-Jacobi equation

$$H\left(q, \frac{\partial S}{\partial q}, \lambda\right) = K(J, \lambda) . \tag{4}$$

The transformation generated by S is such that the transformed Hamiltonian $K(J, \lambda)$ depends only upon the action J (the new momentum) which is thus constant. The new variable (which will be made an angle by imposing that the trajectories close when ψ augments by 2π) is thus a linear function of the time

$$\psi = \frac{\partial K}{\partial J}\, t + \psi_0 . \tag{5}$$

Hence the action-angle variables (ψ, J) solve in principle the one-degree of freedom problem (1). Geometrically, J labels the orbits and ψ is an angular variable along each orbit.

More specifically a solution of (4) can be written under the form:

$$S(q, J, \lambda) = \int_{q_0(h,\lambda)}^{q} P(q', h, \lambda)\, dq' + \int_{h^\star}^{h} \frac{\partial q_0}{\partial h'}\, p_0\, dh' \tag{6}$$

where h is a still undefined function of J,

$$h = K(J, \lambda) , \tag{7}$$

$h^\star$ is an arbitrary value of h and P is the function implicitly defined by

$$H(q, P, \lambda) = h . \tag{8}$$

The quantities $q_0(h, \lambda)$, $p_0(h, \lambda)$ are the coordinates of a curve (parametrized by h) which we shall assume to be transversal to all the periodic orbits of the system. It will be useful later on to consider that (q_0, p_0) are defined implicitly by the analytical function:

$$F(q_0, p_0, \lambda) = 0 . \tag{9}$$

Note that q_0 is usually taken as a constant. This makes (6) simpler as the second term disappears. But this is not always possible as the curve of initial conditions (9) has to intercept all trajectories. Of course this is possible in the few instances (harmonic oscillator, pendulum, $\cdots$) which have been explicitly worked out.

Note also that the second term in (6) could be dropped anyway if we are not interested in the position of the points corresponding to $\psi = 0$. This second term is designed to force $\psi = 0$ on the curve of "initial conditions" (9).

By definition of the canonical transformation (3), generated by S, we have

$$\psi = \left(\frac{\partial K}{\partial J}\right) \int_{q_0}^{q} \frac{\partial P}{\partial h}\, dq' \,. \tag{10}$$

If we now impose that ψ is an angular variable, increasing by 2π after a complete circuit of a periodic orbit, we can extract from (10) that, up to a function independent of h:

$$J(h, \lambda) = \frac{1}{2\pi} \oint P\, dq' \tag{11}$$

where the path integral is taken along the closed periodic orbit. Equation (11) is an implicit definition of the yet undefined function (7).

In the next section, we shall consider a problem where λ is no longer considered as constant but as a function of the time: $\lambda = \varepsilon t$.

In that case the canonical transformation defined by (3) is still useful but the Hamiltonian function of the dynamical system expressed in the action-angle variables is no longer $K(J, \lambda)$ but

$$H'(\psi, J, \varepsilon t) = K(J, \varepsilon t) + \varepsilon\, R(\psi, J, \varepsilon t) \tag{12}$$

where εR is the remainder function defined by

$$R = \frac{\partial S}{\partial \lambda} = \left(\frac{\partial K}{\partial J}\right)^{-1} \int_0^{\psi} \left\{\frac{\partial K}{\partial \lambda} - \frac{\partial H}{\partial \lambda}\right\} d\psi + R(0, J, \lambda) \,. \tag{13}$$

Indeed we have:

$$\frac{\partial H}{\partial \lambda} = \frac{1}{\varepsilon}\frac{dH}{dt} = \frac{1}{\varepsilon}\frac{\partial K}{\partial J}\frac{\partial J}{\partial t} + \frac{\partial K}{\partial \lambda} = -\frac{\partial K}{\partial J}\frac{\partial R}{\partial \psi} + \frac{\partial K}{\partial \lambda} \,. \tag{14}$$

The initial value of R can be evaluated from

$$R(0, J, \lambda) = \int_{h^\star}^{h} \left\{\frac{\partial q_0}{\partial h'}\frac{\partial p_0}{\partial \lambda} - \frac{\partial q_0}{\partial \lambda}\frac{\partial p_0}{\partial h'}\right\} dh' - \left\{\frac{\partial q_0}{\partial \lambda} p_0\right\}_{h=h^\star} \tag{15}$$

and of course vanishes if q_0 is a constant.

Note also that R is a periodic function of ψ, as S depends upon ψ through q, and q is a periodic function of ψ.

Let us mention also that the action J is coordinate-independent. Its definition (11) shows that $2\pi J$ is the (signed) area enclosed by the periodic orbit and areas are preserved by canonical transformations. But the angle ψ is not uniquely defined by the dynamical system. It depends upon the choice of the curve (9) of initial conditions.

If ψ_1 corresponds to a curve $F_1(q_0, p_0, \lambda) = 0$ and ψ_2 to a curve $F_2(q_0, p_0, \lambda) = 0$, the relation between ψ_1 and ψ_2 is easily shown from (10) to be

$$\psi_2 = \psi_1 + \Psi(J, \lambda) \tag{16}$$

where $\Psi(J, \lambda)$ is the value of ψ_2 corresponding to the points on the curve $F_1(q_0, p_0, \lambda) = 0$.

As the momenta conjugated to ψ_1 and ψ_2 are the same, the corresponding remainder functions differ only by a function independent of ψ_1 or ψ_2, leading to the equality:

$$R_1(\psi_1, J, \lambda) = R_2(\psi_1 + \Psi, J, \lambda) + \{R_1(0, J, \lambda) - R_2(\Psi, J, \lambda)\} . \tag{17}$$

The difference between the remainder functions can also be evaluated (up to a function independent of ψ and J) by evaluating the remainder function of the canonical transformation from (ψ_1, J) to (ψ_2, J). We find

$$R_1(0, J, \lambda) = R_2(\Psi, J, \lambda) - \int^J \frac{\partial \Psi}{\partial \lambda} \, dJ . \tag{18}$$

3. Perturbation Theory

We shall now assume that the parameter λ in the Hamiltonian function (1) is actually a function of the time but a "slow" function of the time. By this we mean not only that $d\lambda/dt$ is small but that higher order derivatives of λ are smaller yet, i.e. that there exists a small quantity ε such that

$$\frac{1}{n!} \left| \frac{d^n \lambda}{dt^n} \right| \leq \varepsilon^n \ \text{(uniformly)} . \tag{19}$$

Our results will be valid for ε "sufficiently small", i.e. they will be asymptotic results.

The assumption (19) is often quoted under the form:

$$\lambda = \varepsilon t \tag{20}$$

and we shall follow this practice. To go from assumption (19) to (20) one calls $\lambda' = \varepsilon t$, expresses λ as an analytical function of λ' (radius of convergence of the order of unity), inserts the expression $\lambda\ (\lambda')$ in the Hamiltonian (1) and renames λ' as λ.

The assumption (19) may seem to be strong but is is essential. It is not always quoted in full and is sometime hidden in the naive picture (pleasantly recalled by Arnol'd, 1978) that the "devil" pulling the strings (i.e. making λ a function of time) is not only slow but ignores what the dynamical system does. Well, actually he may know it but condition (19) makes him powerless to adjust to the dynamical system.

Another important assumption we make is that the periods of the periodic orbits of the "frozen" system (i.e. when $\lambda =$ constant) in the domain $D \times I$ of definition are uniformly bounded (independently of ε)

$$T(h, \lambda) \leq T_0 ; \tag{21}$$

we shall come back on this assumption in Sect. 6.

We are now ready to apply a classical perturbation technique to the Hamiltonian system

$$H'(\psi, J, \varepsilon t) = K(J, \varepsilon t) + \varepsilon\, R(\psi, J, \varepsilon t) \tag{22}$$

derived in (12).

The perturbation technique we shall use is the Lie transform method introduced by Hori (1966) and Deprit (1969) in Celestial Mechanics. It has now become a standard technique in perturbation theory (Giacaglia, 1972; Nayfeh, 1973; Kirchgraber and Stiefel, 1978).

We shall first summarize it to introduce the notations we need. We follow Henrard (1970a, 1970b).

The technique is based upon the transformation from the $2n$-dimensional phase space (y) to the $2n$-dimensional phase space (x) given by:

$$x = X(y, \varepsilon)\ . \tag{23}$$

The transformation is close to the identity (i.e. $X(y, 0) = y$) and is defined implicitly as the general solution of the auxiliary Hamiltonian system

$$\frac{dx}{d\varepsilon} = S\, W_x^{\tau}(x, \varepsilon) \quad ; \quad W_x = \frac{\partial W}{\partial x} \tag{24}$$

where $W(x, \varepsilon)$ is the (analytical) generating function, τ means transposition, and S is the principal symplectic matrix

$$S = \begin{pmatrix} 0_n & 1_n \\ -1_n & 0_n \end{pmatrix} \tag{25}$$

where 0_n (resp. 1_n) is the null matrix (resp. the unit matrix) of order n.

The transformation is canonical and the transform $H(X(y, \varepsilon), \varepsilon)$ of any analytical function

$$H(x, \varepsilon) = \sum_{i \geq 0} \frac{\varepsilon^i}{i!} H_i^0(x) \tag{26}$$

is noted $L(W)\, H(x, \varepsilon)$ (the Lie Transform of H) or $\bar{H}(y, \varepsilon)$. It is given by:

$$\bar{H}(y, \varepsilon) = \sum_{i \geq 0} \frac{\varepsilon^i}{i!} H_0^i(y) \tag{27}$$

where the functions H_0^i can be computed from the H_k^0 $(k \leq i)$ by means of the recursive formula

$$H_m^n = H_{m+1}^{n-1} + \sum_{\ell=0}^{m} \binom{\ell}{m} (H_{m-\ell}^{n-1}; W_{\ell+1}) \tag{28}$$

where $(f; g)$ stands for the Poisson Bracket of f and g and W_j are the coefficients of W in the power series expansion

$$W(x, \varepsilon) = \sum_{j \geq 0} \frac{\varepsilon^j}{j!} W_{j+1}(x)\ . \tag{29}$$

The recursive formula (28) leads to the equation

$$H_0^i = \tilde{H}_0^i + (H_0^0; W_i) \tag{30}$$

where $\tilde{H}_0^i$ is a function of the H_j^0 $(j \leq i)$ and the W_j $(j < i)$.

Assume that the function H_j^0 (and recursively the function W_j) belongs to a vector space $\mathbf{F}$ of functions closed with respect to the Poisson Bracket and such that $\mathbf{F}$ is the direct sum of $\mathbf{F}_1$ and $\mathbf{F}_2$ with $\mathbf{F}_1$ belonging to the image of the operator $(H_0^0;\ \cdot\ )$. Then the functions W_i can be recursively defined in such a way that the coefficients H_0^i of the transformed function $L(W)H$ belong to $\mathbf{F}_2$.

This perturbation technique is designed for autonomous systems. It can be extended to non-autonomous systems once we establish that the remainder function of the transformation generated by $W(x, \varepsilon, t)$ is

$$R(y, \varepsilon, t) = -\int_0^\varepsilon L(W)\, \frac{\partial}{\partial t}\, W\, d\varepsilon \tag{31}$$

(see Henrard, 1970b, or Henrard and Roels, 1974). But it is conceptually simpler (and practically equivalent) to transform the non-autonomous system (22) into an autonomous one by introducing a new canonical variable λ and its conjugate momentum Λ. The new autonomous Hamiltonian function reads

$$H''(\psi, J, \lambda, \Lambda) = K(J, \lambda) + \varepsilon\{\Lambda + R(\psi, J, \lambda)\}\ . \tag{32}$$

Let us choose $\mathbf{F}$ as the set of analytical functions of (ψ, J, λ) periodic of period 2π in ψ, $\mathbf{F}_1$ the set of such functions with zero mean value with respect to ψ, and $\mathbf{F}_2$ the set of analytical functions of (J, λ). We recognize that $\mathbf{F}_1$ belongs to the image of $\mathbf{F}$ by the operator:

$$(H_0^0;\ \cdot\ ) = -\left(\frac{\partial K}{\partial J}\right) \frac{\partial}{\partial \psi} \tag{33}$$

if $\partial K/\partial J$ is different from zero. But $\partial K/\partial J = 2\pi/T$ where T is the period of the periodic orbit of the autonomous system (1) which we have assumed to be uniformly bounded in $D \times I$ (see (21)). Hence we can recursively define functions W_i (independent of Λ) such that the coefficients H_0^i of the transformed function $L(W)H$ are independent of the angular variable.

We have to stop the recursion at some finite order n because we cannot make sure that the partial sums

$$W^{(n)} = \sum_{j=0}^{n-1} \frac{\varepsilon^j}{j!} W_{j+1} \tag{34}$$

do converge when n goes to infinity. In any case, defining W in (24) as the partial sum (34), the transformed Hamiltonian function in the "averaged" coordinates $y = (\bar{\psi}, \bar{J})$ reads

$$\bar{H}(\bar{\psi}, \bar{J}, \lambda, \bar{\Lambda}) = K(\bar{J}, \lambda) + \varepsilon\bar{\Lambda} + \sum_{j=1}^{n} \frac{\varepsilon^j}{j!} \bar{H}_j(\bar{J}, \lambda) + \varepsilon^{n+1} H_r(\bar{\psi}, \bar{J}, \lambda, \varepsilon)\ . \tag{35}$$

The interested reader may consult Giorgilli and Galgani (1985) for rigorous estimates of the convergence radius of this Hamiltonian function.

Note that the "averaged" coordinates $(\bar{\psi}, \bar{J})$ depend upon the order n of the procedure. They should ideally be written $(\bar{\psi}_n, \bar{J}_n)$ but this notation is too cumbersome.

The differential equation for the "averaged" momentum $\bar{J}$ is

$$\left|\frac{d\bar{J}}{dt}\right| = \varepsilon^{n+1} \left|\frac{\partial H_r}{\partial \bar{\psi}}\right| \le C_1 \varepsilon^{n+1} \tag{36}$$

(where C_1 is the supremum of $|\partial H_r / \partial \bar{\psi}|$ in the domain $D \times I$). It will be used to evaluate the time-variations of $\bar{J}$.

The non-averaged momentum J, the classical action-variable, can be expressed as a function of the averaged variable $(\bar{\psi}, \bar{J})$ by means of the same recursive formula (28). It leads to

$$J = L(W)\bar{J} = \bar{J} + \varepsilon J_n(\bar{\psi}, \bar{J}, \lambda, \varepsilon) \tag{37}$$

where J_n is an analytical function periodic of period 2π in $\bar{\psi}$. The first order contribution of J_n is easy to compute and will be useful later on. We have

$$\begin{aligned} J &= \bar{J} + \varepsilon(\bar{J}; W_1) + 0(\varepsilon^2) \ , \\ J &= \bar{J} - \varepsilon \left(\tfrac{\partial K}{\partial J}\right)^{-1} \{R(\bar{\psi}, \bar{J}, \lambda) - < R(\bar{\psi}, \bar{J}, \lambda) >\} + 0(\varepsilon^2) \ , \end{aligned} \tag{38}$$

where $< \cdot >$ stands for the averaged value over $\bar{\psi}$. Inverting (38) we obtain:

$$\bar{J} = J + \varepsilon \left(\frac{\partial K}{\partial J}\right)^{-1} \{R(\psi, J, \lambda) - < R(\psi, J, \lambda) >\} + 0(\varepsilon^2) \ . \tag{39}$$

4. The Adiabatic Invariant

From (36) it follows that

$$|\bar{J}(t) - \bar{J}(0)| \le C_1 \varepsilon^{n+1} |t| \tag{40}$$

as long as the trajectory remains in the domain $\bar{D} \times I$ of definition (where $\bar{D}$ is the image of D by the transformation $(q,p) \to (\psi, J) \to (\bar{\psi}, \bar{J})$. It can be reduced to $0 \le \bar{\psi} \le 2\pi$ and $\bar{J}_{\min} \le \bar{J} \le \bar{J}_{\max}$). Of course, the constant C_1 in (40) depends upon the order n and may get very large for large n.

The trajectory can leave this domain only if $\lambda = \varepsilon t$ leaves the interval I of definition (we assume that this never happens) or if $\bar{J}(t)$ leaves the interval $[\bar{J}_{\min}, \bar{J}_{\max}]$. But as we can monitor $\bar{J}(t)$ by (40) itself, the usual bootstrap argument insures that this does not happen as long as

$$\bar{J}_{\min} + C_0 \varepsilon \le \bar{J}(0) \le \bar{J}_{\max} - C_0 \varepsilon \ ; \ \ -\varepsilon^{-n} \le t \le \varepsilon^{-n} \ . \tag{41}$$

These estimates lead to the definition of the Adiabatic Invariant. We shall say that $\bar{J}(\psi, J, \lambda)$ is an adiabatic invariant of order n if

$$|\bar{J}(t) - \bar{J}(0)| \le C_1 \varepsilon^{n+1} |t| \quad \text{for } |t| \le \varepsilon^{-n} \ . \tag{42}$$

This definition implies several different estimations (which can and have been used as different definitions of the adiabatic invariant) according to what we want to stress. If we want to stress the accuracy to which the invariant is preserved, we have

$$|\bar{J}(t) - \bar{J}(0)| \leq C_1 \varepsilon^n \quad \text{for } |t| \leq \varepsilon^{-1} , \tag{43}$$

(this is the most often quoted definition). If we want to stress the length of time for which it is preserved, we have

$$|\bar{J}(t) - \bar{J}(0)| \leq C_1 \varepsilon \quad \text{for } |t| \leq \varepsilon^{-n} , \tag{44}$$

(this may be the most useful one when the Hamiltonian is periodic in λ – see later on (48)). A compromise would be

$$|\bar{J}(t) - \bar{J}(0)| \leq C_1 \varepsilon^{n/2} \quad \text{for } |t| \leq \varepsilon^{-n/2} . \tag{45}$$

This is why we think that (42) which embodies them all should be prefered.

Of special interest, is the estimate (44). Maybe we could reserve the name "Crude Adiabatic Invariant of order n " to a quantity which verifies this estimate. Indeed, the non-averaged momentum J, which has a nice geometrical meaning, does verify this estimate because, by (37), we have

$$\begin{aligned} &|J(t) - J(0)| \leq |J(t) - \bar{J}(t)| + |\bar{J}(t) - \bar{J}(0)| + |\bar{J}(0) - J(0)| \\ &|J(t) - J(0)| \leq C_2 \varepsilon \ \text{ for } |t| \leq \varepsilon^{-n} \end{aligned} \tag{46}$$

Also, Arnol'd (1964) has shown that with the further assumptions

$$\frac{\partial^2 K}{\partial J^2} \neq 0 \quad \text{and} \quad H(x, \lambda) = H(x, \lambda + 2\pi) , \tag{47}$$

this "crude" adiabatic is also "perpetual", i.e.

$$|J(t) - J(0)| \leq C_2 \varepsilon \quad \text{for all } t . \tag{48}$$

We have stressed this point because the much quoted work of Lenard (1959) introduces some confusion about this. Lenard makes the ad-hoc assumption that the function $\lambda(t)$ is constant for $t < 0$ and $t > T$ and varies "slowly" in between for $0 \leq t \leq T$ where T is assumed to be of the order of ε^{-1}.

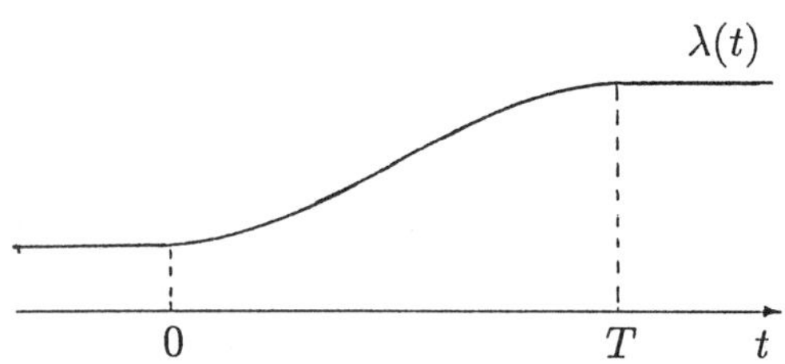

Fig. 1. Variation of λ for the ad-hoc assumption of Lenard.

From this assumption, it follows that $J = \bar{J}$ outside the interval $0 < t < T$ and indeed Lenard is able to show that

$$|J(T) - J(0)| \leq C\varepsilon^n \quad \text{for } T = \varepsilon^{-1} , \tag{49}$$

which looks deceptively akin to (43), although we have just shown that J is only a "crude" adiabatic invariant (see (46)).

The explanation of the paradox is shown in Fig. 2 where it can be seen that the estimate (49) is not valid inside the interval $0 \leq t \leq T$.

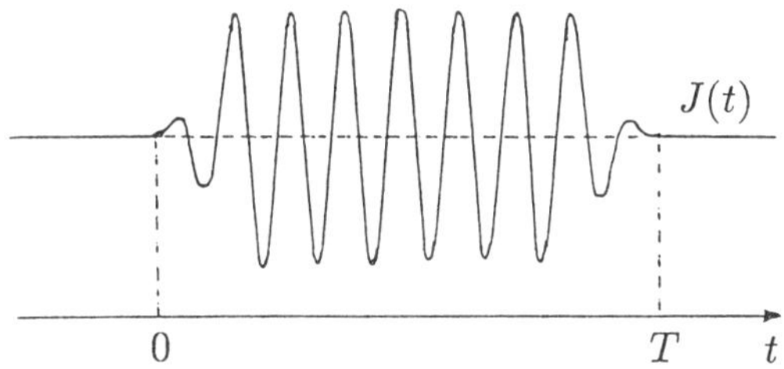

Fig. 2. Resulting variation of J shown in solid line. The variation of $\bar{J}$ is shown as a dashed line.

We should mention here more recent results in the line of Lenard's work. With various hypotheses on the behaviour of $H(x, \lambda)$ when λ goes to infinity, Littlewood (1963), Wasow (1973, 1976), Stengle (1977), Meyer (1976, 1980) evaluate the difference $|J(\infty) - J(-\infty)|$. They find for linear systems (see Wasow, 1973)

$$|J(\infty) - J(-\infty)| = C \exp\left\{-\frac{d}{\varepsilon}\right\} \tag{50}$$

where C and d are analytical functions of ε (not vanishing for $\varepsilon = 0$). A similar result is obtained by Meyer (1976) for some nonlinear systems (see also Chirikov, 1979).

We shall not describe more fully these results as we are not going to use them. Although they are important because they deal with an infinite period of time and because they give precise (and not only asymptotic) estimates of the invariance, they suffer from the same weakness than Lenard's result. They do not tell us what happens between the two limits.

5. Explicit Approach to Action-Angle Variables

The classical definition of action-angle variables recalled in Sect. 2 is not very useful practically if the quadrature (6) cannot be performed by means of elementary functions. Furthermore this mixed variable generating function leads to the same practical problem that Poincaré-von Zeipel methods generate in perturbation theory : functions defined implicitly are not easy to use, estimate, differentiate, $\cdots$

We shall present in this section a new way of defining the same action-angle variables which lends itself better to numerical computation. The same kind of approach has been used successfully in different contexts (Jefferys, 1968; Deprit and Henrard, 1969; Henrard and Lemaître, 1986b).

Let us write $x = (q, p)$ the two-dimensional phase space vector. The Hamiltonian function (1) is thus

$$H(x, \lambda) = h \ . \tag{51}$$

Let us assume that we can compute, if need be numerically, the solution of the differential equations associated with (51). Let us note it:

$$x = X'(t, x_0, \lambda) \tag{52}$$

where the initial condition x_0 lies on a curve (see (9))

$$F(x_0, \lambda) = 0 \ . \tag{53}$$

The solution (52) is periodic of period $T(h, \lambda)$.

Let us now define the action-angle variables as

$$\psi = \frac{2\pi}{T} t \ , \tag{54}$$

$$J = \frac{1}{4\pi} \int_0^T \left(\frac{\partial X'}{\partial t} \right)^\tau SX' \, dt \ , \tag{55}$$

where the superscript τ stands for the transpose of a matrix and where S is the principal symplectic matrix as in (25). The momentum J can also be written

$$J = \frac{1}{4\pi} \oint p \, dq - q \, dp \tag{56}$$

which shows better that it represents the area (divided by 2π) enclosed by the periodic orbit and that it is the same quantity as defined in (11). The geometrical expression (56) is often more useful for theoretical considerations but (55) is the easiest way to actually evaluate the momentum J. This is what we want to stress by taking (55) rather than (56) as a definition.

From the relations $F(x_0, \lambda) = 0$ and $J = J(x_0, \lambda)$ we can extract, numerically if need be,

$$x_0 = X_0(J, \lambda) \ . \tag{57}$$

This can be performed numerically by a fast converging Newton-Raphson procedure as we can evaluate $\partial X_0 / \partial J$ (see equation (61)).

The general solution (52) can now be rewritten as

$$x = X(\psi, J, \lambda) \tag{58}$$

which can be thought of as defining a transformation from the phase space (ψ, J) to the phase space (x).

This transformation is canonical. Indeed ψ and J are the same as the quantities defined in Sect. 2. We can also check directly that

$$(q;p) = \frac{\partial q}{\partial \psi}\frac{\partial p}{\partial J} - \frac{\partial q}{\partial J}\frac{\partial p}{\partial \psi} = 1 \tag{59}$$

which can be verified by differentiating the identity

$$H(q(\psi, J, \lambda), p(\psi, J, \lambda), \lambda) = h(J, \lambda) \tag{60}$$

and using the fact that $\partial J/\partial h = T/2\pi$ which can be derived from (56).

Let us note that there is no need to evaluate directly (58). It can always be evaluated indirectly through (52) and (57) which lend themselves to easy numerical procedures.

But together with the transformation (58) we need to evaluate its partial derivatives.

The derivatives with respect to ψ are easy. They correspond to derivatives with respect to time of (52) with a scale factor $2\pi/T$.

The derivatives with respect to J are implemented by means of the variational equations (see Appendix 1) which can be numerically integrated together with (52) if we can evaluate their initial conditions, the derivatives of (57). These are the solutions of the linear system

$$H_x(x_0) \cdot \frac{\partial X_0}{\partial J} = \frac{2\pi}{T}, \quad F_x(x_0) \cdot \frac{\partial X_0}{\partial J} = 0\,. \tag{61}$$

The first equation comes from the fact that $H(x_0, \lambda) = K(J, \lambda)$ and that $\partial K/\partial J = \dot{\psi} = 2\pi/T$. The second equation comes from $F(x_0, \lambda) = 0$.

The linear system (61) is singular only if the curves $H(x, \lambda) = h$ and $F(x, \lambda) = 0$ are tangent at $x = x_0$. This is not the case as we have assumed that the curve of initial conditions $F(x_0, \lambda) = 0$ is transverse to the trajectories $H(x, \lambda) = h$.

The partial derivatives with respect to λ can also be implemented by means of the variational equation (see Appendix 1). For this we do also need the derivatives of (57) with respect to λ. But before computing them we shall evaluate the remainder function $R(\psi, J, \lambda)$ of the transformation (58) with respect to λ.

A basic property of the remainder function (Wintner, 1941) is that it is the solution of the partial differential equations

$$\frac{\partial R}{\partial \psi} = \left(\frac{\partial X}{\partial \psi}\right)^{T} S \frac{\partial X}{\partial \lambda}\,; \quad \frac{\partial R}{\partial J} = \left(\frac{\partial X}{\partial J}\right)^{T} S \frac{\partial X}{\partial \lambda}\,. \tag{62}$$

The Frobenius conditions of integrability of (62) are verified whenever the transformation $X(\psi, J, \lambda)$ is canonical. The solution of (62) can be written

$$R(\psi, J, \lambda) = \int_0^{\psi} \left(\frac{\partial X}{\partial \psi}\right)^{T} S \left(\frac{\partial X}{\partial \lambda}\right) d\psi + \int_{J^\star(\lambda)}^{J} \left(\frac{\partial X_0}{\partial J}\right)^{T} S \left(\frac{\partial X_0}{\partial \lambda}\right) dJ \tag{63}$$

where $J^\star(\lambda)$ is an arbitrary value of J. Note that if the curve of initial conditions $F(x_0, \lambda) = 0$ does not depend explicitly upon λ, the vectors $\partial X_0/J\partial J$ and $\partial X_0/\partial \lambda$ are colinear and the second integral vanishes.

We know from Sect. 2 that the remainder function $R(\psi, J, \lambda)$ is periodic with respect to ψ. This can be checked directly here by differentiating (55) with respect to λ. We find after an integration by parts

$$\begin{aligned} 0 &= \frac{1}{2\pi}\int_0^{2\pi}\left(\frac{\partial X}{\partial \psi}\right)^{\tau} S\frac{\partial X}{\partial \lambda}d\psi + \frac{1}{4\pi}\left[\left(\frac{\partial X}{\partial \lambda}\right)^{\tau} SX\right]_0^{2\pi} \\ &= \frac{1}{2\pi}\int_0^{2\pi}\left(\frac{\partial X}{\partial \psi}\right)^{\tau} S\frac{\partial X}{\partial \lambda}d\psi = R(2\pi, J, \lambda) - R(0, J, \lambda) \,. \end{aligned} \tag{64}$$

Using the identity (A.6) in Appendix 1, equation (63) becomes

$$R(\psi, J, \lambda) = \left\{H_x(x_0)\cdot\frac{\partial X_0}{\partial \lambda}\right\}\frac{T\psi}{2\pi} + \int_0^{\frac{T\psi}{2\pi}}\left(\frac{\partial X'}{\partial t}\right)^{\tau} SV(t)\,dt + R(0, J, \lambda) \,. \tag{65}$$

From this and from the periodicity of $R(\psi, J, \lambda)$ we conclude that $(\partial X_0/\partial\lambda)$ is the solution of the linear system:

$$\begin{aligned} H_x(x_0)\frac{\partial X_0}{\partial \lambda} &= -\frac{1}{T}\int_0^{T}\left(\frac{\partial X'}{\partial t}\right)^{\tau} SV(t)\,dt \,, \\ F_x(x_0)\frac{\partial X_0}{\partial \lambda} &= -\frac{\partial F}{\partial \lambda}(x_0) \,. \end{aligned} \tag{66}$$

Higher order partial derivatives of the transformation $X(\psi, J, \lambda)$ to action-angle variables, can also be evaluated by means of higher order variational equations. The initial conditions for such higher order variational equations can be evaluated by differentiating (61) and (66). Partial derivatives of $T(x_0, \lambda)$ which appear in the right-hand members of (61) are themselves evaluated from the derivatives of the periodicity condition

$$X'(T, x_0, \lambda) = x_0 \,. \tag{67}$$

6. Extension of Perturbation Theory to the Case of Unbounded Period

The perturbation theory we have briefly described in Sect. 3 is valid only in a domain $D \times I$ where the period of $T(h, \lambda)$ of the "frozen" (λ = constant) dynamical system is uniformly bounded. The domain D must then remain at a finite distance from a critical curve on which T goes to infinity.

As we shall see later (Part 2), the classical adiabatic invariant is no longer conserved if we approach close enough to the critical curve. But how close is close enough?

We shall show in this section that we can approach the critical curve to a distance η vanishing with ε (but such that ε/η goes to zero with ε) and still be able to define an adiabatic invariant to all orders. These estimates will enable us to make the link between the classical adiabatic invariant at a finite distance from the critical curve (see Sect. 4) and the analysis of Part 2 which is valid in a neighborhood of the critical curve, vanishingly small with ε.

As we shall see in more details in Part 2, the transformation to action-angle variables becomes singular on the critical curve. Hence the problem in approaching the critical curve is not only that the period becomes unbounded but also that the radius of convergence of the analytical functions goes to zero (the radius of convergence has to be taken into account because the coefficient C_1 in (36) is a function of the high order derivatives of the original Hamiltonian). We shall have to take care of the two difficulties.

Let us go back to the Hamiltonian (32) and assume that in the domain D of definition, the radius ϱ of convergence of $K(J,\lambda)$ and $R(\psi, J, \lambda)$ as an analytical function of J is bounded by η_1 (we shall estimate it later) and that the period (of the frozen system) is bounded by $2\pi/\eta_2$:

$$\varrho \geq \eta_1, \quad \frac{\partial K}{\partial J} \geq \eta_2 \,. \tag{68}$$

Let us introduce the changes of scale of momenta and time:

$$J = \eta_1 \tilde{J}, \quad \Lambda = \eta_1 \tilde{\Lambda}, \quad \tau = \eta_2 t \,. \tag{69}$$

The new Hamiltonian, describing the dynamical system in the new time variable τ is then

$$\tilde{H}(\psi, \tilde{J}, \lambda, \tilde{\Lambda}) = \tilde{K}(\tilde{J}, \lambda) + \frac{\varepsilon}{\eta_1 \eta_2} \{\eta_1 \tilde{\Lambda} + \tilde{R}(\psi, \tilde{J}, \lambda)\} \tag{70}$$

with

$$\tilde{K}(\tilde{J}, \lambda) = \frac{1}{\eta_1 \eta_2} K(\eta_1 \tilde{J}, \lambda) \,, \quad \tilde{R}(\psi, \tilde{J}, \lambda) = R(\psi, \eta_1 \tilde{J}, \lambda) \,. \tag{71}$$

The radius of convergence of $\tilde{K}$ and $\tilde{R}$ is now not smaller than one and the "period" in the new time is not larger than 2π.

The perturbation method used in Sect. 3 can be reproduced without modification except that the "small parameter" is no longer ε but $\varepsilon/\eta_1\eta_2$.

The first order approximation (39) remains valid except that the second order neglected terms are now of the order $\varepsilon^2/\eta_1\eta_2^2$:

$$\bar{J} = J + \varepsilon \left(\frac{\partial K}{\partial J}\right)^{-1} \{R(\psi, J, \lambda) - < R(\psi, J, \lambda) >\} + 0\left(\frac{\varepsilon^2}{\eta_1 \eta_2^2}\right) \,. \tag{72}$$

There remains to estimate η_1 and η_2. We will show in Part 2 (see (II.20)) than when h approaches the value corresponding to the critical curve (which we shall assume to be zero) the radius of convergence (η_1) and the period ($2\pi/\eta_2$) can be estimated by

$$\eta_1 \geq C_1 h \log |h^{-1}|, \quad \eta_2 \geq C_2 / \log |h^{-1}| \,. \tag{73}$$

Hence an adiabatic invariant to all orders can be defined as long as $\varepsilon \log |h^{-1}|/h$ remains small. We shall show later (see Sect. II.7) that $h \geq \varepsilon^{2/3}$ is in some sense an optimal choice. In that case, the neglected terms in (72) are of the order of $\varepsilon^{4/3} \log \varepsilon^{-1}$.

Part II Transition Through a Critical Curce

1. Introduction

By the virtue of its own success, the classical adiabatic invariant theory was led to a trap. Indeed classical adiabatic invariant theory is able to describe slow but finite deformation of the guiding trajectory (trajectory of the frozen system to which the real trajectory stay close). But simple dynamical systems such as the pendulum possess critical curves formed by homoclinic (or doubly asymptotic) orbits and their limit point. Chances are then that during its deformation the guiding trajectory will bump on a critical curve.

As we have seen in Sect. I.6, we cannot define an n-th order adiabatic invariant when we come close to this critical curve. What happens then?

It was soon realized (Best, 1968) that, except for the obvious discontinuity forced by the definition of the adiabatic invariant, this invariant does not change "much" for "most" of the trajectories crossing the critical curve. This "not much" was apparently estimated as $\varepsilon(\log\varepsilon^{-1})^2$ for the pendulum in an unpublished work of Gardner quoted by Aamodt (1971–1972). A more precise but incorrect estimate (see Cary et al., 1986) had already be given by Chirikov (1959).

As far as we know, Timofeev (1978) was the first to give a precise (and correct) estimate of the change of the adiabatic invariant in the particular case of a pendulum, the restoring torque of which varies linearly with time. Such a result could be gathered also from the estimates of Yoder (1973, 1979) but Yoder was interested in capture probability and not so much in change in the adiabatic invariant.

More recently a very throughout analysis of "separatix crossing" led Cary et al. (1986) (see also Escande, 1985) and independently Neishtadt (1986) to very general estimates of the change in the invariant and of its distribution with respect to the initial phase. The basic ideas for such an analysis can also be found in Hannay (1986).

Indeed estimates of the change in the invariant are not only useful in order to follow precisely the guiding trajectory but mostly because of the fact that it can produce chaotic motion (Menyuk, 1985). Changes in the invariant are very sensitive to the initial phase and so is the "final" (after transition) phase. If the system is forced to go periodically through a transition this is bound to produce the very unstable and unpredictable kind of motion known as "chaotic motion". From the distribution of the changes, estimates of the "diffusion time" or the Lyapunov characteristic number of the motion could be derived.

Celestial Mechanicians were not so worried about changes in the invariant or chaotic motion (although they are now – e.g. Wisdom, 1983, 1985, and Henrard, Lemaître, 1986b) but rather about probability of capture.

Indeed in most instances and specifically in the case of the pendulum, when the guiding trajectory comes close to the critical curve (let us say coming from positive rotation), it can end up in two possible states, either libration or negative circulation. If the pendulum is a model of a resonance, this means a capture (or a non-capture) into resonance and Celestial Mechanics has many of these resonances (either Orbit-Orbit or Spin-Orbit) to explain.

As a matter of fact this problem of "Capture into resonance" was investigated even before its connection with the adiabatic invariant was perceived (Goldreich, 1965, 1966). Formulae based upon a pendulum model were proposed for the probability of capture in the Spin-Orbit case (Goldreich and Peale, 1966; Counselman and Shapiro, 1970, and Burns, 1979) and the Orbit-Orbit case (Yoder, 1973, 1979). Yoder and independently Neishtadt (1975) were apparently the first to make the connection between this problem and the adiabatic invariant theory. Henrard (1982a) proposed a formula to compute the probability of capture for general Hamiltonian systems (with one degree of freedom and slowly varying).

This formula (see (72)) is simple and almost intuitive. It is interesting to notice that it was stated without proof for the anharmonic oscillator by Dobrott and Green (1971) under the name of "Kruskal Theorem". A similar formula applies for a class of dissipative systems and is stated by Arnol'd (1964).

We shall address the two problems, change in the invariant and probability of capture, simultaneously. The analysis will be similar to the one developed in Henrard (1982) and Cary et al. (1986) but several estimates will be improved.

In Sect. 2 we shall define the type of dynamical systems we are considering, state the basic assumptions and introduce some notations.

Section 3 is the key for the improved estimates. We shall describe an exact solution of the autonomous (λ constant) problem in a finite disk around the unstable equilibrium where in previous work an approximate solution in a vanishingly small (with ε) neighborhood was used.

Section 4 deals with a complete neighborhood of the critical curve. Based upon the results of Sect. 3, we shall be able to describe exactly the singularity of the action-angle variables for a general Hamiltonian system (see (18) and (25)). We shall also develop approximate formulae for the action-variables (see (20), (27)) and for the adiabatic invariants (see (34), (35)). These approximate formulae are expressed in terms of quantities ($\omega, J_i^\star, h_i^\star, g_i$) which can readily be computed (if need be numerically) for general one-degreee of freedom systems.

Section 5 is devoted to an estimation of two quantities which will be instrumental in the next sections: the change of energy and the time of flight for a trajectory of the non-autonomous system ($\lambda = \varepsilon t$).

When the trajectory comes close to the unstable equilibrium, this estimation becomes quite delicate as the time spent there can be very long. In Henrard (1982), we were content with an approach of the order of $h \sim \varepsilon^p$ (with $p > 1$) which does not make too much difficulties (the time involved is only of the order of $\log \varepsilon^{-1}$). Most other authors claim that their estimates are valid up to $h \sim \exp\{-1/\varepsilon\}$ (involving a time of the order of $1/\varepsilon$ and thus a change in λ of the order of unity). Most of these claims are not supported.

A very careful analysis was made by Cary et al. (1986) and supports this claim but only in part. They arrive at the conclusion that valid estimates can be obtained as long as h remains much larger than $\exp\{-1/\varepsilon\}$. Our analysis follows closely theirs. Our estimates of the error terms are somewhat sharper mainly due to the fact that we use a better description of the autonomous system in the vicinity of the unstable equilibrium (see Sect. 3). Our conclusion is that an optimum choice (in a sense that will be made more precise in Appendix 6) is to take $h \geq \exp\{-1/\varepsilon^{1/3}\}$ still retaining the exponential character but allowing a change in λ of the order of $\varepsilon^{2/3}$ only. The corresponding estimate in Cary et al. (1986) would be $h \geq \exp\{-1/\varepsilon^{4/5}\}$.

Formulae (58) and their approximate counterparts (59) may be regarded as the main results of this second part. They encapsulate our knowledge of the behaviour of the dynamical system under the form of a Poincaré's mapping. Sects. 6, 7 and 8 are then an exploitation of the approximate formulae (59) but the more accurate formulae (58) may hopefully lead in the future to a refined description of the separatrix crossing.

In Sect. 6, we compute the probability of transition from one of the domains of phase space defined by the separatrix to one of the other two domains and in Sects. 7 and 8, we evaluate the change in the adiabatic invariant due to this transition.

The adiabatic invariant change depends very critically upon the value of h at one of the closest approaches to the unstable equilibrium. As we show in Sect. 6 that this value can be considered as a random variable (with uniform distribution on a small interval of order ε), we can compute in Sect. 7 the root mean square of the distribution of the adiabatic invariant.

This is the instrumental quantity in the description of the diffusion process resulting in chaotic motion.

2. Neighborhood of an Homoclinic Orbit

We assume that the one-degree of freedom dynamical system described by the Hamiltonian function (I.1)

$$H(x, \lambda) = h \tag{1}$$

(we use the notation of Sect. I.5) possesses in its domain D of definition one and only one non-degenerate unstable equilibrium $x^\star(\lambda)$, limit point of two homoclinic trajectories $\Gamma_1(\lambda)$ and $\Gamma_2(\lambda)$.

The global topology of the two homoclinic curves $\Gamma_1(\lambda)$ and $\Gamma_2(\lambda)$ may be of various types as shown in Fig. 3.

All these dynamical systems are equivalent on an open neighborhood of the critical curve and we shall use the bow-tie model which is easier to draw to illustrate our analysis.

Notice that the angle-action variables introduced in Sect. II.2 cannot be defined on the full domain D as it contains a critical curve on which they are singular.

But we can define three subdomains on which they are well-defined. The domain D_1 (resp. D_2) is the open set of D touching Γ_1 (resp. Γ_2) and D_3 is the open set of D touching both curves (see Fig. 4). Most of our analysis in Sects. II.3 and II.4 will be devoted to the estimation of limits when the periodic trajectories defined in one of these domains approach its boundary Γ_1, Γ_2 or $\Gamma_3 = \Gamma_1 \cup \Gamma_2$.

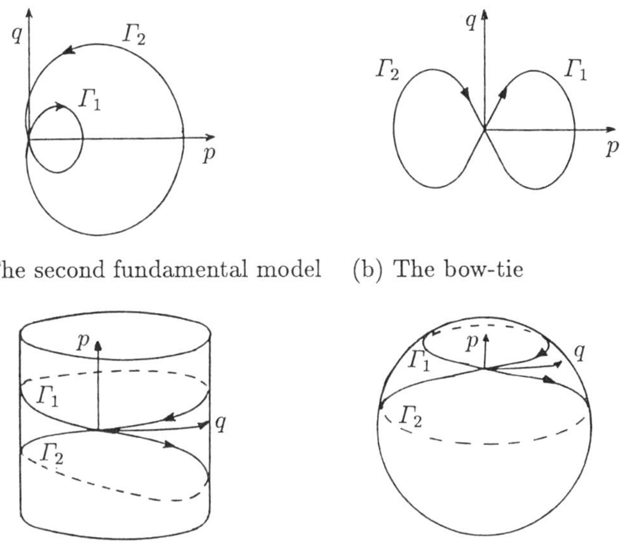

(a) The second fundamental model (b) The bow-tie

(c) The pendulum on the cylinder (d) The "Colombo's top" on the sphere

Fig. 3. Different types of global topology of the Homoclinic orbits.

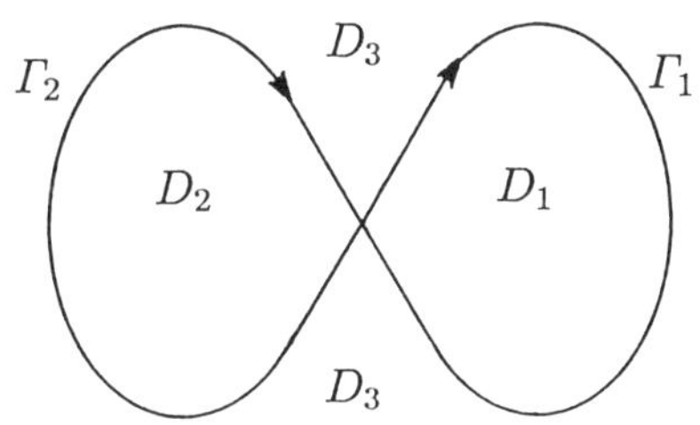

Fig. 4. The three subdomains defined by the homoclinic orbits Γ_1 and Γ_2.

In order to simplify the subsequent analysis, and without loss of generality, we shall make three assumptions.

First, and this is trivial, we shall assume that $h = 0$ corresponds to the critical curve formed by $x^\star(\lambda)$, $\Gamma_1(\lambda)$ and $\Gamma_2(\lambda)$. This can always be achieved by subtracting $H(x^\star(\lambda), \lambda)$ from the Hamiltonian. In the same spirit, we shall assume that the value of h is positive in the domain D_3 and negative in D_1 and D_2. This can always be achieved by changing, if need be, the sign of one of the canonical coordinates q or p in order to change the sign of the Hamiltonian function.

Let us remark that this assumption imposes the direction of the time arrow on the orbits (Figure 3 has been drawn accordingly) and subsequently the sign of the action-variables J_i $(1 \leq i \leq 3)$ in each domain D_i. The sign will be positive if the orbits are travelled clockwise and negative if they are travelled counter-clockwise.

Secondly, we shall assume that $x^\star(\lambda) \equiv 0$. At first it seems that it can be achieved by a translation $x = x^\star(\lambda) + x'$ but then when we shall consider $\lambda = \varepsilon t$, the translation will be time dependent and the corresponding Hamiltonian function will no longer be $H(x(\varepsilon t) + x', \varepsilon t)$. We show in the Appendix 2 that it can nevertheless be achieved at the price of a slight modification of the dynamical system. This modification does not affect the generality of our analysis or the various estimates we are going to make.

Finally, we shall assume that the time scale $\omega(\lambda)$ defined in (5) is actually independent of λ and equal to its value $\omega(\tau)$ for a fixed τ (the pseudo-crossing time) to be defined later. We show in Appendix 2 how this can be achieved by a change in the parametrization of the independent variable. This change does not affect the final results.

As mentioned earlier, an important part of our analysis will consist in defining and estimating quantities (actually functions of the parameter λ) which describe the dynamical system in a small neighborhood of the homoclinic orbit. At the lowest order, these quantities are the critical values of the action-variables (defined below), the time scale at the unstable equilibrium $\omega(\lambda)$ (defined in (5)), the "steepness" parameters $h_i^\star$ (defined in (21)) and the "out-of-symmetry" parameters g_i (defined in (31)) introduced by Cary et al. (1986).

Let us start by defining the critical values of the action-variables $J_i^\star(\lambda)$ $(1 \leq i \leq 3)$ as the limits of the action-variable in the domain D_i when the periodic curves tend towards the homoclinic curves. They are

$$\begin{aligned} J_i^\star(\lambda) &= \frac{1}{4\pi} \int_{-\infty}^{+\infty} \left(\frac{\partial X_i^\star}{\partial t}\right)^T S\, X_i^\star \, dt\,, \quad 1 \leq i \leq 2\,, \\ J_3^\star(\lambda) &= J_1^\star(\lambda) + J_2^\star(\lambda)\,, \end{aligned} \tag{2}$$

where $X_i^\star(t)$ is the homoclinic orbit Γ_i. The integrals are of course finite as they are the area (divided by 2π) of the domain D_i $(1 \leq i \leq 2)$:

$$J_i^\star(\lambda) = \frac{1}{4\pi} \oint_{\Gamma_i} p\, dq - q\, dp\,, \quad 1 \leq i \leq 2\,. \tag{3}$$

Formulae (2) and (3) are equivalent for the same reasons that (I.55) and (I.56) are equivalent. Formula (3) stresses the geometrical character of $J_i^\star$, while (2) is more useful in the computations.

3. The Autonomous Problem Close to the Equilibrium

It is well-known that, in the vicinity of an equilibrium, one can "normalize" an Hamiltonian system (Birkhoff, 1927). This normalization is in general only asymptotic but the formal power series can be shown in some cases to be convergent (e.g. Siegel and Moser, 1971) and thus the normalization to be analytical. This is the case for a one-degree of freedom system such as the one we are analysing.

Curiously, we did not find a formal proof of this when the equilibrium is unstable. We are thus forced to refer to a much more powerful result of Moser (1958) concerning two-degrees of freedom systems. Indeed we can always "supplement" our one-degree of freedom system with another degree of freedom (let us say an harmonic oscillator) completely independent from it. It is easy to show that the normalization can be chosen to keep the two degrees of freedom independent. We can then drop after normalization the artificial second degree of freedom and end up with the analytical normalization we are seeking.

Actually, Moser (1958) does show that there exists a group of canonical normalization and a group of analytical normalization but does not show that their intersection is not empty. Such a result could be gathered from the work of Rüssmann (1964) but again in an indirect way. Some readers may be interested in a more direct proof. This if why we give, in Appendix 8, a complement to Moser's paper to show that indeed one can define a normalization which is both canonical and analytic.

From the previous discussion, we know that there exists a disk of radius δ around the unstable equilibrium $x = 0$ of the system (1) in which is defined an analytical canonical transformation from the phase space $x = (q, p)$ to the phase space $z = (z_1, z_2)$:

$$x = X_N(z, \lambda) \tag{4}$$

which transforms the Hamiltonian function $H(x, \lambda)$ into the normalized Hamiltonian

$$h = H_N(Z, \lambda) = \omega(\lambda)Z + 0(Z^2) \tag{5}$$

where Z is the product of the two coordinates

$$Z = z_1 z_2 \; . \tag{6}$$

The function $\omega(\lambda)$ is one of the eigen-values (the other one is $-\omega(\lambda)$) of the matrix $SH_{xx}(0, \lambda)$. It is bounded away from zero as we have assumed that the equilibrium $x = 0$ is non-degenerate and it can always be chosen as positive, if need be by exchanging z_1 and z_2. Furthermore, as we have mentioned in the previous section, it can be made independent of λ by a change of the time variable. We assume, from Sect. II.5 on, that it is indeed independent of λ and equal to the value $\omega(\tau)$ reached at some particular value τ of λ to be defined later.

We shall also consider the inverse of the function (5):

$$Z = Z(h, \lambda) = \frac{h}{\omega} + 0(h^2) \; . \tag{7}$$

In the plane (z_1, z_2) the trajectories are given by the branches of the hyperbola $z_1 \cdot z_2 = Z(h, \lambda)$ as shown in Fig. 5. In the domain D_3 the two branches belong to the same trajectory while in the domain D_1 and D_2 the two branches belong to different orbits (one in D_1 and the other one in D_2).

In each of the three open domains D_i, we can define angle-action variables (ψ_i, J_i). For this we have to choose a curve of "initial conditions" (see (I.9)). We shall choose $z_1 = z_2 = \sqrt{Z}$ for D_3 and $z_1 = -z_2 = \pm\sqrt{|Z|}$ for D_1 and D_2.

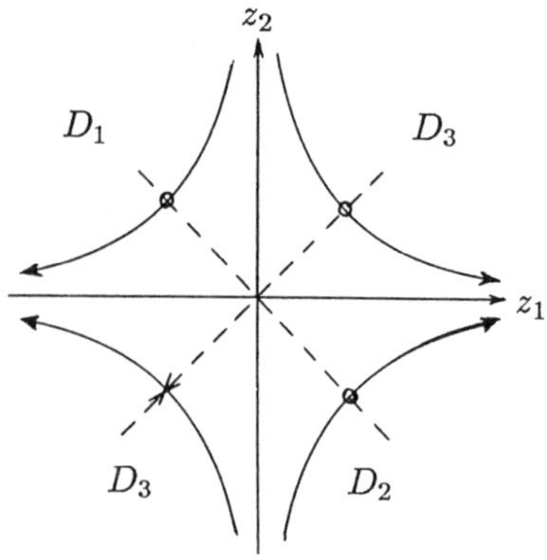

Fig. 5. Trajectories in normalized coordinates (z_1, z_2) showing the apices ($\circ$) and the anti-apex ($\times$).

This initial point along a trajectory will be called its "apex". The return point $z_1 = z_2 = -\sqrt{Z}$ along a trajectory in D_3 will be called the "anti-apex". Apex and anti-apex are called "vertex", by Cary et al. (1986).

The normalizing transformation (4) is not uniquely defined although the normalized Hamiltonian (5) is uniquely defined. This makes the definition of the apices coordinate-dependent. We shall come back on this later.

The transformation from the normalizing coordinates (z_1, z_2) to the angle-action variables (ψ_i, J_i) of each domain D_i are easy to define according to Sect. I.2. The generating functions $S'(z_1, J_i, \lambda)$ is (see (I.6)):

$$S_i'(z_1, J_i, \lambda) = \int_{\pm\sqrt{|Z_i|}}^{z_1} \frac{Z_i}{z_1} dz_1 + \frac{1}{2} Z_i = \pm \frac{1}{2} Z_i \log \frac{z_1^2 e}{|Z_i|} \tag{8}$$

where Z_i is a yet unknown function of J_i and λ. Its definition depends upon the global properties of the trajectory and it cannot be determined by the analysis of this section which is purely local being confined to the disk of radius δ around the origin. We shall determine this function or rather its inverse $J_i = \mathcal{J}_i(Z, \lambda)$ in the next section.

Note that a "$\pm$" sign is inserted in (8) and in what follows, to indicate that the sign of the function should be ajusted (in an obvious way) in accordance with the quadrangle of the plane (z_1, z_2) to which the domain D_i belongs.

The angular variables ψ_i are given by

$$\psi_i = \frac{\partial S_i'}{\partial J_i} = \pm \frac{1}{2} \frac{\partial Z_i}{\partial J_i} \log \frac{z_1^2}{|Z_i|} \tag{9}$$

and the transformation from angle-action variables to normalized variables is:

$$z_1 = \pm\sqrt{|Z_i|} \exp\left\{ \left(\frac{\partial Z_i}{\partial J_i} \right)^{-1} \psi_i \right\} \quad , \quad z_2 = \pm\sqrt{|Z_i|} \exp\left\{ - \left(\frac{\partial Z_i}{\partial J_i} \right)^{-1} \psi_i \right\} . \tag{10}$$

When we consider that $\lambda = \varepsilon t$, the normalizing transformation (4) and the transformation to action-angle variables (10) are time-dependent. The remainder function to be added to H_N in order to produce the "new Hamiltonian" of the dynamical system can be evaluated as we did in Sect. I.5 and is the sum of the remainder function of the normalizing transformation $R_N(z, \lambda)$ and the remainder function of the transformation to action-angle variables, $R_i'(\psi_i, J_i, \lambda)$. We have from (I.63):

$$R_N(z,\lambda) = \int_0^{z_1} \left(\frac{\partial X_N}{\partial z_1}\right)^T S \left(\frac{\partial X_N}{\partial \lambda}\right) dz_1 + \int_0^{z_2} \left\{\left(\frac{\partial X_N}{\partial z_2}\right)^T S \left(\frac{\partial X_N}{\partial \lambda}\right)\right\}_{z_1=0} dz_2 . \tag{11}$$

The equilibrium $x = 0$ being sent on $z = 0$, the normalizing transformation has no independent term and the remainder function R_N has no linear term

$$R_N(z, \lambda) = 0(||z||^2) . \tag{12}$$

On the other hand, we have from (I.13):

$$R_i'(\psi_i, J_i, \lambda) = \frac{\partial S'}{\partial \lambda} = \pm\frac{1}{2}\left(\frac{\partial Z_i}{\partial \lambda}\right) \log \frac{z_1^2}{|Z_i|} = \left(\frac{\partial Z_i}{\partial \lambda}\right)\left(\frac{\partial Z_i}{\partial J_i}\right)^{-1} \psi_i . \tag{13}$$

Summing the two contributions (11) and (13) and defining the function $J_i = \mathcal{J}_i(Z, \lambda)$ as the inverse of the yet unknown function mentioned earlier, we have

$$R_i(\psi_i, J_i, \lambda) = -\left(\frac{\partial \mathcal{J}_i}{\partial \lambda}\right) \psi_i + R_N(z, \lambda) . \tag{14}$$

As mentioned earlier, the normalizing transformation is not uniquely defined, although the product $Z = z_1 z_2$ is uniquely defined. This makes the definition of the apices coordinate-dependent and, therefrom, the angular variables ψ_i and the remainder functions R_i coordinate-dependent. But the dependence is rather tame. If we note $\tilde{\psi}_i$ and $\tilde{R}_i$ the angular variables and the remainder functions corresponding to a different choice of normalizing transformation, we find, with the help of (I.10) for the angular variables and of (I.17) for the remainder functions, that

$$\tilde{\psi}_i = \psi_i + \left(\frac{\partial K}{\partial J_i}\right) \Psi(Z, \lambda) , \quad \tilde{R}_i(\psi_i) = R_i(\psi_i) - \frac{\partial \mathcal{J}_i}{\partial \lambda}\frac{\partial K}{\partial J_i} \Psi(Z, \lambda) \tag{15}$$

with

$$K(J_i, \lambda) = H_N(Z_i(J_i, \lambda), \lambda) . \tag{16}$$

The analytical function $\Psi(Z, \lambda)$ measures the shift in angular variable between the apices. Notice that the formula (15) depends upon the same function $\Psi(Z, \lambda)$ in the three domains D_i.

4. The Autonomous Problem Close to the Homoclinic Orbit

In this section we shall evaluate the unknown functions $\mathcal{J}_i(Z,\lambda)$ we just mentioned and the first order correction to the adiabatic invariant expressed by (I.39) or (I.72). These evaluations will make it necessary to introduce the steepness parameters $h_i^\star$ (see (21) and (28)) and the out-of-symmetry parameters g_i (see (31) and (33)) mentioned earlier.

Let us remember that $2\pi J_i$ is the area enclosed by the closed curve $H(x,\lambda) = h = H_N(Z,\lambda)$. It can be evaluated as the difference between the area $2\pi J_i^\star(\lambda)$ (see (2)) enclosed by the critical curve and an area that can be divided into two parts A_1 and A_2 as shown in Fig. 6 (which is drawn for the domain D_2).

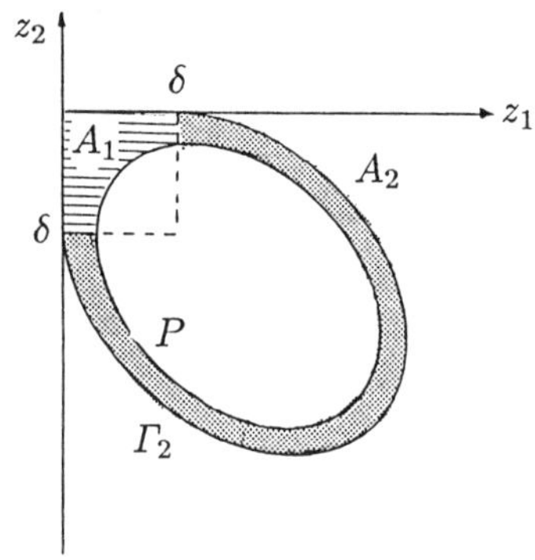

Fig. 6. Evaluation of the area enclosed by the curve P.

The area A_1 is equal to

$$A_1 = \int_{|Z/\delta|}^{\delta} \frac{|Z|}{z_1} dz_1 + \delta \left(\frac{|Z|}{\delta} \right) = |Z| \log \frac{\delta^2 e}{|Z|} . \tag{17}$$

The area A_2 is an analytical function of h (and thus of Z) vanishing with h (and thus with Z). Collecting those results and remembering the sign convention we made in Sect. II.1, we obtain

$$\mathcal{J}_i(Z,\lambda) = J_i^\star(\lambda) + \frac{Z}{2\pi} \log \frac{\Phi_i(Z,\lambda)}{|Z|} \qquad (1 \le i \le 2) \tag{18}$$

where the functions $\Phi_i(Z,\lambda) = \delta^2 e \exp(A_2/|Z|)$ are analytical functions. These two functions are invariants of the dynamical system and, together with the function $Z(h,\lambda)$, they characterize it completely. These are the functions we introduced in the previous section (see (14)). The functional dependences of J_i with respect to h:

$$J_i(h,\lambda) = \mathcal{J}_i(Z(h,\lambda),\lambda) \qquad (1 \le i \le 2) \tag{19}$$

will also be most useful. They also characterize completely the (autonomous) dynamical system as they give implicitly the relationship between the Hamiltonian h and the momentum J_i in each of the domains D_i $(1 \leq i \leq 2)$.

Their approximations close to the homoclinic orbits and the approximations of their derivatives are given by

$$\begin{aligned} J_i(h, \lambda) &= J_i^\star(\lambda) + \frac{h}{2\pi\omega} \log \frac{e h_i^\star}{|h|} + 0(h^2 \log |h^{-1}|) , \\ \frac{\partial J_i}{\partial h} &= \frac{1}{2\pi\omega} \log \frac{h_i^\star}{|h|} + 0(h \log |h^{-1}|) \end{aligned} \tag{20}$$

The parameters

$$h_i^\star(\lambda) = \frac{\omega(\lambda)}{e} \, \Phi_i(0, \lambda) \quad (1 \leq i \leq 2) \tag{21}$$

which we call the "steepness parameters" measure the rates at which J_i approach $J_i^\star$ when h goes to zero. As such they will enter in many of our estimates.

For simple dynamical systems, the functions J_i and thus the parameters $h_i^\star$ can be evaluated analytically (see the "paradigms" in Sect. 3). For more general dynamical systems, they may have to be evaluated numerically. The following scheme may be useful in this connection.

From (19), the period $T_i(h, \lambda)$ of a periodic trajectory in the domain D_i can be expressed as

$$T_i(h, \lambda) = 2\,\pi \, \frac{\partial J_i}{\partial h} = \left(\frac{\partial Z}{\partial h} \right) \left\{ \log \frac{\Phi_i}{|Z| e} + \frac{Z}{\Phi_i} \frac{d\Phi_i}{dZ} \right\} . \tag{22}$$

On the other hand, the period is the sum of the time spent in the disk around the origin from $z_1 = |Z/\delta|$ to $z_1 = \delta$ and the time spent outside, $T_\delta^i(h, \lambda)$, which can be evaluated numerically. We find

$$T_i(h, \lambda) = \frac{\partial Z}{\partial h} \log \frac{\delta^2}{|Z|} + T_\delta^i(h, \lambda) . \tag{23}$$

Comparing (22) and (23), we find

$$h_i^\star(\lambda) = \omega(\lambda) \, \delta^2 \, \exp\{\omega(\lambda) \cdot T_\delta^i(0, \lambda)\} . \tag{24}$$

Formula (18) is valid for the two domains D_1 and D_2. In order to evaluate the area enclosed by a trajectory in the domain D_3, we have to add twice the area A_1 plus the two areas of the type A_2 corresponding to each lobe along Γ_1 and Γ_2. We find

$$\mathcal{J}_3(Z, \lambda) = J_3^\star(\lambda) + \frac{Z}{\pi} \log \frac{\Phi_3(Z, \lambda)}{|Z|} \tag{25}$$

with

$$\Phi_3(Z, \lambda) = [\Phi_1(Z, \lambda) \, \Phi_2(Z, \lambda)]^{1/2} \quad , \quad J_3^\star(\lambda) = J_1^\star(\lambda) + J_2^\star(\lambda) . \tag{26}$$

Here also we introduce for further references, the functional dependence of J_3 with respect to h, its approximation and the approximation of its derivative:

$$\begin{aligned} J_3(h,\lambda) &= \mathcal{J}_3(Z(h,\lambda),\lambda) \ , \\ J_3(h,\lambda) &= J_3^\star(\lambda) + \frac{h}{\pi\omega} \ \log \frac{eh_3^\star}{|h|} + 0(h^2 \ \log|h^{-1}|) \\ \frac{\partial J_3}{\partial h} &= \frac{1}{\pi\omega} \ \log \frac{h_3^\star}{|h|} + 0(h \log|h^{-1}|) \ , \end{aligned} \tag{27}$$

with

$$h_3^\star(\lambda) = [h_1^\star \ h_2^\star]^{1/2} = \frac{\omega}{e} \ \Phi_3(0,\lambda) \ . \tag{28}$$

We turn now to the evaluation of the first order corrections to the adiabatic invariants for small (but not too small) values of h. From (I.72), we have

$$\bar{J_i} = J_i + \varepsilon \left(\frac{\partial J_i}{\partial h}\right) \ \{R(\psi_i, J_i, \lambda) - < R(\psi_i, J_i, \lambda) >\} + 0(\varepsilon^2 h^{-1} \log^2|h^{-1}|) \ . \tag{29}$$

In estimating the error term, we made use of (18) and (25) to estimate the derivatives of h with respect to J_i and thus to estimate the quantities η_1 and η_2 entering (I.72). This result was already announced in (I.73).

We shall evaluate the adiabatic invariant (29) at the apex ($\psi_i = 0$). From (14) we have that

$$R_i(0, J_i, \lambda) = R_N\left(\pm\sqrt{|Z|}, \pm\sqrt{|Z|}\right) = 0(|Z|) = 0(h) \ . \tag{30}$$

It remains to evaluate the mean value of the remainder functions. This is done in Appendices 3 and 4 where we find that

$$\begin{aligned} < R_i > &= \left(\frac{\partial J_i}{\partial h}\right)^{-1} \ \{-g_i(\lambda) + 0(h\log|h^{-1}|)\} \quad 1 \le i \le 2 \ , \\ < R_3 > &= \left(\frac{\partial J_3}{\partial h}\right)^{-1} \ \{-g_3(\lambda) + \pi\Delta_{12} + 0(h\log|h^{-1}|)\} \ , \end{aligned} \tag{31}$$

with

$$\Delta_{12} = \frac{\partial J_1}{\partial \lambda}\frac{\partial J_2}{\partial h} - \frac{\partial J_2}{\partial \lambda}\frac{\partial J_1}{\partial h} \tag{32}$$

and

$$g_3(\lambda) = g_1(\lambda) + g_2(\lambda) \ . \tag{33}$$

The functions $g_i(\lambda)$, which are important because they measure the first order corrections to the adiabatic invariants (see (34) and (35) below), vanish when the functions R_i ($1 \le i \le 2$) are odd in ψ. This is the case when the dynamical system possesses the right type of symmetry and when the apices are chosen accordingly. This is why we have called these functions the "out of symmetry" parameters. All the simple dynamical systems considered in Part III have the right type of symmetry and the corresponding functions $g_i(\lambda)$ vanish. For more general systems, the functions can be evaluated numerically by means of the numerical integration of the variational equations (see Appendix 3).

Gathering these results, we find that the adiabatic invariants in each of the three domains D_i are given by

$$\bar{J}_i = J_i + \varepsilon g_i + 0(\varepsilon h \log|h^{-1}|, \varepsilon^2 h^{-1} \log^2 |h^{-1}|) \quad 1 \le i \le 2\,, \tag{34}$$

$$\bar{J}_3 = J_3 + \varepsilon\{g_3 - \pi \Delta_{12}\} + 0(\varepsilon h \log|h^{-1}|, \varepsilon^2 h^{-1} \log^2 |h^{-1}|) \tag{35}$$

where the J_i $(1 \le i \le 2)$ are evaluated at the apices.

As the values of the J_i depend upon the definition of the apices (and thus upon the choice of the particular normalizing transformation (4)) so are the corrections g_i. But the combinations

$$\left(\frac{\partial J_i^\star}{\partial \lambda}\right)^{-1} g_i - \left(\frac{\partial J_j^\star}{\partial \lambda}\right)^{-1} g_j \qquad 1 \le i, j \le 3\,, \tag{36}$$

are independent of this choice. Indeed, we have from (31) that

$$g_i = -\lim_{h \to 0} \left(\frac{\partial J_i}{\partial h}\right) < R_i > \qquad 1 \le i \le 2\,. \tag{37}$$

Hence, if we denote by $\tilde{g}_i$ the value corresponding to a different choice of normalizing transformation, we have by (15)

$$\left(\frac{\partial J_i^\star}{\partial \lambda}\right)^{-1} (\tilde{g}_i - g_i) = -\Psi(0, \lambda) \qquad 1 \le i \le 2\,, \tag{38}$$

where the function $\Psi(Z, \lambda)$ is the same for both domains D_i $(1 \le i \le 2)$. From (38) and the fact that $g_3 = g_1 + g_2$ and $J_3^\star = J_1^\star + J_2^\star$, it is easy to conclude that the combinations (36) are independent upon the choice of the apices.

5. Traverse from Apex to Apex

We shall now be concerned with solutions of the non-autonomous system described by the Hamiltonian function:

$$H(x, \varepsilon t) = h(t)\,. \tag{39}$$

As we have seen in Sect. II.3, this system is equivalent in a disk of radius δ around the origin, to the system described by

$$H'(z, \varepsilon t) = H_N(Z, \varepsilon t) + \varepsilon\, R_N(z_1, z_2, \varepsilon t) \tag{40}$$

in the normalizing coordinates (z_1, z_2).

We shall be concerned more specifically with a "traverse" from apex to apex (or from apex to anti-apex) close to one of the homoclinic orbits Γ_i.

Let us assume that at time t_0, a trajectory of (39) is at an apex (or anti-apex) with $h = h_0$, $\lambda = \lambda_0$, $z_1^2 = z_2^2 = \zeta_0^2 = |Z(h_0, \lambda_0)|$, and that the following apex (or anti-apex) corresponds to $t = t_1$, $h = h_1$, $\lambda = \lambda_1$, $z_1^2 = z_2^2 = \zeta_1^2 = |Z(h_1, \lambda_1)|$.

We plan to evaluate the difference in "energy" and in time between those two consecutive apices:

$$\Delta h = h_1 - h_0 \quad , \quad \Delta\lambda = \lambda_1 - \lambda_0 = \varepsilon\,\Delta T = \varepsilon\,(t_1 - t_0)\,. \tag{41}$$

In order to obtain these estimates, we compare the solution of the non-autonomous system (39) (or (40)) which passes through an apex at time λ_j/ε with the energy h_j:

$$x(t, h_j, \lambda_j) \quad \text{or} \quad z(t, h_j, \lambda_j) \qquad (1 \le j \le 2\,)\,. \tag{42}$$

with the solution of the autonomous system described by $H(x, \lambda_j)$ or $H_N(Z, \lambda_j)$ which we denote

$$x^\star(t, h_j, \lambda_j) \;\; \text{or} \;\; z^\star(t, h_j, \lambda_j) = (\zeta_j \exp\{\Omega_j t\}\;,\; \zeta_j \exp\{-\Omega_j t\}) \tag{43}$$

where $\Omega_j = \Omega(\zeta_j^2, \lambda_j)$, the function Ω being the derivative $\partial H_N/\partial Z$.

The comparison is quite delicate if one wishes to reach very small values of h_0 (of the order of $\exp\{-1/\varepsilon\}$) which implies very long periods of time (of the order of $1/\varepsilon$).

The main step of this comparison is developed in detail in Appendix 5. It compares, in the disk of radius δ around the origin, the solution $z(t, h_j, \lambda_j)$ of the non-autonomous system with the solution

$$u^\star = (\zeta_j \exp\{\mu_j(t)\}, \zeta_j \exp\{-\mu_j(t)\}) \quad , \; \mu_j = \int_{t_j}^{t} \Omega(\zeta_j^2, \varepsilon t)\, dt\;, \tag{44}$$

of the intermediary system described by $H_N(Z, \varepsilon t)$.

We find the estimate

$$||z - u^\star|| \le c_6\varepsilon \;\; \text{for} \quad c_7 \exp\left\{-\frac{1}{c_1\varepsilon}\right\} \le h_j \le c_5 \tag{45}$$

where c_1, c_5, c_6, c_7 are constants (independent of ε) evaluated in the Appendix.

It is for the comparison of $u^\star$ and $z^\star$ that the assumption we have made in Sect. II.2 that ω is independent of λ is useful. Indeed this assumption makes the estimate, in the disk of radius δ around the origin:

$$|\mu_j - \Omega_j t| \le c_8\varepsilon h_j \log|h_j^{-1}| \tag{46}$$

sharper than the corresponding estimate ($|\mu_j - \Omega_j t| \le c_8\varepsilon \log|h_j^{-1}|$) to which one would be led if ω, the leading term in Ω, were indeed a function of time. The estimate (46) leads to a total estimate

$$||z - z^\star|| \le c_9\varepsilon \tag{47}$$

in the disk of radius δ around the unstable equilibrium.

The value of ΔT, the time spent from apex (h_0, λ_0) to the next one (h_1, λ_1) is then estimated as follows. Let

$$\Delta T = T_0^\varepsilon + T_1^\varepsilon + T_\delta^\varepsilon \tag{48}$$

where $T_0^\varepsilon(h_0, \lambda_0)$ and $T_1^\varepsilon(h_1, \lambda_1)$ are the values of the time spent in the disk of radius δ in the neighborhood of the two apices and T_δ^ε the value of the time spent outside this disk. The superscript ε is there to recall that we are considering the non-autonomous system with $\lambda = \varepsilon t$.

From Appendix 5 (see A.59), we get an estimate for T_j^ε:

$$T_j^\varepsilon = \frac{1}{2\Omega_j} \log \frac{\delta}{\zeta_j} + 0(\varepsilon h_j \log^2 |h_j^{-1}|) \, . \tag{49}$$

As the time T_δ^ε spent outside the disk is uniformly bounded with respect to ε and h_j, we estimate from (49) that $\lambda_j - \varepsilon t$ is of the order of $\varepsilon \log \mid h_j^{-1} \mid$ outside the disk. Hence $x(t)$ and $x^\star(t)$ which are respectively solutions of the dynamical systems $H(x, \varepsilon t)$ and $H(x, \lambda_j)$ for initial conditions the difference of which are of the order of ε (by virtue of (47)) differ at most by $\varepsilon \log \mid h_j^{-1} \mid$. We can then approximate T_δ^ε, the time spent along $x(t)$, by T_δ (defined in (23)), the time spent along $x^\star(t)$. We obtain

$$T_\delta^\varepsilon = \frac{1}{2}\{T_\delta(h_0, \lambda_0) + T_\delta(h_1, \lambda_1)\} + 0(\varepsilon \log h_m^{-1}) \tag{50}$$

where h_m is the smaller of $|h_0|$ and $|h_1|$ (and h_M which will be introduced later is the larger of $|h_0|$ and $|h_1|$).

From (48), (49) and (50), we estimate ΔT as a mean value of periods of trajectories of the autonomous system (see (23))

$$\begin{aligned} \Delta T &= \frac{1}{2}\{T(h_0, \lambda_0) + T(h_1, \lambda_1)\} + 0(\varepsilon \log h_m^{-1}) \\ &= \pi \left\{ \left(\frac{\partial J}{\partial h} \right) (h_0, \lambda_0) + \left(\frac{\partial J}{\partial h} \right) (h_1, \lambda_1) \right\} + 0(\varepsilon \log h_m^{-1}) \, . \end{aligned} \tag{51}$$

We now proceed by estimating Δh:

$$\Delta h = h_1 - h_0 = \varepsilon \int_{t_0}^{t_1} \frac{\partial H}{\partial \lambda}(x, \varepsilon t) \, dt \, . \tag{52}$$

The integral can be split into two parts: One starting from t_0 on an interval of $T(h_0, \lambda_0)/2$ and the other one on an interval $T(h_1, \lambda_1)/2$ ending at t_1. Each of these integrals is then compared with the corresponding integrals with $x^\star(t, h_i, \lambda_i)$ substituted to x and λ_i substituted to εt. We obtain

$$\Delta h = \varepsilon \int_0^{T(h_0,\lambda_0)/2} \frac{\partial H}{\partial \lambda}(x_0^\star, \lambda_0) \, dt + \varepsilon \int_{-T(h_1,\lambda_1)/2}^{0} \frac{\partial H}{\partial \lambda}(x_1^\star, \lambda_1) \, dt + 0(\varepsilon^2 \log h_m^{-1}) \, . \tag{53}$$

On the other hand, the second integral may be approximated by:

$$\varepsilon \int_{-T(h_1,\lambda_1)/2}^{0} \frac{\partial H}{\partial \lambda}(x_1^\star, \lambda_1) \, dt = \varepsilon \int_{-T(h_0,\lambda_0)/2}^{0} \frac{\partial H}{\partial \lambda}(x_0^\star, \lambda_0) \, dt + 0(\varepsilon^2 \log h_m^{-1}) \, . \tag{54}$$

This is obtained after some lenghty but straightforward calculations by splitting the integrals into a part inside the disk of radius δ where most of the time is spent but where we have an exact and simple representation of the solutions $x_0^\star$ and $x_1^\star$, and a part outside this disk where the time spent is uniformly bounded and the estimate $(x_1^\star - x_0^\star) \sim 0(\varepsilon \log h_m^{-1})$ is enough.

We obtain eventually (the subscript j may be taken indifferently as 0 or 1)

$$\Delta h = \varepsilon \int_0^{T(h_j,\lambda_j)} \frac{\partial H}{\partial \lambda}(x_j^\star, \lambda_j)\, dt + 0(\varepsilon^2 \log h_m^{-1})\,. \tag{55}$$

It remains to compute the integral in the right-hand member of (55). Using (I.13) and the fact that the remainder function is periodic, we obtain

$$0 = R(2\pi) - R(0) = \left(\frac{\partial K}{\partial J}\right)^{-1} \int_0^{2\pi} \left\{\frac{\partial K}{\partial \lambda} - \frac{\partial H}{\partial \lambda}\right\} d\psi = \int_0^T \left\{\frac{\partial K}{\partial \lambda} - \frac{\partial H}{\partial \lambda}\right\} dt \tag{56}$$

and from (55) and (56)

$$\begin{aligned}\Delta h &= 2\pi\varepsilon \left(\frac{\partial K}{\partial J}\right)^{-1} \frac{\partial K}{\partial \lambda} + 0(\varepsilon^2 \log |h_m^{-1}|) \\ &= -2\pi\varepsilon \left(\frac{\partial J}{\partial \lambda}\right)(h_j, \lambda_j) + 0(\varepsilon^2 \log |h_m^{-1}|)\,. \end{aligned}\tag{57}$$

The last equality is obtained by differentiating $K(J(h,\lambda),\lambda) = h$ with respect to λ.

The approximations (51) and (57) define mappings from (h_0, λ_0) to (h_1, λ_1), from apex to apex in each of the domains D_i. These mappings reproduce (approximately) the behaviour of the non-autonomous dynamical system in the vicinity of the homoclinic orbit. They are Poincaré's mappings with the apices defining the surfaces of section. To make these mappings precisly area-preserving, we can modify them without increasing the error terms and write them under the form:

$$\begin{aligned}\lambda_1 &= \lambda_0 + \varepsilon\pi \left[\left(\frac{\partial J_i}{\partial h}\right)(h_0, \lambda') + \left(\frac{\partial J_i}{\partial h}\right)(h_1, \lambda')\right]\,, \\ h_1 &= h_0 - \varepsilon\pi \left[\left(\frac{\partial J_i}{\partial \lambda}\right)(h_0, \lambda') + \left(\frac{\partial J_i}{\partial \lambda}\right)(h_1, \lambda')\right]\,. \end{aligned}\tag{58}$$

where

$i = (1,2,3)$ is the index of the domain D_i

and

$$\lambda' = \lambda_0 + \varepsilon\pi \left(\frac{\partial J_i}{\partial h}\right)(h_0, \lambda')\,.$$

Such a mapping plays the same role in the case of slowly varying Hamiltonian systems we are investigating as the standard mapping (Chirikov 1979) plays in case of rapidly varying Hamiltonian systems. It encapsulates our knowledge of the behaviour of the system.

In what follows we shall use only an approximation of this mapping which is easier to handle. It is obtained by substituting the approximations (20) and (27) for the functions J_i and taking $\lambda' = \lambda_0$

$$\Delta\lambda_i = \frac{\varepsilon}{2\omega}\left\{\log\frac{h_i^\star}{|h_0|} + \log\frac{h_i^\star}{|h_1|}\right\} + 0(\varepsilon^2 \log|h_m^{-1}|, \varepsilon h_M \log h_M^{-1}) \; ;$$
$$\Delta h_i = -2\pi\varepsilon\left(\frac{\partial J_i^\star}{\partial\lambda}\right) + 0(\varepsilon^2 \log|h_m^{-1}|, \varepsilon h_M \log h_M^{-1}) \; . \qquad (59)$$

The subscripts (i) in Δh_i, $\Delta\lambda_i$ have been inserted to recall that the mapping is different in each of the domains D_i. The functions $\omega(\lambda)$, $h_i^\star(\lambda)$ and $J_i^\star(\lambda)$ are evaluated at $\lambda = \lambda_0$. We recall also that h_m (resp. h_M) stands for the minimum (resp. maximum) of the absolute values of h_0 and h_1.

Formulae (59) are not meaningful if $\varepsilon \log|h_m^{-1}|$ is not small. We thus make the assumption

$$h_m \geq \varepsilon\eta \gg \exp\{-\varepsilon^{-1}\} \; . \qquad (60)$$

Later, we shall be led to the choice (see A.83)

$$\eta = \frac{1}{\varepsilon}\exp\{-\varepsilon^{-1/3}\} \qquad (61)$$

in order to minimize the error terms on the final results.

6. Probability of Capture

We are now in a position where we can analyse the transition from one domain to another one. We shall investigate in this section the basic question: where does the trajectory go when, from inside the domain D_i (which is shrinking), it is pushed towards the critical curve? Does it stay indefinitely close to the critical curve? Does it end up eventually well inside one of the other domains D_j where the adiabatic invariant can again inform us about its ultimate fate, and which one of the other domains?

We shall show that, except for a set of initial conditions, the measure of which is exponentially small with ε, the trajectory does end up in one of the other two domains after a time such that the parameter λ has not changed significantly.

In some cases we shall be able to say which one of the other domains is visited. In other cases, it depends very sensitively upon the initial conditions. So sensitively that the accuracy on the initial conditions needed to decide which one it is, is not physically meaningful and, as a consequence, we shall resort to a probabilistic argument.

Let us first investigate the case where the trajectory is initially in domain D_3 and approaches the critical curve close enough so that the formula (59) becomes meaningful. As we are approaching the critical curve and not going away from it, Δh_3, the increment of h (see (59)), is negative and, at each turn, from apex to apex, the value of h decreases by an amount proportional to ε (we assume of course that the $\partial J_i^\star/\partial\lambda$ are bounded away from zero). Eventually, h takes on a value h_0 such that

$$0 < h_0 \leq -\Delta h_3 \; . \tag{62}$$

This is the last time the trajectory goes through the apex in domain D_3. We shall call it the main apex. As we use the approximation (59), we have to exclude from our consideration, initial conditions such that h_0 comes closer to one of its limiting values (0 and $-\Delta h_3$) than $\varepsilon\eta$ (see (60)).

This is part of the set of initial conditions we were mentioning earlier and for which our analysis fails. The corresponding trajectories could stay for a very long time, possibly forever, close to the "unstable equilibrium". Note that the "unstable equilibrium" is an equilibrium of the "frozen" system with λ constant. In the system we are analysing, with $\lambda = \varepsilon t$, the equilibrium may be replaced by a very complicated invariant set.

Let us assume now that the two domains D_i $(1 \leq i \leq 2)$ increase in size. It means that Δh_1 and Δh_2 (see (59)) are also negative and that $\Delta h_3 = \Delta h_1 + \Delta h_2$ is larger in absolute value than either of them. If h_0 happens to be in the interval

$$\varepsilon\eta \leq h_0 \leq -\Delta h_1 - \varepsilon\eta \; , \tag{63}$$

the first traverse along Γ_1 after that will bring the trajectory inside the domain D_1 with a negative value of h. From there-on, the trajectory will loose energy at the rate of Δh_1 for each turn in D_1 and will end up well inside this domain.

On the other hand, if h_0 is in the interval

$$-\Delta h_1 + \varepsilon\eta < h_0 < -\Delta h_3 - \varepsilon\eta \; , \tag{64}$$

the trajectory will arrive at the anti-apex in domain D_3 with a value of the energy equal to $h_0' = h_0 + \Delta h_1$, with

$$\varepsilon\eta < h_0' < -\Delta h_2 - \varepsilon\eta \; . \tag{65}$$

The traverse along Γ_2 after this will bring the trajectory inside D_2 and from there-on it will go deeper and deeper in D_2.

If we do not know the exact value of h_0 but assume that the distribution of possible values is uniform on the interval of definition (62) (we shall come back on this assumption later), the probability of the trajectory ending up in D_i is proportional to the length of the interval (63) or (65). As a consequence, we have

$$P_i = \frac{\Delta h_i}{\Delta h_3} = \frac{\partial J_i^\star}{\partial\lambda} \Big/ \frac{\partial J_3^\star}{\partial\lambda} + 0(\varepsilon\log(\varepsilon\eta)^{-1}) \tag{66}$$

where P_i is the probability of the trajectory ending up in domain D_i. With the assumption (61), the error term is $\varepsilon^{2/3}$.

If, on the other hand, the two domains D_i $(1 \leq i \leq 2)$ do not increase in size but only one of them does, say D_1, the trajectory will certainly end up in that domain. Indeed only Δh_1 is negative and the trajectory can only leave D_3 along Γ_1 and enters then D_1. Once it has entered it, it will remain in it, decreasing its energy by Δh_1 at each traverse.

Let us investigate now the case where the trajectory is initially in one of the two domains D_i $(1 \leq i \leq 2)$. Let us choose D_2 to simplify the notations.

The value of Δh_2 is then positive (as we are approaching the critical curve) and we shall eventually enter the domain D_3 (except possibly for an exponentially small set of initial conditions). At the first apex in Domain D_3, which we shall call the main apex, the energy is h_0 with

$$\varepsilon\eta < h_0 < \Delta h_2 - \varepsilon\eta \tag{67}$$

as we have crossed $h = 0$ in the last traverse along Γ_2 just before this apex.

Again we have to consider two cases according to the sign of Δh_1. If it is positive, the energy will increase at each successive traverse and the trajectory will end up in D_3.

If Δh_1 is negative and if h_0 happens to fall in the interval

$$\varepsilon\eta \leq h_0 \leq -\Delta h_1 - \varepsilon\eta \ , \tag{68}$$

the trajectory will enter the domain D_1 on its first traverse along Γ_1 and will remain there loosing energy at each successive traverse in D_1.

If h_0 does not belong to the interval (68), it means that it belongs to

$$-\Delta h_1 + \varepsilon\eta \leq h_0 \leq \Delta h_2 - \varepsilon\eta \tag{69}$$

and that $-\Delta h_1 < \Delta h_2$ or that $\Delta h_3 = \Delta h_2 + \Delta h_1 > 0$. Hence the trajectory does not enter D_1 on its first traverse along Γ_1 and increases its energy by Δh_3 on the total trip from apex to apex. This will be true for the successive trips from apex to apex until formula (59) is no longer valid and we are deep enough in domain D_3.

Again if we do not know the exact value of h_0 but assign a uniform distribution of probability on its value in the domain of definition (67), the probability P_i of the trajectory ending up in D_i ($i = 1$ or 3) is proportional to the length of the intervals (68) or (69):

$$P_1 = -\frac{\partial J_1^\star}{\partial \lambda} \Big/ \frac{\partial J_2^\star}{\partial \lambda} + 0(\varepsilon \log(\varepsilon\eta)^{-1}) \quad , \quad P_3 = \frac{\partial J_3^\star}{\partial \lambda} \Big/ \frac{\partial J_2^\star}{\partial \lambda} + 0(\varepsilon \log(\varepsilon\eta)^{-1}) \ . \tag{70}$$

The various cases we have analysed may be summarized in a unique formula. We may consider a jump from domain D_i to domain D_j ($1 \leq i, j \leq 3$) if the trajectories are leaving D_i, i.e.

$$\text{leaving } D_i : \qquad \text{sgn}(h_i) \cdot \left(\frac{\partial J_i^\star}{\partial \lambda} \right) > 0 \ . \tag{71}$$

In that case, the probability of the jump from D_i to D_j is given by:

$$P_r(i,j) = -\text{sgn}(h_i h_j) \frac{\partial J_j^\star}{\partial \lambda} \Big/ \frac{\partial J_i^\star}{\partial \lambda} + 0(\varepsilon \log(\varepsilon\eta)^{-1}) \ , \tag{72}$$

where $\text{sgn}(h_i)$ is the sign of h in the domain D_i. Written in this way, Formulae (71) and (72) are independent of the assumption on the sign of h made in Sect. II.2.

Of course, formula (72) should be understood with the following convention. If the right-hand member is negative, the probability is actually zero and if the right-hand member is larger than one, the probability is actually one. We shall call the function $P_r(i,j)$ the probability function. It is equal to the probability of transition when its value lies between zero and one.

We ought to come back now on the assumption that h_0, the value of the Hamiltonian at the main apex, is a random variable uniformly distributed on its interval of definition (see (62) or (67)).

When the probabilistic argument is introduced simply by our lack of knowledge about the precise initial conditions (or for that matter the precise modelization of the dynamical system) of a unique "test particle" as it happens in most problems of capture into resonance in Celestial Mechanics, then the assumption is as good as another one. After all, we do not have that many planets or satellites to make precise statistics about them. The situation is different with asteroids ... but this is another story!

On the other hand, if we are thinking in terms of distribution of many test particles in a dynamical system as it is natural in problems involving charged particles in a plasma, then it becomes important to relate the distribution on the values of h_0 (at the main apex) with the distribution of initial conditions far from the transition.

It appears to be the reason why the uniform distribution on h_0 was taken for granted by most investigators in Celestial Mechanics (Goldreich, 1965, 1966; Goldreich and Peale, 1966; Allan, 1969, 1970; Sinclair, 1972, 1974; ...). Relations between the distribution of h_0 and the distribution of initial conditions far from the transition were nevertheless investigated by Counselman and Shapiro (1970), Yoder (1973, 1979) and Neishtadt (1975). Neishtadt found that it is a simple consequence of Liouville's Theorem.

We shall paraphrase Neishtadt's argument by using Poincaré-Cartan Integral Invariant.

Let us take two small sets of points P_i $(1 \leq i \leq 2)$ in the extended phase space (q, p, t). We take them at the main apex, centered respectively around a value $h_0(i)$ of h_0:

$$P_i : (q,p,t) \text{ such that } \big\{z_1 = z_2(\text{apex}),\, h_0 \in [h_0(i) - \delta h, h_0(i) + \delta h], \\ t \in [t_0 - \delta t, t_0 + \delta t]\big\} . \tag{73}$$

The values of the Integral Invariant for these sets of points are then

$$\int\!\!\int_{P_i} dq\, dp - dH\, dt = -\int\!\!\int_{P_i} dH\, dt = -4\delta h \cdot \delta t . \tag{74}$$

Let us define Q_i $(1 \leq i \leq 2)$ as the sets of points in phase space translated from P_i along the trajectories back to a time $t = \tau$ when they are far away from the transition. As the integrals in (74) remain invariant, we have

$$\int\!\!\int_{Q_i} dq\, dp - dH\, dt = \int\!\!\int_{Q_i} dq\, dp = -4\delta h \cdot \delta t . \tag{75}$$

The areas of the two sets of initial conditions Q_i are then equal. Hence if the distribution of test particles in the phase-space far from transition is uniform, so is the distribution of values of h_0 for test particles crossing the apex per unit of time.

We have assumed in the argument above a uniform distribution of test particles in the full phase-space far from transition because it is the simplest assumption. Actually the set of points Q_i can be shown to be very narrow strips along the guiding trajectories far from transition (see Escande, 1985). This is a direct consequence of the estimates (83) and (88) in the next section where it is shown that during a transition, the difference in time (which translates into a difference "along" the guiding trajectory) is much larger than the difference in h for two neighboring trajectories.

Hence a distribution of initial points (far from transition) uniform along the guiding trajectory (in the angular coordinate of the action-angle variables) and smooth across them (in the action coordinate) would produce an "almost" uniform distribution of h_0 at the main apex.

Liouville's Theorem about the conservation of area in phase-space can be used also to make the formula (72) about probability of capture, almost intuitive.

Indeed, if we assume that together, the three domains D_i ($1 \leq i \leq 3$) form an invariant set and that at time zero there is a uniform distribution of particles all over the phase-space, then formula (72) is just a paraphraze of Liouville's Theorem as it insures that the density of particles remains the same in each of these domains D_i. Indeed, it states that the number of particles entering (or leaving) a domain is proportional to its increase (or decrease) in area (see Henrard and Lemaître, 1983).

It is possible that the above argument could be the base of a simplified proof of formula (72), but alone it would not show how intricate and unstable is the phenomenon of capture in one of the domains D_i, how a tiny (of the order of ε) difference in initial conditions can decide the fate of the test particle, whether it enters one domain or another one.

This instability is of course the reason why transitions through a critical curve can lead to "chaotic motion". This is what we shall investigate in detail in the next two sections.

7. Time of Transit

The results of Sect. 5 concerning the changes in h and λ in one traverse must be combined to obtain the total changes in h and λ (and from there in the adiabatic invariant) for a trajectory leaving one domain D_i ($1 \leq i \leq 3$) and entering another one D_j ($1 \leq j \neq i \leq 3$). This makes six cases to analyse.

But, as shown in Fig. 7, these various cases may be reduced to only two of them (a) and (e) by an exchange in the indices 1 and 2 of the inside region and/or a reversal of time.

Even the two cases (a) and (e) may be analysed together if we break the trajectories into two halves: one mostly in domain D_2 up to the main apex A_0

in Fig. 7, and the other one from this apex on either in D_3 or back in D_1. But this last half trajectory is similar to the first one (in domain D_2) with exchange of the indices (1) and (2) and time reversal.

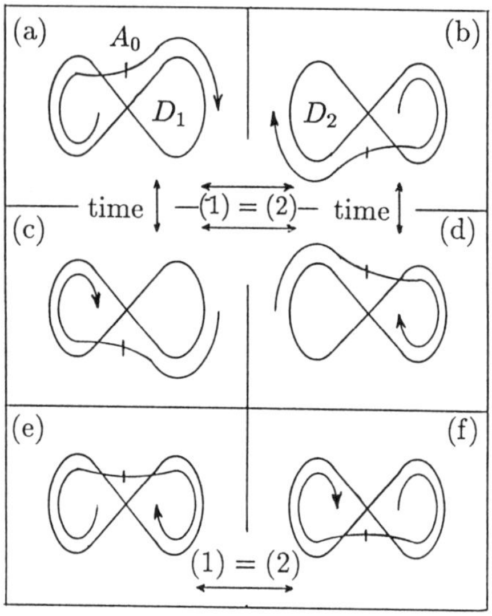

Fig. 7. The six cases of transition from one domain to another one.

In this section, we shall analyse case (a). The results of this section will then be assembled in the next section to produce a general formula valid for each subcase (a)–(f).

We start the trajectory (a) at an apex A_N corresponding to N complete traverses (apex to apex) along the curve Γ_2 in the domain D_2 before reaching the main apex A_0. The value of h and λ at each apex A_k $(0 \leq k \leq N)$ in between will be denoted h_k and λ_k (see Fig. 8).

At each apex A_k, the value of the adiabatic invariant is given by $\bar{J}_2(h_k, \lambda_k)$ where the function $\bar{J}_2(h_k, \lambda_k)$ can be deduced recursively from (59).

Far from the critical curve, the value of $\bar{J}_2$ should remain constant from apex to apex but, close to the critical curve, it is no longer true and it is precisely these differences that we wish to evaluate.

From (59), we know that from apex to apex, the difference $\Delta h = h_{k+1} - h_k$ remains more or less constant. But this is not the case for the difference $\Delta\lambda = \lambda_{k+1} - \lambda_k$ which depends sensitively upon the value of h_k.

It is thus the successive values of λ_k and their dependence upon the "final" state (h_0, λ_0) at the main apex that will be the key to the variation of the adiabatic invariant. Put otherwise the rate of change of h per traverse remains constant but the time spent in a traverse is very sensitive to initial conditions. From this it follows that the guiding trajectory (the trajectory of the autonomous

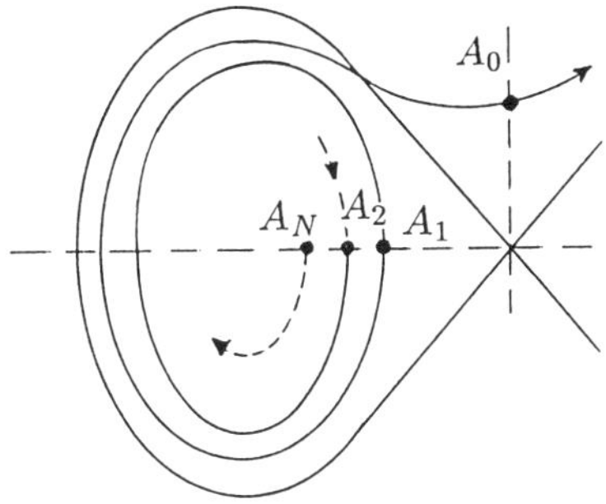

Fig. 8. First half of the trajectory from A_N to A_0.

system defined by $J_2(h, \lambda)$) and the true trajectory lose synchronization when we approach the critival curve as the true trajectory may spend a variable amount of time close to the unstable equilibrium.

It is this default of synchronization that we shall evaluate by comparing the "true time of transit":

$$\Lambda_2 = \lambda_0 - \lambda_N \tag{76}$$

with the "pseudo time of transit":

$$\Lambda_2^\star = \tau_2 - \lambda_N \tag{77}$$

where τ_2 is the value of λ where transition from D_2 to D_3 would take place if the adiabatic invariant were conserved. The "pseudo crossing time" τ_2 is thus defined by

$$J_2^\star(\tau_2) = \bar{J}_2(\lambda_N, h_N) \ . \tag{78}$$

The loss of synchronization will be the difference $\Lambda_2 - \Lambda_2^\star$ and it will be evaluated as a function of the value h_0 of the Hamiltonian at the main apex and of the pseudo crossing time τ_2.

The error terms in the evaluation of Λ_2 and $\Lambda_2^\star$ are functions of N, $\log(\varepsilon\eta)^{-1}$ and ε. The analysis conducted in the Appendix 6 shows that they are minimized if we take

$$N \sim \log(\varepsilon\eta)^{-1} \sim \varepsilon^{-1/3} \ . \tag{79}$$

Because $h_N \sim N\Delta h \sim N\varepsilon \sim \varepsilon^{2/3}$, we are, for this value of h_N, deep enough in the domain D_2 for the adiabatic invariant to be preserved (see Sect. I.6).

From there-on, we shall assume that N and η have the order of magnitude specified in (79) in order to simplify the expressions of the error terms, and for the ease of notation, we shall momentarily drop the index (2) affecting the quantities Λ_2, J_2, $h_2^\star$, g_2, ... We shall reestablish it in the final result.

Inserting (20) and (34) in (78), we find

$$J^\star(\tau) = J^\star(\lambda_N) + \frac{h_N}{2\pi\omega} \log \frac{h^\star e}{|h_N|} + \varepsilon g + 0(\varepsilon^{4/3} \log^2 \varepsilon^{-1}) \tag{80}$$

where the quantities ω, $h^\star$, g are evaluated at $\lambda = \lambda_N$.

According to (80) itself, the difference $\tau - \lambda_N$ is of the order of $\varepsilon^{2/3}\log\varepsilon^{-1}$. Hence this equation can be solved for τ by linearizing $J^\star(\tau) - J^\star(\lambda_N)$ and evaluating the quantities ω, $h^\star$, g, $\partial J^\star/\partial\lambda$ at $\lambda = \tau$ rather than at $\lambda = \lambda_N$, without increasing the error term:

$$\left(\frac{\partial J^\star}{\partial\lambda}\right)\Lambda^\star = \frac{h_N}{2\pi\omega}\log\frac{h^\star e}{|h_N|} + \varepsilon g + 0(\varepsilon^{4/3}\log^2\varepsilon^{-1}) . \tag{81}$$

In order to evaluate Λ, we have to sum the individual increments (59). This is done with some details in the Appendix 6, and gives

$$\left(\frac{\partial J^\star}{\partial\lambda}\right)\Lambda = \frac{h_N}{2\pi\omega}\log\frac{h^\star e}{|h_N|} - \frac{\varepsilon}{\omega}\left|\frac{\partial J^\star}{\partial\lambda}\right|\left\{\log\frac{\Gamma(1-x)}{x\sqrt{2\pi}} + x\log\frac{h^\star}{b}\right\} + 0(\varepsilon^{4/3}\log^2\varepsilon^{-1}) \tag{82}$$

where b is the absolute value of the increment Δh along a traverse and x the relative size of h_0 (see (84)).

The term depending on the value of h at the apex A_N drops from the difference $\Lambda - \Lambda^\star = \lambda_0 - \tau$ and we obtain eventually for the loss of synchronization in the domain D_i $(1 \le i \le 2)$

$$\left(\frac{\partial J_i^\star}{\partial\lambda}\right)(\tau_i - \lambda_0) = \varepsilon g_i + \frac{b_i}{2\pi\omega}\left\{\log\frac{\Gamma(1-x)}{\sqrt{2\pi x}} + x\log\frac{h_i^\star}{b_i}\right\} + 0(\varepsilon^{4/3}\log^2\varepsilon^{-1}) \tag{83}$$

where

$$\begin{aligned} b_i &= 2\pi\varepsilon\left|\frac{\partial J_i^\star}{\partial\lambda}\right| , \\ x &= h_0/b_i \in \,]\eta, 1-\eta[\, , \\ \eta &= 0(1/\varepsilon\exp\{-1/\varepsilon^{1/3}\}) . \end{aligned} \tag{84}$$

Formulae (83) and (84) are valid for all configurations of Fig. 7, whether the trajectory leaves or enters the domain D_i $(1 \le i \le 2)$.

The analysis of the second half of the trajectory in the domain D_3 proceeds along the same lines. Starting from the main apex A_0, we follow the trajectory from apex to apex (from A'_k to A'_{k+1}) until we reach an apex A'_M well inside the domain D_3. We choose again $M = 0(\varepsilon^{-1/3})$. We then compare the "true time of transit" in domain D_3:

$$\Lambda_3 = \lambda_M - \lambda_0 \tag{85}$$

with the "pseudo-time of transit"

$$\Lambda_3^\star = \lambda_M - \tau_3 \qquad \text{with} \qquad J_3^\star(\tau_3) = \bar{J}_3(\lambda_M, h_M) . \tag{86}$$

The pseudo-time of transit is computed from (27) and (35):

$$\begin{aligned} \left(\frac{\partial J_3^\star}{\partial\lambda}\right)\Lambda^\star = &-\frac{h_M}{\pi\omega}\log\frac{h_3^\star e}{h_M} - \frac{\varepsilon}{2\omega}\left[\frac{\partial J_1^\star}{\partial\lambda}\log\frac{h_2^\star}{h_M} - \frac{\partial J_2^\star}{\partial\lambda}\log\frac{h_1^\star}{h_M}\right]\operatorname{sgn}\left(\frac{\partial J_3^\star}{\partial\lambda}\right) \\ &- \varepsilon g_3 + 0(\varepsilon^{4/3}\log^2\varepsilon^{-1}) . \end{aligned}$$

The computation of Λ by summation of the individual increments (59) is carried out in Appendix 6 and produces:

$$\begin{aligned}\left(\frac{\partial J_3^\star}{\partial\lambda}\right)\Lambda = &-\frac{h_M}{\pi\omega}\log\frac{h_3^\star e}{h_M} - \frac{\varepsilon}{2\omega}\left[\frac{\partial J_1^\star}{\partial\lambda}\log\frac{h_2^\star}{h_M} - \frac{\partial J_2^\star}{\partial\lambda}\log\frac{h_1^\star}{h_M}\right]\operatorname{sgn}\left(\frac{\partial J_3^\star}{\partial\lambda}\right)\\ &+\frac{\varepsilon}{2\omega}\left[\frac{\partial J_1^\star}{\partial\lambda}\log\frac{h_2^\star}{b_3} - \frac{\partial J_2^\star}{\partial\lambda}\log\frac{h_1^\star}{b_3}\right]\operatorname{sgn}\left(\frac{\partial J_3^\star}{\partial\lambda}\right)\\ &-\frac{b_3}{2\pi\omega}\left[\log\left[\frac{\Gamma(y)\Gamma(\alpha+y)\sqrt{y}}{2\pi}\right] - 2y\log\frac{h_3^\star}{b_3}\right]\\ &+0(\varepsilon^{4/3}\log^2\varepsilon^{-1})\end{aligned} \tag{87}$$

where $y = h_0/b_3$ and the values of b_3, α are defined in (89). The quantity y is related to x in (84) by $y = \alpha x$.

The dependence upon the last apex A'_M drops out from the difference $\Lambda - \Lambda^\star = \tau_3 - \lambda_0$ and we obtain for the loss of synchronization in the domain D_3:

$$\begin{aligned}\left(\frac{\partial J_3^\star}{\partial\lambda}\right)(\tau_3-\lambda_0) = \varepsilon g_3 &- \frac{1}{2\pi\omega}\left[b_i\log\frac{h_3^\star}{b_3} - \frac{1}{2}b_3\log\frac{h_i^\star}{b_3}\right]\\ &-\frac{b_3}{2\pi\omega}\left\{\log\left[\frac{\Gamma(\alpha x)\Gamma(1-\alpha+\alpha x)\sqrt{\alpha x}}{2\pi}\right] - 2\alpha x\log\frac{h_3^\star}{b_3}\right\}\\ &+0(\varepsilon^{4/3}\log^2\varepsilon^{-1})\end{aligned} \tag{88}$$

where

$$\begin{aligned} b_3 &= 2\pi\varepsilon\left|\tfrac{\partial J_3^\star}{\partial\lambda}\right| \quad , \qquad \alpha = \tfrac{\partial J_i^\star}{\partial\lambda} \Big/ \tfrac{\partial J_3^\star}{\partial\lambda} \geq 0\,,\\ x &= \tfrac{h_0}{b_i} \in \left]\beta+\eta, 1-\eta\right[\quad , \beta = \max(0, \tfrac{\alpha-1}{\alpha})\,,\end{aligned} \tag{89}$$

$$\eta = 0\left(\frac{1}{\varepsilon}\exp(-1/\varepsilon^{1/3})\right)\,. \tag{90}$$

Formulae (88) and (89) are valid for all configurations of Fig. 7, whether the trajectory leaves or enters the domain D_i $(1 \leq i \leq 2)$.

8. Change in the Invariant

The loss of synchronization between the real trajectory and the guiding trajectory is instrumental in computing the change in the adiabatic invariant during a transition. Indeed, from the definition of the pseudo-crossing-time (equation (78) or (86)), we have, for a transition from domain D_i to domain D_j $(1 \leq i \neq j \leq 3)$

$$\begin{aligned}\Delta\bar J &= \bar J_j(h_M,\lambda_M) - \bar J_i(h_N,\lambda_N)\\ &= J_j^\star(\tau_j) - J_i^\star(\tau_i)\\ &= J_j^\star(\tau_i) - J_i^\star(\tau_i) + \left(\frac{\partial J_j^\star}{\partial\lambda}\right)(\tau_j-\tau_i) + 0(\varepsilon^{4/3}\log^2\varepsilon^{-1})\,.\end{aligned} \tag{91}$$

The first difference in the right-hand member of (91) is simply the jump resulting from the definition of the action variable as an area. It would exist even if the action J were a perfect invariant during transition. The third term involves the loss of synchronization on both sides of the crossing of the critical curve, which has been evaluated in Sect. 7. The error term comes from the neglected terms of the order of $(\tau_i - \tau_j)^2$.

Combining (83) and (88), we find

$$\begin{aligned}
\Delta\bar{J} &= \Delta_1(i,j) + \Delta_2(i,j) + 0(\varepsilon^{4/3}\log^2\varepsilon^{-1}) \ , \\
\Delta_1(i,j) &= J_j^\star - J_i^\star + \varepsilon\left(\frac{\partial J_j^\star}{\partial\lambda}\right)\left\{\left(\frac{\partial J_j^\star}{\partial\lambda}\right)^{-1} g_j - \left(\frac{\partial J_i^\star}{\partial\lambda}\right)^{-1} g_i\right\} \\
\Delta_2(i,j) &= \frac{\varepsilon}{\omega}\left(\frac{\partial J_j^\star}{\partial\lambda}\right) G_{ij}(z) \ .
\end{aligned} \tag{92}$$

The first terms Δ_1 depends only upon the pseudo-crossing-time τ_i and not upon the value of h_0 at the main apex. Due to its symmetry, its contribution to the change in the adiabatic invariant is not cumulative but cancels out when we consider periodic jumps back and forth between the two domains.

Notice also that while the quantities g_i and g_j do depend upon the choice of the particular normalizing transformation used in defining the apices (see Sect. II.4), the combination of them which appears in (92) is independent of this choice as explained at the end of Sect. II.4.

The second term Δ_2 contains the meaningful part of the adiabatic invariant change. The analytical expression of the function G_{ij} depends upon whether one of the domains involved in the jump is the "double" domain D_3 or not. We have

$$\begin{aligned}
G_{k3} = G_{3k} &= \frac{1}{2}(1-2z)(1-2\alpha)\log\varepsilon^{-1} \\
&\quad + \frac{1}{2}(1-2z)\left[\log\frac{h_k^\star\varepsilon}{b_k} - 2\alpha\log\frac{h_3^\star\varepsilon}{b_3}\right] \\
&\quad + \log\left\{\frac{\Gamma(\alpha-\alpha z)\Gamma(1-\alpha z)\Gamma(z)}{(2\pi)^{3/2}}\right\}
\end{aligned} \tag{93}$$

where

$$\alpha = \frac{b_k}{b_3} \geq 0 \quad , \quad \eta \leq z = \frac{b_k - h_0}{b_k} \leq \min(1, \alpha^{-1}) - \eta \ ,$$

and when neither (i) or (j) is equal to 3:

$$\begin{aligned}
G_{ij} &= z(1+\alpha)\log\varepsilon^{-1} \\
&\quad + z\left[\log\frac{h_i^\star\varepsilon}{b_i} + \alpha\log\frac{h_j^\star\varepsilon}{b_j}\right] \\
&\quad + \log\left\{\frac{\Gamma(1-z)\Gamma(1-\alpha z)}{2\pi z\sqrt{\alpha}}\right\}
\end{aligned} \tag{94}$$

where

$$\alpha = \frac{b_i}{b_j} \geq 0 \quad , \quad \eta \leq z = \frac{h_0}{b_i} \leq \min(1, \alpha^{-1}) - \eta \ .$$

Notice that, in (94), the function G_{ij} is symmetric. It is invariant for the permutation $(\alpha, z) \to (\alpha^{-1}, \alpha z)$ resulting from the exchange of the indices (i) and (j).

Formulae (92), (93) and (94) summarize the effect of the separatrix crossing upon the value of the adiabatic invariant. They are equivalent to the formulae obtained by Cary et al. (1986) except for the error terms. We have displayed them somewhat differently in order to isolate in Δ_2 the terms depending upon the value of h_0 at the principal apex.

Also, in displaying the functions G_{ij} $(1 \le i \ne j \le 3)$, we have isolated on the first lines the leading terms in $\log \varepsilon^{-1}$. The other terms are of the order of unity except for a very small range of values of z near the limit of definition where they can reach at most (for $z = \eta$ or $1 - \eta$) the order of $\varepsilon^{-1/3}$.

If ε is small enough, these leading terms dominate the expressions and the other terms can be ignored. This is what we shall do here. A more refined analysis taking those other terms into account can be found in Appendix 7.

As we have seen in Sect. 6, the value of h_0 at the main apex can be considered as a random variable the distribution of which is uniform on its interval of definition. Hence Δ_2 is also a random variable, the distribution of which is characterized mainly by its mean value and its second moment:

$$< \Delta_2 >= \frac{1}{z_{\max}} \int_0^{z_{\max}} \Delta_2 \, dz \ , \quad \sigma^2(\Delta_2) = \frac{1}{z_{\max}} \int_0^{z_{\max}} [\Delta_2 - < \Delta_2 >]^2 \, dz \ . \tag{95}$$

These are easy to compute when G_{ij} is reduced to its leading term. For the mean values, we find

$$\begin{aligned} &(i,j) = (3,k) \text{ or } (k,3) : \alpha = b_k/b_3 \ , \\ &\begin{cases} \alpha < 1 : < \Delta_2 >= 0 \ , \\ \alpha > 1 : < \Delta_2 >= \frac{\varepsilon}{2\omega} \left(\frac{\partial J_j^\star}{\partial \lambda} \right) \frac{(1-2\alpha)(\alpha-1)}{\alpha} \log \varepsilon^{-1} \ , \end{cases} \\ &i \ne 3 \text{ and } j \ne 3 : \alpha = \min \left\{ \frac{b_i}{b_j}, \frac{b_j}{b_i} \right\} \ , \\ &\quad < \Delta_2 >= \frac{\varepsilon}{2\omega} \left(\frac{\partial J_j^\star}{\partial \lambda} \right) (1+\alpha) \log \varepsilon^{-1} \ . \end{aligned} \tag{96}$$

For the same reasons of symmetry than in the case of Δ_1, the mean value of Δ_2 does not contribute to changes in the adiabatic invariant that can be cumulative. Here again, if a test particle jumps from domain D_i to domain D_j, and then back to D_i, the contributions of the mean value of Δ_2 for each jump cancel each other.

The real key to the diffusive change in the adiabatic invariant is then the second moment which we call the diffusion parameter:

$$\sigma_{ij}(\Delta_2) = \frac{b_j}{\max\{b_i, b_j\}} \left| \frac{\partial J_1^\star}{\partial \lambda} - \frac{\partial J_2^\star}{\partial \lambda} \right| \frac{\varepsilon \log \varepsilon^{-1}}{2\omega\sqrt{3}} \tag{97}$$

for a jump from domain D_i to domain D_j. We recall that the quantities b_m are given by

$$b_m = 2\pi\varepsilon \left| \frac{\partial J_m^\star}{\partial \lambda} \right| . \tag{98}$$

The diffusion parameter is thus a function of the derivatives of the critical areas: $\frac{\partial J_m^\star}{\partial \lambda}$ and of the time scale at the equilibrium ω, evaluated at the pseudo-crossing-time τ_i.

The leading term (97) in the diffusion parameter disappears in a special but important case, the symmetric case when

$$\frac{\partial J_1^\star}{\partial \lambda} = \frac{\partial J_2^\star}{\partial \lambda} . \tag{99}$$

In that case, there can be no transition between domains D_1 and D_2. According to the sign of $h_3(\partial J_3^\star/\partial\lambda)$, we can have a transition from both D_1 and D_2 to D_3 or a transition from D_3 to either D_1 or D_2 with equal probability.

In order to compute the diffusion parameter in the symmetric case, we ought to go back to the complete formula (93). Fortunately, this formula can be much simplified as we have $\alpha = \frac{1}{2}$. We find:

$$G_{k3} = G_{3k} = \frac{1}{2}(2x-1)\log\frac{h_k^\star}{h_3^\star} - \log\{2\sin\pi x\} \tag{100}$$

where

$$\eta \le x = \frac{h_0}{b_k} \le 1 - \eta .$$

The mean value of G_{k3} is zero and the quadrature involved in the computation of the diffusion parameter can be performed analytically (see Gradshteyn and Ryzhic 1965, p. 540). We find

$$\sigma_{ij}(\Delta_2) = \left| \frac{\partial J_j^\star}{\partial \lambda} \right| \left[1 + \frac{1}{\pi^2}\log^2\frac{h_i^\star h_j^\star}{h_3^\star h_3^\star} \right]^{1/2} \frac{\pi\varepsilon}{2\omega\sqrt{3}} . \tag{101}$$

When the geometric symmetry (99) is accompanied by a time symmetry such that

$$h_1^\star = h_2^\star \tag{102}$$

(which implies that $h_3^\star = h_1^\star = h_2^\star$), the second term of the square root in (101) disappears and we obtain:

$$\sigma_{ij}(\Delta_2) = \left| \frac{\partial J_j^\star}{\partial \lambda} \right| \frac{\pi\varepsilon}{2\omega\sqrt{3}} . \tag{103}$$

This last formula is the one given by Timofeev (1978) in the case of the pendulum with varying amplitude and by Cary et al. (1986) in the general symmetric case.

Part III The Paradigms

1. Introduction

Most of the applications of the classical adiabatic invariant theory have dealt with a very few "paradigm" problems, essentially the harmonic oscillator and the pendulum.

The concept itself of an adiabatic invariant has been inspired by the harmonic oscillator described by

$$H = \frac{1}{2}(p^2 + \omega^2 q^2) \tag{1}$$

for which the action-variable is given by

$$J = \frac{1}{2\omega}(p^2 + \omega^2 q^2) = \frac{h}{\omega} . \tag{2}$$

The first order invariance of J when ω is a slow function of the time may have helped shape the old quantum theory. At the Solvay Congress of 1911, Lorentz asked how the amplitude of an oscillator would vary if its period were slowly changed. Would the number of quanta of its motion change? Einstein (Langevin and de Broglie, 1912) immediately gave the answer that the action $J = h/\omega$ would remain constant and thus the number of quanta would remain unchanged if $d\omega/dt$ were small enough.

The first proof of adiabatic invariance to all order was given by Kulsrud (1957) again for the harmonic oscillator giving hope that this would be the case for more general problems. In quick succession after this breakthrough, Kruskal (1957) proved an analogous result for the gyrating particle and Lenard (1959) for the anharmonic oscillator. Explicit formulae for the third order adiabatic invariant of the harmonic oscillator can be found in Stern (1971).

Another example of "easy" action-variable is provided by the two body problem. Its Hamiltonian function expressed in polar coordinates in the plane of motion, is given by

$$H = \frac{1}{2}(R^2 + \frac{G^2}{r^2}) - \frac{\mu}{r} \tag{3}$$

where R is the radial velocity conjugated to the radial variable r and G is the angular momentum, conjugated to the angular variable θ:

$$G = \sqrt{\mu a(1-e^2)} . \tag{4}$$

The quantities a and e are respectively the semi-major axis and the eccentricity of the elliptic trajectory.

The action-variable associated with the motion in r is:

$$\begin{aligned} J &= \frac{1}{2\pi}\oint R\,dr = \frac{1}{\pi}\sqrt{\mu/a}\int_{a(1-e)}^{a(1+e)} \{(a(1+e)-r)\cdot(r-a(1-e))\}^{1/2}\frac{dr}{r} \\ &= \frac{2}{\pi}\{\sqrt{\mu a} - G\} . \end{aligned} \tag{5}$$

As G is constant, so is the Delaunay's action:

$$L = \sqrt{\mu a} \; . \tag{6}$$

The actions L and G have been used consistently in Celestial Mechanics since the time of Delaunay (1867), but curiously enough their role as "adiabatic invariant" when μ is a slow function of the time has been overlooked although it yields immediately the interesting result that the eccentricity remains constant and the semi-major axis scales as $1/\mu$ when the parameter μ changes slowly with time.

This result was already known for particular laws of variation of μ by Poincaré (1911), Jeans (1924) and others and has been applied to the problem of evolution of binary stars (see Dommanget, 1963, for further references). But the fact that it is a consequence of the adiabatic invariant theory shows that it is a quite general property of the two body problem.

By far the most popular "paradigm" problem has been the pendulum which we shall review in Sect. 2. It is almost impossible to find a reference about resonances or adiabatic invariant which does not use implicitly or explicitly the pendulum as a model.

Its role as the "first fundamental model for resonance" is certainly well deserved. It is simple enough that it can be solved analytically (albeit by the use of unwieldy elliptic integrals) but complex enough that it contains the delicate type of motions which are the homoclinic orbits. Furthermore it presents itself under the very convivial face of a simple mechanical device rich in intuitive pictures. The legend has it that under the disguise of a swinging chandelier in Pisa cathedral it distracted Galileo from its devotion ...

Notwithstanding its importance, the pendulum is not the only useful model and sometimes its usefulness has been overemphasized. We describe in Sect. 3 what we have called "the second fundamental model for resonance" (Henrard and Lemaître, 1983a). In many instances of first order resonance, this model is a better representation than the pendulum. It is also richer qualitatively as it possesses a non-degenerate bifurcation. Borderies and Goldreich (1984) have shown how to compute analytically the most significant quantities related to the transitions through the critical curves.

This second fundamental model for resonance is the "eldest brother" of a countable family of models representing resonances of n-th order. The most significants of them have been analyzed by Borderies and Goldreich (1984) and Lemaître (1984). We shall not reproduce their analysis here but refer interested reader to the original papers.

Section 4 is devoted to a brief description of the "Colombo's top" model. This simple dynamical system has been studied in relation to the problem of the Spin-Orbit interaction of Planets and Satellites (Colombo, 1966; Peale, 1969, 1974; Ward, 1975; Ward et al., 1979; Borderies, 1980). For small values of the obliquity, it can be approximated by the second fundamental model. It has the very nice feature of being defined upon a compact manifold (a sphere). The most significant quantities related to the transitions through the critical curves can also be expressed analytically.

In Sect. 5, we generalize a remark made in Sect. 2 concerning the pendulum problem and show how in some circumstances a dissipative force can be modelized by a slow variation of the parameters of an Hamiltonian system.

2. The Pendulum

Various physical problems of interest can be modelized by a pendulum with slowly varying parameters.

The most obvious one is, of course, the pendulum itself with variable length $\lambda(t)$:

$$H = \frac{1}{2} I^2 - b(t) \cos\phi \qquad \text{with} \quad b = gL^3 . \tag{7}$$

Notice that we did not use the usual normalization $y = p_\phi / L^2$ which, from the Hamiltonian

$$H' = \frac{1}{2} p_\phi^2 \; / \; L^2 - gL \cos\phi \tag{8}$$

leads to the Hamiltonian

$$H'' = \frac{1}{2} y^2 - \frac{g}{L} \cos\phi . \tag{9}$$

Indeed, when λ is a function of time, the usual normalization is no longer canonical. Instead, we have used a change of time scale $\tau = t/\lambda^2$.

More generally, many resonance problems with variable restoring torque can be modelized by (7).

On top of the (slow) variation of the restoring torque $b(t)$ of the pendulum, one can take into account (see Yoder, 1979) a (small) outside torque $(-\dot{c})$ by considering the differential equation

$$\ddot{\phi} = -b \sin\phi - \dot{c} . \tag{10}$$

Defining the momentum $I = \dot{\phi} + c$, one is led to the Hamiltonian function

$$H = \frac{1}{2} (I - c)^2 - b \cos\phi . \tag{11}$$

In plasma physics, one considers particles moving in a wave field with slowly varying amplitude and phase velocity leading to the equation (see, for instance, Cary et al., 1986)

$$\ddot{\phi} = -b(t) \sin(\phi - d(t)) . \tag{12}$$

The equation for the angular variables $\psi = \phi - d$ is similar to (10) with $\dot{c}$ replaced by $\ddot{d}$. Hence we are led to (11) with c replaced by $\dot{d}$.

Menyuk (1985) prefers to consider a two modes system

$$H = \frac{1}{2} I^2 - \alpha \cos(\phi - \varepsilon t) - \beta \cos(\phi + \varepsilon t) \tag{13}$$

which can be put under the form (12) if we take

$$\begin{cases} b\sin d = (\alpha - \beta)\sin \varepsilon t \ , \\ b\cos d = (\alpha + \beta)\cos \varepsilon t \ . \end{cases} \tag{14}$$

The equation for the synchroneous motor:

$$\ddot{\phi} = -b_0 \sin\phi + a\dot{\phi} \tag{15}$$

with a small dissipative term ($a\dot{\phi}$) has been studied by Andronov et al. (1966) and Urabe (1954, 1955). Burns (1978) applied their results to the rotation of Mercury. This problem can also be modelized by the Hamiltonian (7). If we define the "momentum" I by

$$I = \dot{\phi} e^{-at} \ , \tag{16}$$

and the new time variable τ by:

$$a\tau = e^{at} \ , \tag{17}$$

we are led to the differential equations:

$$\frac{d\phi}{d\tau} = I \quad , \quad \frac{dI}{d\tau} = -e^{-2at} b_0 \sin\phi = -(a\tau)^{-2} b_0 \sin\phi \ , \tag{18}$$

which corresponds to the Hamiltonian (7) with $b = b_0 e^{-2at} = (a\tau)^{-2} b_0$.

It may seem strange that a dissipative problem like (15) is mapped onto an Hamiltonian problem like (7). This apparent paradox is explained when one considers that (16) is time dependent and thus that a conservation of area in the phase space (ϕ, I) corresponds to an exponential decrease in area in the phase space $(\phi, \dot{\phi})$. We shall come back on this representation of dissipative forces by Hamiltonian systems in Sect. 5.

From this brief review, it is obvious that the slowly varying pendulum can modelize a large variety of interesting phenomena which all can be described by the Hamiltonian function (11). We shall slightly modify its expression to have $h = 0$ on the stable equilibrium and study the Hamiltonian function:

$$h = \frac{1}{2}(I - c)^2 + 2b\sin^2\frac{\phi}{2} \tag{19}$$

where b and c are slow functions of the time.

First we have to clarify a point about the global topology of this dynamical system. It is defined on the cylinder and the cylinder is a two-dimensional manifold such that a closed curve does not always define a finite area. The connection between the action-variable and area should be modified accordingly.

In order to avoid unnecessary complications, we prefer to perform the following "surgery" on the manifold. Pick a negative value I_0 of I large enough (in absolute value) that it is outside the domain D of interest. Take $\sqrt{2(I - I_0)}$ and ϕ as polar coordinates in the plane to obtain a mapping from D on the cylinder to an open set D' in the plane.

The area enclosed by a level curve of (19) in the domain D' is equal to the usual integral

$$2\pi J = \oint I \, d\phi \,, \tag{20}$$

except for an additive constant $\pm 2\pi I_0$ in case of circulation (positive or negative). We can then use the definition (20) for the action-variable and think of it as the area enclosed by the level curves of (19) in D'.

This being settled, the action-angle variables (ψ, J) of the frozen system (19) are well-known. They are related to the variable (ϕ, I) by means of

In case of libration	In case of circulation
$\alpha = h/2b < 1$ $\sin\phi/2 = \sqrt{\alpha}\sin\ell$	$\beta^{-1} = h/2b > 1$ $\sin\phi/2 = \sin\theta$
$S = 4\sqrt{b}\{(\alpha-1)\mathbf{F}(\ell,\alpha) + \mathbf{E}(\ell,\alpha)\} + c\phi$ $\psi = \mathbf{F}(\ell,\alpha)\pi/2\mathbf{K}(\alpha)$ $J = 8\sqrt{b}\{(\alpha-1)\mathbf{K}(\alpha) + \mathbf{E}(\alpha)\}/\pi$ $\partial J/\partial h = 2\mathbf{K}(\alpha)/(\pi\sqrt{b})$	$S = 4\sqrt{b/\beta}\mathbf{E}(\theta,\beta)\cdot\mathrm{sgn}(I-c) + c\phi$ $\psi = \mathbf{F}(\theta,\beta)(\pi/\mathbf{K}(\beta))\cdot\mathrm{sgn}(I-c)$ $J = \frac{4}{\pi}\sqrt{b/\beta}\mathbf{E}(\beta) + c\,\mathrm{sgn}(I-c)$ $\partial J/\partial h = \sqrt{\beta/b}\mathbf{K}(\beta)/\pi$

The functions $\mathbf{F}(\ell,\alpha)$, $\mathbf{E}(\ell,\alpha)$, ... are the usual elliptic integrals (see Abramowitz and Stegun, 1965, for the notations) and $\mathbf{Z}(\ell,\alpha)$ which appears later is the Jacobi's zeta function.

What is a little less known, although it can be found under various forms, more or less explicit in Best (1968), Timofeev (1978), Menyuk (1985), are the formulae for the slowly varying pendulum. One finds that the remainder function of the canonical transformation going from (ϕ, I) to (ψ, J) is in case of
Libration:

$$\varepsilon R = \dot{c}\phi + \frac{2\dot{b}}{\sqrt{b}}\mathbf{Z}(\ell,\alpha) \,, \tag{21}$$

Circulation:

$$\varepsilon R = \frac{\dot{c}}{\mathbf{K}(\beta)}\{2\mathbf{K}(\beta)\theta - \pi\mathbf{F}(\theta,\beta)\} + \frac{2\dot{b}}{b}\sqrt{b/\beta}\mathbf{Z}(\theta,\beta) \,. \tag{22}$$

In both cases, the mean value of the remainder function vanishes

$$< R > = \frac{1}{2\pi}\int_0^{2\pi} R \, d\psi = 0 \,. \tag{23}$$

From this, it is easy to compute the first order correction to the adiabatic invariant. From (I.72), we obtain, in case of:
Libration:

$$\bar{J} = J + \frac{2}{\pi}\mathbf{K}(\alpha)\{\frac{2\dot{b}}{b}\mathbf{Z}(\ell,\alpha) + \frac{\dot{c}}{\sqrt{b}}\phi\} + 0(\frac{\varepsilon^2\log^2(\alpha-1)}{\alpha-1}) \,,$$

Circulation:

$$\bar{J} = J + \frac{1}{\pi}\mathbf{K}(\beta)\{\frac{2\dot{b}}{b}\mathbf{Z}(\theta,\beta) + 2\sqrt{\beta/b}\dot{c}\theta\} - \dot{c}\sqrt{\beta/b}\mathbf{F}(\theta,\beta) + 0(\frac{\varepsilon^2\log^2(\beta-1)}{\beta-1}) \,. \tag{24}$$

The invariants attached to the critical curves are then easy to compute. We have, with D_1 corresponding to positive rotation (see Fig. 9):

$$\begin{array}{ll} \omega = \sqrt{b} & \text{see (II.5)}\,, \\ J_1^\star = \frac{4}{\pi}\sqrt{b} + c\,;\; J_2^\star = \frac{4}{\pi}\sqrt{b} - c\,;\; J_3^\star = \frac{8}{\pi}\sqrt{b} & \text{see (II.2)}\,, \\ h_1^\star = h_2^\star = h_3^\star = 32b & \text{see (II.21), (II.28)}\,, \\ g_1 = g_2 = g_3 = 0 & \text{see (II.31)}\,, \end{array} \tag{25}$$

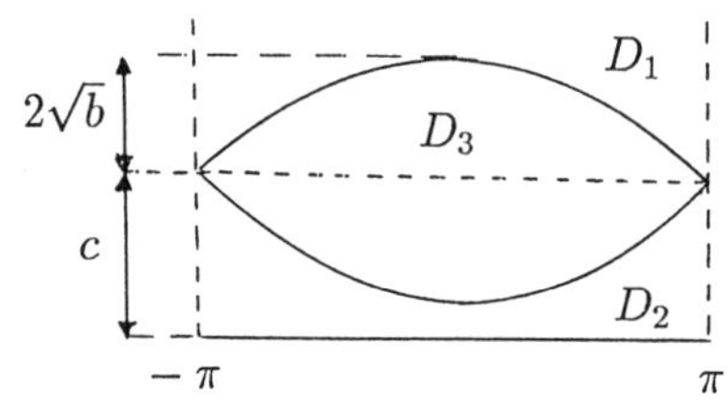

Fig. 9. The phase-space of the pendulum with the three domains.

In order to compute the probability of transitions, we define the parameters

$$B = \frac{2}{\pi}\frac{\dot{b}}{\sqrt{b}} \quad , \quad C = \frac{\pi}{2}\sqrt{b}\frac{\dot{c}}{\dot{b}} \,. \tag{26}$$

in such a way that the derivatives of the critical areas are

$$\varepsilon\frac{\partial J_1^\star}{\partial\lambda} = B(1+C) \quad , \quad \varepsilon\frac{\partial J_2^\star}{\partial\lambda} = B(1-C) \quad , \quad \varepsilon\frac{\partial J_3^\star}{\partial\lambda} = 2B \,. \tag{27}$$

The probabilities of transition take then the simple form illustrated in Fig. 10. It is computed from the probability function (see (II.72))

$$Pr(i,j) = -\mathrm{sgn}(h_i h_j)\frac{\partial J_j^\star}{\partial\lambda} \Big/ \frac{\partial J_i^\star}{\partial\lambda} \,. \tag{28}$$

Let us recall that, when the right-hand side of (28) is negative, the probability should be taken as zero and, when the right-hand side is larger than one, the probability should be taken as one.

The main term in the diffusion parameter (see (II.97) becomes, for a jump from domain D_i to domain D_j:

$$\sigma_{ij} = \frac{\dot{c}\log|b/\dot{c}|}{\sqrt{3b}} Pr(i,j) \,, \tag{29}$$

where $Pr(i,j)$ is the probability function of the corresponding jump. In the symmetric case, when $\dot{c} = 0$, we obtain (see (II.103)):

$$\sigma_{ij} = \left|\frac{2\dot{b}}{b\sqrt{3}}\right| Pr(i,j) \,. \tag{30}$$

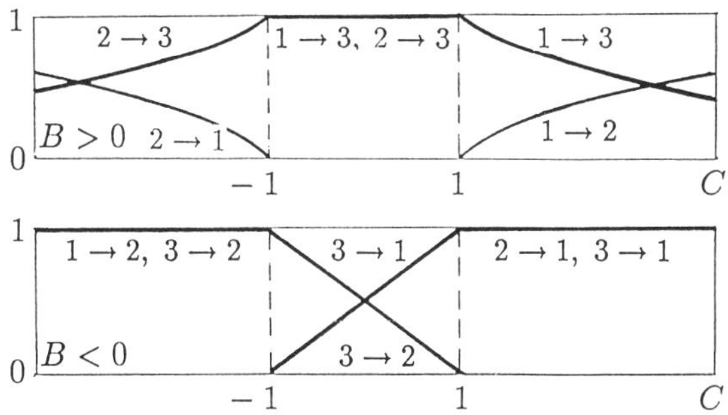

Fig. 10. Probabilities of transition: (a): for $B > 0$; (b) for $B < 0$.

In this case, the probability is 1 if $j = 3$ and 1/2 if $j = 1$ or 2.

Equivalent formulae have been proposed by Timofeev (1978) for the symmetric case ($\dot{c} = 0$) and by Cary et al. (1986) for the non-symmetric case ($\dot{c} \neq 0$).

3. The Second Fundamental Model

If we look carefully at the way a resonance problem is reduced to the pendulum, we find most of the time the following steps.

First appropriate action-angle variables are introduced in such a way that the problem is reduced (by averaging over the "fast" frequencies) to a one-degree of freedom Hamiltonian system:

$$H = H_0(S) + \varepsilon H_1(S, s) \tag{31}$$

where $H_1(S, s)$ is 2π-periodic in s (the resonant angle) and $\partial H_0/\partial S$ vanishes for a particular value $S^\star$ of S (the exact resonance). Indeed, when ε is small, $\partial H_0/\partial S$ is the zero-order approximation of the resonant frequency.

In a second step, H_0 and H_1 are expanded in Taylor series with respect to S and only the most significant terms are retained.

Here we come to a cross-point. In some cases, $H_1(S, s)$ is a non-vanishing function of s for any S in the vicinity of $S^\star$. Its simplest form is $\varepsilon H_1(S, s) = b(S) \cos s$ and we obtain the pendulum Hamiltonian

$$H = \beta(S - S^\star)^2 + b(S^\star) \cos s \ . \tag{32}$$

But, in many instances, the Hamiltonian H is actually defined as an analytic function of the cartesian coordinates $x = \sqrt{2S} \cos s$, $y = \sqrt{2S} \sin s$. This means that its expression in the polar coordinates (S, s) possesses the d'Alembert characteristic in the couple $(\sqrt{S}, s)$ (see Henrard, 1974). In such cases, the analogous simplest form for the truncated Hamiltonian is

$$H = \beta(S - S^\star)^2 + \varepsilon \cdot \sqrt{2S} \cos s \ . \tag{33}$$

Of course, if $S^\star$ is not too small (larger than the typical variation of S at least), the square root $\sqrt{2S}$ can be expanded around $S^\star$ to yield after truncation a pendulum with a restoring torque $b = \varepsilon\sqrt{2S^\star}$. But when $S^\star$ is allowed to be small, the dynamical system (33) is essentially different from the pendulum.

The dynamical system (33) itself or some of its avatars has been described and analysed by various authors especially in connection with the orbit-orbit resonance in Celestial Mechanics (e.g. Poincaré, 1899, 1902; Schubart, 1966; Message, 1966; Yoder, 1973; Neishtadt, 1975; Sessin, 1981; Sessin and Ferraz-Mello, 1984). Its significance as a general model has been pointed out by Henrard and Lemaître (1983a). Indeed, our first application of this model (see Sect. IV.2) has nothing to do with Celestial Mechanics.

The Hamiltonian (33) depends upon three parameters $(\beta, S^\star, \varepsilon)$. In most applications, β is of the order of unity, $S^\star$ measures the closeness to the exact resonance and can be vanishingly small, while ε measures the strength of the restoring torque and can also be small (but not vanishingly small).

The three parameters are not really independent and, by scaling the time and the action, we can consider a model with only one truly independent parameter.

Let us define

$$\begin{aligned} r &= s \cdot \operatorname{sgn}(\beta) \qquad \text{if } \beta\varepsilon < 0 \, , \\ r &= s \cdot \operatorname{sgn}(\beta) + \pi \text{ if } \beta\varepsilon > 0 \, , \end{aligned} \tag{34}$$

and the scaled time and momentum

$$\tau = \left|\frac{\beta\varepsilon^2}{4}\right|^{1/3} \cdot t \quad , \quad R = \left|\frac{2\beta}{\varepsilon}\right|^{2/3} \cdot S \, . \tag{35}$$

The Hamiltonian function (33) is replaced by

$$K(r, R; \delta) = -3(\delta + 1)R + R^2 - 2\sqrt{2R}\cos r \tag{36}$$

where δ is the parameter of the model and is given by

$$3(\delta + 1) = 2\left|\frac{2\beta}{\varepsilon}\right|^{2/3} S^\star \, . \tag{37}$$

The Hamiltonian function (36) is not differentiable at $R = 0$. This is because the polar coordinates (r, R) are not a regular map of the phase space at the origin. But the cartesian coordinates

$$y = \sqrt{2R}\sin r \quad , \quad x = \sqrt{2R}\cos r \, , \tag{38}$$

where x is the momentum conjugated to y, transforms (36) into

$$K(y, x; \delta) = -\frac{3}{2}(\delta + 1)(x^2 + y^2) + \frac{1}{4}(x^2 + y^2)^2 - 2x \tag{39}$$

which is regular at the origin.

The level curves $K(y, x; \delta) = h$ of the Hamiltonian (39) are illustrated in Figs. 11 and 12 for various values of the parameter δ. In the three dimensional

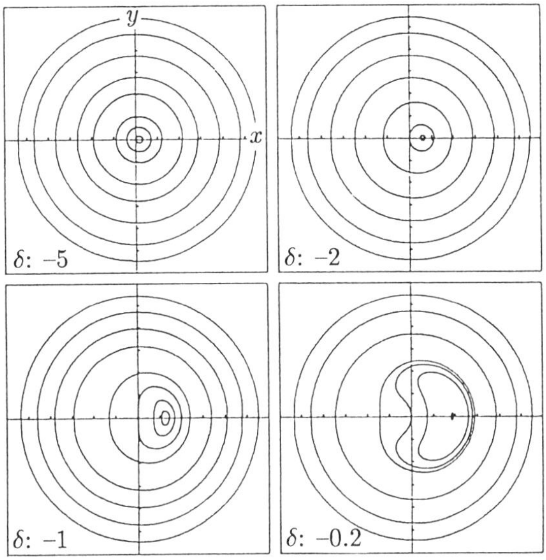

Fig. 11. Level curves of the second fundamental model for negative values of δ.

space (x, y, z), they are the intersections of the plane $h + 2x = z$ and the surface with a cylindrical symmetry:

$$z = -\frac{3}{2}(\delta + 1)(x^2 + y^2) + \frac{1}{4}(x^2 + y^2)^2 . \tag{40}$$

When $(\delta + 1)$ is negative, the surface is a kind of paraboloid and the intersections are topologically equivalent to circles. But when $(\delta + 1)$ is positive, the surface is a kind of "sombrero" (Ferraz-Mello: private communication).

The intersections of the sombrero (40) and an inclined plane has interesting features as can be seen in Fig. 12. Perhaps the best way to visualize it is to walk with an inclined sombrero under the rain ...

When δ is larger than zero, there appears an unstable equilibrium associated with two homoclinic orbits (see Fig. 13). In between those two curves (where the water would start to gather if you walk under the rain) lies a domain (D_3), the resonance domain, which corresponds to the libration domain of the pendulum. Inside the smallest of the homoclinic orbits and outside the largest one, we have an "internal" (D_1) and "external" (D_2) domain corresponding to the positive or negative circulations.

When δ is large enough in absolute value, the vicinity of the origin is occupied with curves very close to circles. This corresponds to a "non-resonance" situation for which S is almost constant and s circulates almost linearly.

One interesting feature of the second fundamental model is, of course, the generic bifurcation which happens at $\delta = 0$ with the apparition of the unstable equilibrium. This does not happen in the pendulum model.

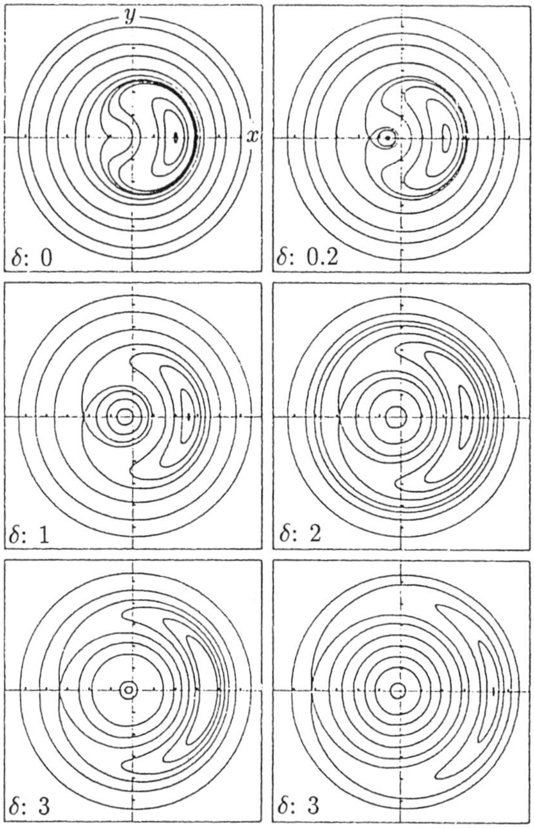

Fig. 12. Level curves of the second fundamental model for positive values of δ (The sombrero).

In principle, the action-angle variables of the second fundamental model could also be evaluated by means of elliptic integrals. But the expressions would be very involved. Fortunately, Borderies and Goldreich (1984) have shown how to evaluate simply the quantities related to the transitions through the critical curves. More information about this evaluation can be found in Lemaître (1984b).

We first observe that the equilibria of the dynamical system described by (39) are given by $y_i^\star = 0$ and $x_i^\star$ roots of the equation:

$$x^3 - 3(\delta + 1)x - 2 = 0 \, . \tag{41}$$

For any $\delta > 0$, they are three roots (see Figs. 13 and 14) given by:

$$\begin{aligned} x_1^\star &= 2(\delta + 1)^{1/2} \cos \varDelta \, , \\ x_2^\star &= -(\delta + 1)^{1/2} (\cos \varDelta - \sqrt{3} \sin \varDelta) \, , \\ x_3^\star &= -(\delta + 1)^{1/2} (\cos \varDelta + \sqrt{3} \sin \varDelta) \, , \end{aligned} \tag{42}$$

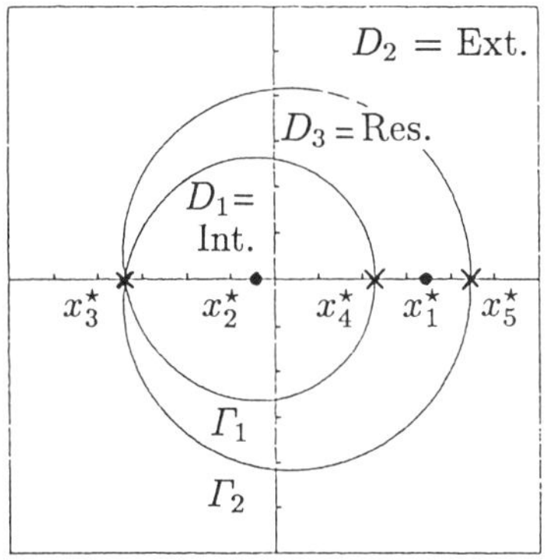

Fig. 13. Notations relative to the second fundamental model.

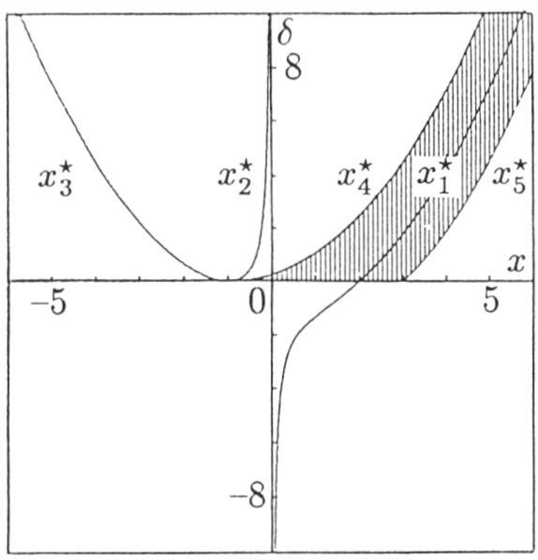

Fig. 14. Location of the equilibria ($x_i^\star$: $1 \le i \le 3$) and the intersections of the homoclinic orbits with the x axis ($x_4^\star$ and $x_5^\star$).

where Δ is defined implicitly by

$$\cos 3\Delta = (\delta + 1)^{-3/2} . \tag{43}$$

The unstable equilibrium corresponds to the third root. For the ease of notation in what follows, we shall define the quantity:

$$x^\star = -x_3^\star > 1 . \tag{44}$$

When we linearize (39) in the vicinity of the unstable equilibrium, we find

$$K = \frac{x^\star(4 - x^{\star 3})}{4} + \frac{2(x^{\star 3} - 1)^{1/2}}{x^\star} \left\{ \frac{\delta x^2 - \delta y^2}{2} \right\} + \cdots \tag{45}$$

from which it is obvious that the time scale of the unstable equilibrium is

$$\omega = \frac{2(x^{\star 3} - 1)^{1/2}}{x^\star} . \tag{46}$$

The homoclinic orbits intercept the x-axis at the location (see Figs. 13 and 14):

$$x_4^\star = x^\star - 2(x^\star)^{-1/2} \quad , \quad x_5^\star = x^\star + 2(x^\star)^{-1/2} \,. \tag{47}$$

The critical areas $J_i^\star$ are given by the integrals

$$\begin{aligned} J_i^\star &= \frac{1}{2\pi}\oint R\,dr = \frac{1}{2\pi}\oint R\frac{\dot r}{\dot R}dR \\ J_i^\star &= \frac{3}{2\pi}\oint \frac{(2R+2x^{\star 2}-\delta-1)}{[(2R-x_4^{\star 2})(x_5^{\star 2}-2R)]^{1/2}}dR \,, \end{aligned} \tag{48}$$

where the path integrals are taken along the homoclinic orbit Γ_i.

After some algebra, we find

$$\begin{aligned} J_1^\star &= \frac{3}{\pi}\{(\delta+1)[\arcsin((x^\star)^{-3/2})-\frac{\pi}{2}]+\frac{(x^{\star 3}-1)^{1/2}}{x^\star}\}<0\,, \\ J_2^\star &= \frac{3}{\pi}\{(\delta+1)[\arcsin((x^\star)^{-3/2})+\frac{\pi}{2}]+\frac{(x^{\star 3}-1)^{1/2}}{x^\star}\}>0\,, \\ J_3^\star &= J_1^\star + J_2^\star \,, \end{aligned} \tag{49}$$

$$\begin{aligned} \frac{\partial J_1^\star}{\partial\delta} &= \frac{3}{\pi}[\arcsin((x^\star)^{-3/2})-\frac{\pi}{2}]\,, \\ \frac{\partial J_2^\star}{\partial\delta} &= \frac{3}{\pi}[\arcsin((x^\star)^{-3/2})+\frac{\pi}{2}]\,, \\ \frac{\partial J_3^\star}{\partial\delta} &= \frac{6}{\pi}\arcsin((x^\star)^{-3/2})\,. \end{aligned} \tag{50}$$

The probabilities of transition are easily computed from (50). When δ decreases ($\dot\delta < 0$), the transitions are always from D_1 or D_3 towards D_2 (all transitions are towards the external domain). When δ increases ($\dot\delta > 0$), the transitions are from D_2 towards either D_1 or D_3 (all transitions come from the external domain). The probability of capture into resonance (transition from D_2 to D_3) is given by

$$P_c = 2\Bigg/\left[1+\frac{\pi}{2\arcsin((x^\star)^{-3/2})}\right] \tag{51}$$

a formula first proposed by Yoder (1973) and independently by Neishtadt (1975). The function (51) is illustrated in Fig. 15. For large values of δ this function behaves like

$$P_c \sim \frac{4}{\pi}\,[3(\delta+1)]^{-3/4} \,. \tag{52}$$

The other invariants of the transitions are computed similarly. We find:

$$\begin{aligned} h_1^\star = h_2^\star = h_3^\star &= 64(x^\star)^{-8}(x^{\star 3}-1)^3 \quad \text{(see (II.21))}, \\ g_1 = g_2 = g_3 &= 0 \qquad\qquad\qquad\quad \text{(see (II.31))}. \end{aligned} \tag{53}$$

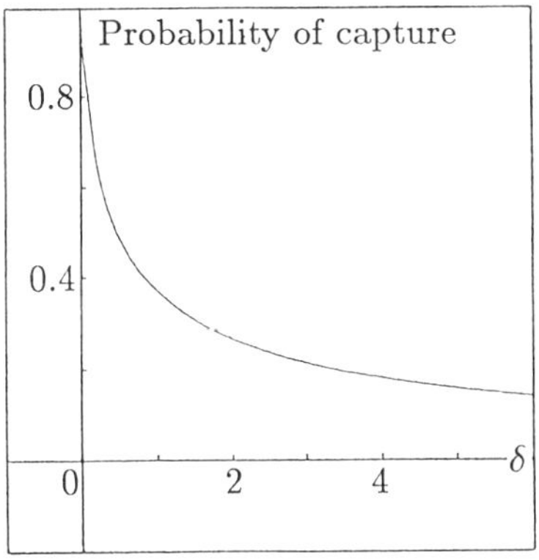

Fig. 15. Probability of capture into resonance when δ increases.

The principal terms in the diffusion parameter (see (II.97) for a jump from domain D_i to domain D_j is given by

$$\sigma_{ij} = -\dot{\delta} \log \dot{\delta} \frac{\sqrt{3}}{2} \frac{x^\star}{(x^{\star 3} - 1)^{1/2}} P_r(i,j) \tag{54}$$

where $P_r(i,j)$ is the probability function (see (II.72)) of the corresponding jump.

4. The Colombo's Top

In 1693, Cassini described the rotation of the Moon by his three famous empirical laws. The first one notes that the rotation rate is synchroneous with the orbital mean motion such that one side always faces the Earth. The other two describe the position of the spin axis of the Moon with respect to the normal to the orbit plane and the normal to the eccliptic plane.

Colombo (1966) showed that Cassini's second and third laws were generalizable to any axially symmetric planet with an arbitrary spin angular velocity. They correspond to an equilibrium (called Cassini's state by Peale, 1969) of the following dynamical system.

Consider an axially symmetric oblate planet, the spin axis of which is directed along its symmetry axis. The averaged effect of the Sun on its equatorial bulge can be modelized as the effect of a ring of material at distance r from the center of mass of the planet. Let us assume further that the plane of motion of the Sun (or equivalently the plane of the ring of material) is not fixed in space but precess uniformly. Such a dynamical system may be described by the Hamiltonian function

$$H = -\frac{1}{2}(Z-b)^2 - a\sqrt{1-Z^2} \cos h \tag{55}$$

where $h - \pi/2$ and $\arccos Z$ are the spherical coordinates of the angular momentum of the planet (in a frame moving with the ring of material). The parameters

a and b are given by

$$a = \frac{2}{3}\frac{\varrho\mu}{n^2}\frac{C}{C-A}\sin i \quad , \quad b = \frac{2}{3}\frac{\varrho\mu}{n^2}\frac{C}{C-A}\cos i \tag{56}$$

where ϱ is the spin angular velocity of the planet, μ the precession angular velocity of the ring and n the orbital angular velocity of the planet. $A = B < C$ are the principal moments of inertia of the planet, and i is the angle between the normal to the plane of the ring and the precession axis. More information about the geometry of the problem and the derivation of the Hamiltonian (55) can be found in Colombo (1966), Peale (1969) or Henrard (1987).

The parameters a and b have been defined in such a way that they are positive for most of the applications in the Solar System. If they were not, it would be easy to make them so either by a translation $(h \to h + \pi)$ which would reverse the sign of a, or by the involution $(h, Z) \to (-h, -Z)$ which reverses the sign of b. We shall thus assume in what follows that a and b are positive.

Observe also that, when a is small $(a = \varepsilon^{3/2}a')$, and b close to one $(b = 1 - b'\varepsilon)$, the dynamical system can be approximated in the neighborhood of $Z = 1$ (with $Z = 1 - \varepsilon I$) by

$$\frac{1}{\varepsilon^2}H = -\frac{1}{2}(I - b')^2 - a'\sqrt{2I}\cos h + 0(\varepsilon) \tag{57}$$

which corresponds to the second fundamental model we have described in the previous section.

The second fundamental model of resonance is a good model for physically interesting systems only when I is small enough, while the Colombo's top is a valid approximation for any values of Z in its domain of definition $(-1 \le Z \le 1)$. In this sense, the Colombo's top may be thought of as an extension of the second fundamental model on a nice compact manifold: the sphere.

Inded, the topology of the unit sphere is the proper topology associated with the phase space of the Colombo's top. If we introduce the cartesian coordinates of the unit vector along the angular momentum of the planet:

$$X = \sqrt{1 - Z^2}\sin h \quad , \quad Y = -\sqrt{1 - Z^2}\cos h \quad , \quad Z = Z \ , \tag{58}$$

the Hamiltonian (55) reads

$$H = -\frac{1}{2}(Z - b)^2 + aY \tag{59}$$

and it is then apparent that the curves $H =$ constant (which are the trajectories of the dynamical system described by (55) are the intersections of the parabolic cylinder (59) and the unit sphere in the three dimensional space (X, Y, Z).

The singularities $Z = \pm 1$ introduced in (55) by the use of spherical coordinates are removed in (59). As shown in Fig. 16 (A), as many as four parabolic cylinders of the family (59) can be tangent to the sphere. The points of tangency correspond to equilibria of the dynamical system.

When there are four equilibria, one of them is unstable as the intersection of the corresponding parabolic cylinder and the sphere are not reduced to the

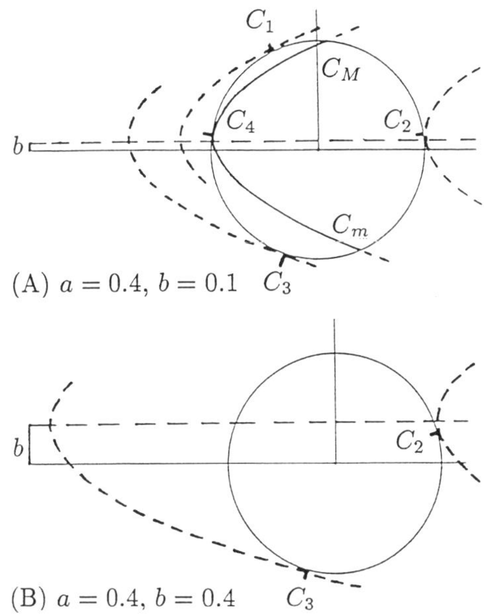

Fig. 16. The parabolic cylinders tangent to the sphere and the Cassini's states.

tangency point but contains the stable and unstable manifolds of the equilibrium which form two closed loops on the sphere, the homoclinic orbits.

The location of the equilibria, called "Cassini's states" by Peale (1969) and numbered 1 to 4 as in Fig. 16, are given by

$$X_i = 0 \quad , \quad Y_i = \frac{-aZ_i}{(Z_i - b)} \tag{60}$$

and Z_i, root of

$$(Z_i - b)^2(1 - Z_i^2) = a^2 Z_i^2 \; . \tag{61}$$

One checks easily that the fourth degree polynomial (61) possesses a double root $Z_D = Z_1 = Z_4$ (i.e. $C_1 = C_4$ in Fig. 16 (A)) whenever a and b are such that

$$a = -Y_D^3 \quad , \quad b = Z_D^3 \; . \tag{62}$$

Hence the Colombo's top possesses four Cassini's states whenever

$$a^{2/3} + b^{2/3} < 1 \tag{63}$$

and only two for larger values of a and b as shown in Fig. 16 (B).

We show in Fig. 17 the projections of the trajectories on the plane (Y, Z) for various values of the parameters a and b. Also shown in Fig. 18 are three other projections of the trajectories for one typical set of values of these parameters.

When a and b are small enough (see condition (63)), the two homoclinic orbits associated with the unstable Cassini's state (no. 4) divide the sphere into

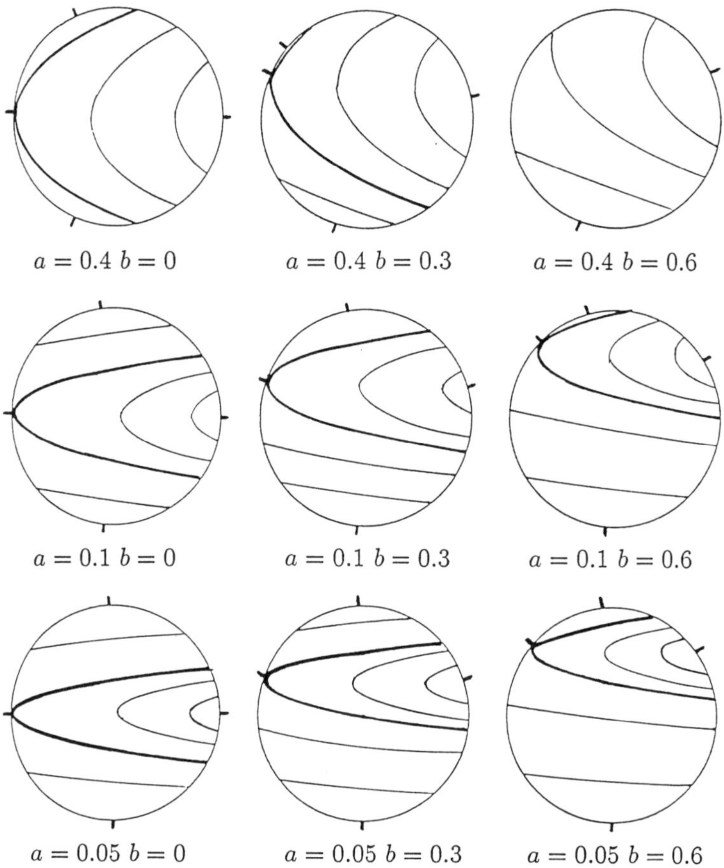

Fig. 17. Projection upon the plane (Y, Z) of the trajectories of the dynamical system (55) for various values of the parameters.

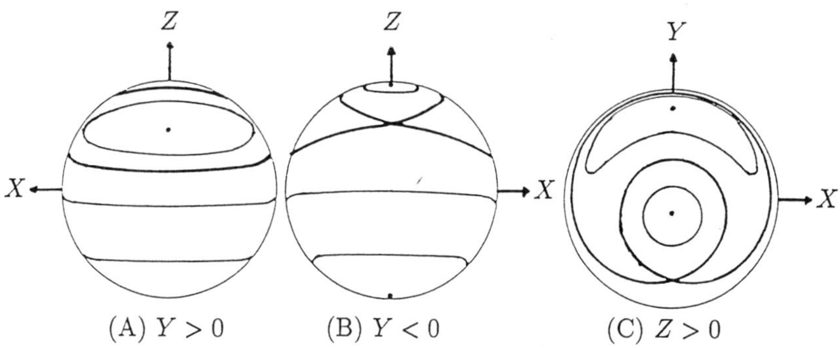

Fig. 18. Different projections of the last case ($a = 0.05$, $b = 0.06$) of the Fig. 16: (a) Projection upon the plane (X, Z) of the half sphere $Y > 0$. (b) Projection upon the plane (X, Z) of the half sphere $Y < 0$. (c) Projection upon the plane (X, Y) of the half sphere $Z > 0$.

three domains, each of them associated with one of the stable Cassini's states (see Fig. 19).

Notice that the numbering of the domains and the numbering of the Cassini's states they contain, do not correspond. This comes from a conflict of "traditions": The traditional numbering of the Cassini's states and the numbering we have adopted here of the domains of definition of the action-angle variables (number 3 being reserved for the domain touching both homoclinic orbits).

When we linearize the dynamical system described by (55) in the vicinity of the unstable equilibrium (located at $X^\star = 0, Y^\star, Z^\star$), we find that it corresponds to the Hamiltonian function

$$H_L = -\frac{Y^{\star 3} + a}{2Y^{\star 3}}(\delta Z)^2 - \frac{Y^\star a}{2}(\delta h)^2 \tag{64}$$

from which one finds that the time scale of the unstable equilibrium is

$$\omega = \left[-\frac{a(Y^{\star 3} + a)}{Y^{\star 2}}\right]^{1/2} . \tag{65}$$

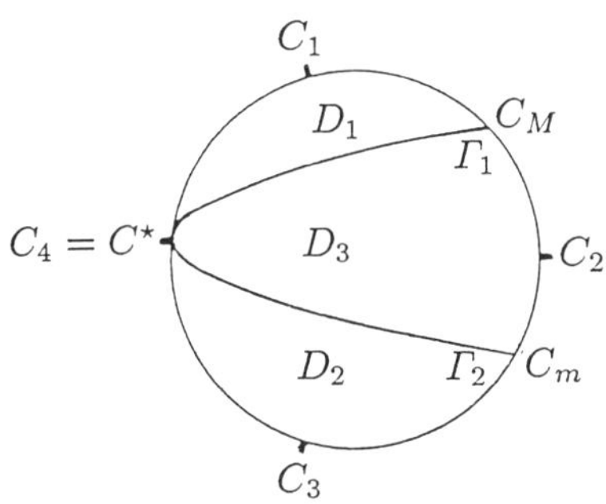

Fig. 19. The three domains defined on the sphere by the homoclinic orbits Γ_1 and Γ_2.

The homoclinic orbits cross the plane $X = 0$ at the points C_M and C_m (see Fig. 19) the location of which are given by

$$\begin{aligned} Z_m &= \frac{Z^\star}{Y^\star}(Y^\star + 2a) - 2[-a(Y^\star + a)]^{1/2} , \\ Z_M &= \frac{Z^\star}{Y^\star}(Y^\star + 2a) + 2[-a(Y^\star + a)]^{1/2} . \end{aligned} \tag{66}$$

The critical values of the action variables $J_i^\star$ are given by

$$J_i^\star = \frac{1}{2\pi}\oint Z\, dh \tag{67}$$

where the path integrals are taken along the homoclinic orbits Γ_i.

They are related to the (unsigned) areas A_i of the three domains D_i by

$$\begin{aligned} A_1 &= 2\pi(1 + J_1^\star) , \\ A_2 &= 2\pi(1 + J_2^\star) , \\ A_3 &= -2\pi J_3^\star = -2\pi(J_1^\star + J_2^\star) . \end{aligned} \tag{68}$$

Transforming the integrals in dh into integrals in dZ as it was done in the previous section, leads, after a considerable amount of algebra, to the following:

$$J_1^\star = -b + J_3^\star \;/\; 2 \quad , \quad J_2^\star = b + J_3^\star \;/\; 2 \tag{69}$$

and

$$J_3^\star = \frac{4b}{\pi} \arcsin S - \frac{4}{\pi}\omega - \frac{2}{\pi} \arctan T \tag{70}$$

with

$$S = \left[-\frac{aZ^{\star 3}}{bY^{\star 3}} \right]^{1/2} \quad , \quad T = \frac{2\omega Y^\star}{Y^{\star 3} + 2a} \tag{71}$$

and ω is the time constant at the unstable equilibrium given by (65). The value of "arctan" in (70) should be taken between 0 and π. Some details of the computation are given in (Henrard and Murigande, 1987).

It is remarkable that the derivatives of the critical area $J_3^\star$ with respect to the two parameters are very simple. After some algebra, we find

$$\frac{\partial J_3^\star}{\partial a} = -\frac{4\omega}{\pi a} \quad , \quad \frac{\partial J_3^\star}{\partial b} = \frac{4}{\pi} \arcsin S \; . \tag{72}$$

The other invariants of the transitions are computed similarly. We find:

$$h_1^\star = h_2^\star = h_3^\star = -32 \tfrac{a(Y^{\star 3} + a)^3}{Y^{\star 7}(Y^\star + a)} \quad \text{see (II.21)}, \tag{73}$$

$$g_1 = g_2 = g_3 = 0 \qquad \text{see (II.31)}. \tag{74}$$

It can be checked that, at the limit for a small and b close to one, these quantities tend to the equivalent invariants of the second fundamental model of resonance (see (53)).

5. Dissipative Forces

In some problems of interest, the slow dependence upon the time of the dynamical system cannot be directly modelized by a slowly varying parameter but rather by the addition of a small non-conservative force. Let the dynamical system be described by the differential equations

$$\dot{q} = \frac{\partial H}{\partial p} + \varepsilon f_1(q,p) \quad , \quad \dot{p} = -\frac{\partial H}{\partial q} + \varepsilon f_2(q,p) \tag{75}$$

where $H(q,p)$ is the Hamiltonian function of the unperturbed (or "frozen") system and $(\varepsilon f_1, \varepsilon f_2)$ the small non-conservative force.

The following trick can reduce the dynamical system described by (75) to an Hamiltonian system with slowly varying parameter.

Let the functions

$$\tilde{q}(Q,P,\varepsilon t) \quad , \quad \tilde{p}(Q,P,\varepsilon t) \tag{76}$$

be defined as the general solution (with the initial conditions (Q,P)) of the auxiliary system of differential equations

$$\frac{d}{dt}\tilde{q} = \varepsilon f_1(\tilde{q},\tilde{p}) \quad , \quad \frac{d}{dt}\tilde{p} = \varepsilon f_2(\tilde{q},\tilde{p}) \tag{77}$$

and let us define the new time variable τ by

$$\frac{dt}{d\tau} = \left[\frac{\partial\tilde{q}}{\partial Q}\frac{\partial\tilde{p}}{\partial P} - \frac{\partial\tilde{q}}{\partial P}\frac{\partial\tilde{p}}{\partial Q}\right] = W(Q,P,\varepsilon t)\ . \tag{78}$$

If we interpret (76) as a change of variables from (q,p) to (Q,P) by means of $q=\tilde{q}$ and $p=\tilde{p}$, the differential equations corresponding to (75) reads

$$\frac{d}{d\tau}Q = \frac{\partial}{\partial P}K(Q,P,\varepsilon t)\ , \quad \frac{d}{d\tau}P = -\frac{\partial}{\partial Q}K(Q,P,\varepsilon t) \tag{79}$$

with

$$K(Q,P,\varepsilon t) = H(\tilde{q}(Q,P,\varepsilon t),\tilde{p}(Q,P,\varepsilon t))\ . \tag{80}$$

The change of variables and of time scale has effectively reduced the system (75) to an Hamiltonian system, the Hamiltonian function of which is (80).

Of course, the time transformation (78) has to be evaluated along a trajectory of (79) which means that the parameter εt in (80) may not have its second derivative (with respect to the new time τ) of the order ε^2. We have actually

$$\frac{d^2}{d\tau^2}(\varepsilon t) = \varepsilon\left\{\frac{\partial W}{\partial Q}\frac{\partial K}{\partial P} - \frac{\partial W}{\partial P}\frac{\partial K}{\partial Q}\right\} + \varepsilon^2 W^2\left[\frac{\partial f_1}{\partial\tilde{q}} + \frac{\partial f_2}{\partial\tilde{p}}\right]\ . \tag{81}$$

When the additive forces (f_1,f_2) are linear (as it is the case in (15)) such that

$$\begin{pmatrix} f_1 \\ f_2 \end{pmatrix} = \begin{pmatrix} b_1 \\ b_2 \end{pmatrix} + A\begin{pmatrix} q \\ p \end{pmatrix}\ , \tag{82}$$

the wronskian $W(Q,P,\varepsilon t)$ in (78) is given by

$$W(Q,P,\varepsilon t) = \exp\{\mathrm{trace}(\varepsilon A t)\} \tag{83}$$

and does not depend upon the variables (Q,P). Hence the second derivative (81) is proportional to ε^2 and, for that matter, the n-th derivative $d^n(\varepsilon t)/d\tau^n$ is proportional to ε^n (as assumed in (I.19)). The change of time variable (78) does not preclude the application of the adiabatic invariant theory to the Hamiltonian function (80). Furthermore, the change of variables (76) is readily computed as being

$$\begin{pmatrix} \tilde{q} \\ \tilde{p} \end{pmatrix} = e^{\varepsilon A t}\begin{pmatrix} Q \\ P \end{pmatrix} + (e^{\varepsilon A t} - 1)A^{-1}\begin{pmatrix} b_1 \\ b_2 \end{pmatrix}\ . \tag{84}$$

For the application of the classical adiabatic invariant to such problems, it may be simpler to shortcut the change of variables (84) and the time transformation (78) in the following way.

Comparing the action variable corresponding to the Hamiltonian $H(q, p)$,

$$J_H = \frac{1}{4\pi} \oint p \, dq - q \, dp \tag{85}$$

and the action variable corresponding to the transformed Hamiltonian (80),

$$J_K = \frac{1}{4\pi} \oint P \, dQ - Q \, dP \, , \tag{86}$$

we find, taking into account that the curves (along which the path integrals are computed) are closed, that

$$J_H = \det(e^{\varepsilon At}) J_K \, . \tag{87}$$

Indeed, εt should be considered as constant along these path integrals defining the actions.

When the dissipative forces (f_1, f_2) are taken into account, the action variable J_K remains on the mean constant (see Sect. I.4) and the action variable J_H changes exponentially

$$J_H(t) = \exp\{\text{trace}(\varepsilon At)\} J_H(0) \, . \tag{88}$$

Part IV Applications

1. Introduction

The history of the development of the Adiabatic Invariant Theory is witness to the fact that it has applications in many fields of Applied Mathematics. Obviously, we shall not be able to cover any one of them in any depth. This chapter should thus be taken more as a series of suggestions about the type of problems that can be approached and about the type of answers that can be expected rather than a review of the applications of the Adiabatic Invariant Theory.

Systems with slowly varying parameters arise quite naturally in Mechanical or Electronical Engineering. Bogoliubov and Mitropolsky (1961) make a point in applying their method of perturbation to such systems. Stoker (1950) and Andronov et al. (1966) describe in detail the jumps and hysteresis phenomena stemming from the slow sweep of a system through a resonance. Such phenomena are used in some mechanical or electronical systems to produce desired effects or should be carefully analysed in other systems to prevent, by design, undesired effects.

Most of the literature on this topic does not use as such the adiabatic invariant theory but especially designed perturbation theories (Bogoliubov and Mitropolsky, 1961) or an heuristic approach (Stoker 1950).

In Sect. 2, we apply the adiabatic invariant theory to a well-known problem in the theory of oscillation: the Duffing's equation. The aim of this application if to show on a simple example that a qualitative and quantitative description of the jump can be developed as a direct application of the adiabatic invariant theory. Furthermore, the adiabatic invariant theory gives information about the behaviour of the free oscillations and thus the transient (for damped systems) of the solution during its passage through the resonance. This transient is usually not analysed although it may be of interest in some applications.

The motion of a charged particle in an electromagnetic field is, of course, a privileged topic for the applications of the adiabatic invariant theory. After all, as we have already mentioned, the theory itself was mostly developed with these applications in mind: acceleration of cosmic rays (Alfven, 1950, and Helwig, 1955), energy and momentum balance for waves in collisionless plasmas (Best, 1968), containment of plasma (Kruskal, 1952, 1957, 1962), in mirror machines (Aamodt, 1971–1972) or Toroïdal devices (Dobrott and Green 1971), high energy accelerators and colliding beam storage rings (Tennyson, 1979), ...

To represent this kind of applications, we have chosen two particular topics. The first one considers a simple model of particle motion in a slowly modulated field and aims at describing the slow chaotic motion which can be generated by repeated separatrix crossings. It is, in some sense, an "academic" application but it connects nicely the adiabatic invariant theory with the large and fast growing literature on chaotic motions in Hamiltonian systems. We follow mostly the work of Escande (1987) but we are able to add to his description an evaluation of the diffusion coefficient (see (35)).

In the second application, we show how the adiabatic invariance of the magnetic momentum is at the root of the analysis of containment devices for plasmas (be it the magnetic field of the Earth for the Van Allen belts or the magnetic bottle for controlling hot plasmas). We have described in some detail how the "guiding center" coordinates can be introduced naturally by a canonical transformation simplifying the zero-th order Hamiltonian. This may be of interest in itself as the guiding center coordinates are usually introduced in a non-canonical way (e.g. Littlejohn, 1979) prohibiting the use of Hamiltonian perturbation methods.

In the last two sections, we come to the applications of the adiabatic invariant theory to Celestial Mechanics. These applications are witnesses of a rather new and exciting type of problems in this field: evolutionary problems in the dynamic of the Solar System.

The analysis of these problems has not been connected straight away with the adiabatic invariant theory. Hence the reader will not find any mention of adiabatic invariants in a large part of the literature cited. It is not difficult however to see that the basic phenomena described there are easily followed and understood with the help of the adiabatic invariant theory (see Peale, 1986).

For obvious reasons, we content ourselves in all these applications with very simple minded physical models which are nicely described by one of the "paradigms" we have analysed in the previous chapter.

This does not mean that the adiabatic invariant theory should be confined to the analysis of such paradigms, giving only qualitative answers to real world problems for which they are only a qualitative approximation.

We have shown in Sect. I.5 how the angle-action variables can be defined by explicit quadratures that can be easily implemented numerically. Hence, by mixing analytical and numerical approaches, the adiabatic invariant theory can produce quantitative answers in more realistic problems for which no analytical expression can be derived for quantities such as the critical values of the adiabatic invariant (see, for instance, Henrard-Lemaître, 1986b, and Lemaître and Henrard, 1988).

2. Passage Through Resonance of a Forced Anharmonic Oscillator

Let us consider the forced oscillations of a pendulum described by the differential equation

$$\ddot{\theta} + \sin\theta = \mu \sin\omega t \; . \tag{1}$$

The frequency ω of the forcing term passes slowly through the resonance value $\omega = 1$

$$\omega = 1 - \varepsilon t \tag{2}$$

We shall consider that the amplitude μ of the forcing term is small and study the small oscillations with $\theta = \eta x$ where η is a small parameter. Expanding $\sin\theta$ in powers of η and keeping only the first two terms, we obtain the undamped Duffing's equation:

$$\ddot{x} + x = \frac{\eta^2}{3!}x^3 + \frac{\mu}{\eta}\sin\omega t \; . \tag{3}$$

The corresponding Hamiltonian reads:

$$H = \frac{1}{2}(P_x^2 + x^2) - \frac{\eta^2}{4!}x^4 - \frac{\mu}{\eta}x\sin\omega t \; .$$

Introducing the action-angle variables for the harmonic oscillator,

$$x = \sqrt{2I}\sin\phi \quad , \quad P_x = \sqrt{2I}\cos\phi \; , \tag{4}$$

we obtain

$$H = I - \frac{\eta^2}{3!}I^2\sin^4\phi - \frac{\mu}{2\eta}\sqrt{2I}\{\cos(\phi - \omega t) - \cos(\phi + \omega t)\} \; . \tag{5}$$

The angular variable $\phi - \omega t$ has a small frequency. We make it stand forward by the canonical transformation

$$\psi = \phi - \omega t \quad , \quad I = J \; , \tag{6}$$

the remainder function of which is $R = -J(1 - 2\varepsilon t)$.

The new Hamiltonian function K is then

$$\begin{aligned} K = (2\varepsilon t)J - \frac{\eta^2}{16}J^2 - \frac{\mu}{2\eta}\sqrt{2J}\cos\psi \\ + \frac{\eta^2}{12}J^2\cos 2(\psi + \omega t) - \frac{\eta^2}{48}J^2\cos 4(\psi + \omega t) \\ + \frac{\mu}{2\eta}\sqrt{2J}\cos(\psi + 2\omega t)\ . \end{aligned} \tag{7}$$

If we take $\eta \sim \mu^{1/3}$, the last two lines are small short periodic perturbations which can be averaged to yield an averaged Hamiltonian function

$$\begin{aligned} \bar{K} = (2\varepsilon t)J - \frac{\eta^2}{16}J^2 - \frac{\mu}{2\eta}\sqrt{2J}\cos\psi \\ - \frac{1}{1 + 2\varepsilon t}\left\{\frac{17}{2304}\eta^4 J^3 + \frac{1}{32}\eta\mu J\sqrt{2J}\cos\psi\right\} \\ + 0(\eta^6)\ . \end{aligned} \tag{8}$$

The action-angle variables (J, ψ) in (8) are the averaged action-angle variables and should be written $(\bar{J}, \bar{\psi})$. We spare ourselves this burden remarking that the differences $J - \bar{J}$ and $\psi - \bar{\psi}$ are small (η^2) and short periodic.

The first line of (8) reproduces the "second fundamental model" we have analysed in Sect. III.3. The second line represents a small correction to this "paradigm" no larger than the errors we have already accepted by taking Duffing's equation rather than the pendulum itself.

We shall thus drop this second line and deal with the second fundamental model by introducing the proper scaling:

$$R = \left(\frac{\eta^3}{4\mu}\right)^{2/3} J \quad , \quad r = \pi - \psi \quad , \quad \tau = \frac{1}{4}\left(\frac{\mu}{2}\right)^{2/3} t\ . \tag{9}$$

With respect to the new time τ, the Hamiltonian is now:

$$K' = -3(\delta + 1)R + R^2 - 2\sqrt{2R}\cos r \tag{10}$$

with the parameter δ given by

$$(\delta + 1) = \frac{32}{3}\varepsilon t \left(\frac{\mu}{2}\right)^{-4/3}\ . \tag{11}$$

The derivative of δ with respect to the new time τ will thus be small if $|\varepsilon| << \mu^{4/3}$.

We are now in a position to use the description of the second fundamental model developed in Sect. III.3.

We shall first assume that well before the passage through resonance (for $\delta + 1$ large in absolute value), the energy of the oscillator is at its minimum; any free oscillation has been damped out by some process.

If $\varepsilon < 0$ (when the frequency of the forcing term is increasing), we start in region D_1 (see Fig. 13) at the equilibrium $x_2^\star$. The value of the adiabatic invariant is $J_1 = 0$.

When t increases and reaches the value $t = 0$ (at the exact resonance), the critical area $2\pi|J_1^\star|$ of domain D_1 decreases and goes to zero. The oscillator experiences a transition to the domain D_2, and a jump in his adiabatic invariant from zero to $J_2^\star = 3$.

Later on, when t is large enough that $(\delta + 1)$ is large in absolute value again but negative, the oscillation is around the equilibrium $x_1^\star$ (very close to zero – see Figs. 11 or 14) with an amplitude in R of about 3.

Indeed the solution curve is almost exactly a circle the area of which is $2\pi J_2 = \pi(2R)$.

Hence the amplitude of oscillation in θ is

$$\theta_{\max} = \eta x_{\max} = 2\sqrt{6}\left(\frac{\mu}{2}\right)^{1/3} = 3.89\mu^{1/3} \,. \tag{12}$$

The passage through the resonance excites a *free oscillation* the amplitude of which is constant and given by (12).

On the other hand, if the passage through the resonance goes the other way, with ε positive, (the frequency of the forcing term is decreasing), we start at the equilibrium $x_1^\star$ for $(\delta + 1)$ large and negative. When we reach $\delta = 0$, the homoclinic orbits appear, divide the phase space into three domains, but we know, as we are following the equilibrium $x_1^\star$, that we enter domain D_3 and this without a transition. The adiabatic invariant remains at its original value $J_3 = 0$ and the oscillator experiences a *forced* oscillation corresponding to the position of $x_1^\star$.

When $(\delta + 1)$ becomes large, the position of $x_1^\star$ is approximately $x_1^\star \sim [3(\delta + 1)]^{1/2}$ and the amplitude of the forced oscillation is

$$\theta_{\max} \simeq 4[2\varepsilon t]^{1/2} = 4[2(1-\omega)]^{1/2} \;; \tag{13}$$

of course this estimates is valid only while the oscillations remain small.

The passage through resonance has excited a *forced oscillation* the amplitude of which is increasing with time.

This much of the analysis could have been gathered from an intelligent interpretation of the classical analysis of the stable periodic orbits of the Duffing's equation in function of the parameter ω. These periodic orbits correspond to the equilibrium points $x_1^\star$, $x_2^\star$ of the second fundamental model (see Fig. 13). But we can go further.

Let us assume, for instance, that ε is positive but that well before the passage through resonance, the free oscillation is not vanishing and has an amplitude $\theta_{\max}$ larger than $3.89\mu^{1/3}$. The adiabatic invariant J is then larger than 3 and, when the oscillator reaches the critical value $\delta = 0$, we enter the external domain D_2. Later on, for $\delta > 0$, the critical area $2\pi J_2^\star$ increases and there will come a time when $J = J_2^\star$. At that time, the oscillator will experience a transition either towards the resonance zone D_3 or the internal zone D_1. The probabilities of such transition can be evaluated as function of δ_T (value of δ at the transition) or equivalently as function of $\theta_{\max}$ far from resonance. Indeed, δ_T is a function of J through $J = J_2^\star(\delta_T)$ and J is a function of $\theta_{\max}$.

Let us concentrate on the case of a transition from D_2 to D_1 corresponding to cases where the oscillator does not lock into resonance. The probability of such an event is given in Fig. 20. In this case, the adiabatic invariant experiences a jump from $J = J_2^\star(\delta_T)$ to $J_1^\star(\delta_T)$ which is smaller.

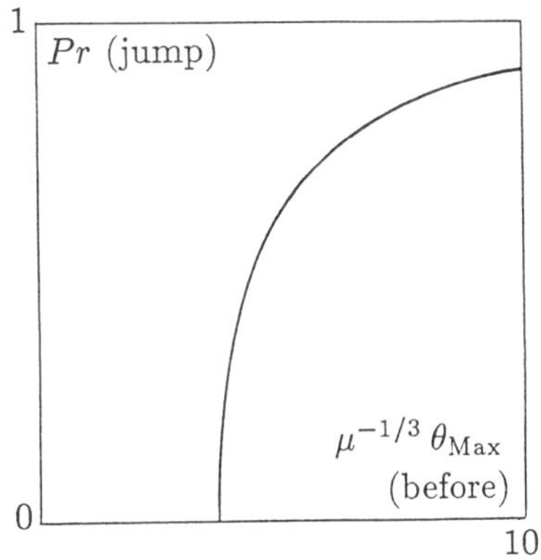

Fig. 20. Probability of transition without locking in resonance as a function of the amplitude of free oscillation before the resonance.

Later on, when δ is large and positive, the oscillation will be around the equilibrium $x_2^\star$ (very close to zero, see Figs. 13 or 14) with an amplitude in R of about $J_1^\star(\delta_T)$. Indeed, the solution curve will be almost exactly a circle the area of which is $2\pi J_1^\star(\delta_T) = \pi(2R)$.

Hence the amplitude of *free oscillation* in θ is given by

$$\theta_{\max}\ (\text{after jump}) = 2\left(\frac{\mu}{2}\right)^{1/3}\sqrt{2J_1^\star(\delta_T)} \tag{14}$$

which is smaller than $\theta_{\max}$ (before jump) given by:

$$\theta_{\max}\ (\text{before jump}) = 2\left(\frac{\mu}{2}\right)^{1/3}\sqrt{2J_2^\star(\delta_T)}\,. \tag{15}$$

We arrive thus at the conclusion that this kind of passage through resonance actually dampens the free oscillation whenever the oscillator is not captured by the resonance. The damping factor is a function of $\theta_{\max}$ (before jump) and is illustrated in Fig. 21.

We have analysed here the forced resonant oscillation of an undamped pendulum because it leads directly to the "second fundamental model" by means of the averaged undamped Duffing's equation, and we could then use the formulae developed in the preceeding section. Let us remark that a linearly damped pendulum could be analysed in much the same way by using the "trick" developed in Sect. III.5.

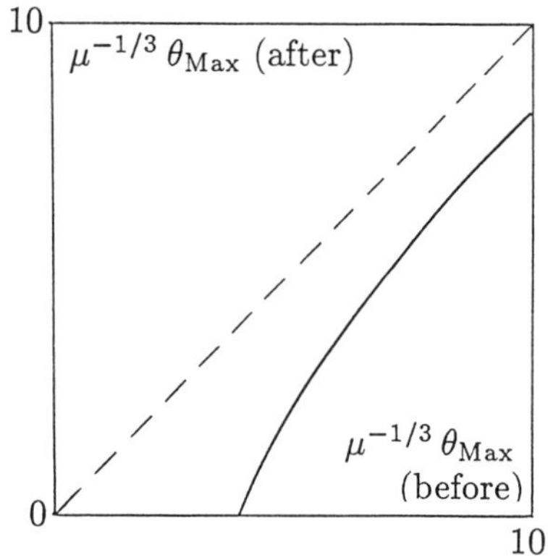

Fig. 21. Amplitude of free oscillation after the passage through resonance as a function of the amplitude of free oscillation before the passage through resonance.

3. Particle Motion in a Slowly Modulated Wave

We consider here the motion of a single charged particle in the field of a modulated electrostatic wave. Setting the particle charge, particle mass and wave number all equal to 1, we find that the system is described by the Hamiltonian

$$H = \frac{1}{2}I^2 - b(\varepsilon t)\cos(\phi - d(\varepsilon t)) \tag{16}$$

where b is the wave amplitude, d the wave phase and ε is a parameter indicating the time scale on which the modulation occurs. The Hamiltonian (16) is equivalent to within a scaling of the time to the Hamiltonian

$$H = \frac{1}{2}I^2 - \frac{b(\tau)}{\varepsilon^2}\cos(\phi - d(\tau))\ . \tag{17}$$

This makes it apparent that a large modulation time (ε small in (16)) is equivalent to a large amplitude of the wave (ε small in (17)).

The Hamiltonian (16) is also representative of the two-mode system described by

$$H = \frac{1}{2}I^2 - a_1\cos(\phi - \alpha(\varepsilon t)) - a_2\cos(\phi + \beta(\varepsilon t)) \tag{18}$$

with

$$\begin{aligned} b(\varepsilon t) &= [a_1^2 + a_2^2 + 2a_1a_2\cos(\alpha+\beta)]^{1/2}\ , \\ \tan d(\varepsilon t) &= \frac{a_1\sin\alpha - a_2\sin\beta}{a_1\cos\alpha + a_2\cos\beta}\ . \end{aligned} \tag{19}$$

This form with $\alpha = \beta = \varepsilon t$, studied for instance by Menyuk (1985), makes more apparent the relationship of our problem with the problem of the interaction of two resonances which has been much studied in relation with chaotic motion (see, for instance, Lichtenberg and Lieberman (1983) and Escande (1985)).

When ε is large, the system (18) with $\alpha = \beta = \varepsilon t$ has two resonances centered at $I = \pm\varepsilon$ with widths $2a_1^{1/2}$ and $2a_2^{1/2}$ respectively. As ε becomes smaller, these two resonances move closer and eventually overlap resulting in chaotic motion. As ε continues to decrease, the orbits do not become increasingly chaotic. The two resonances are almost superposed. The slow change in phase between them induces a very slow stochasticity.

As discussed in Sect. III.2, we shall study the system (16) under its avatar described by the Hamiltonian

$$H = \frac{1}{2}(I - c)^2 + 2b\sin^2\frac{\phi}{2} \tag{20}$$

with $c(\varepsilon t) = \dot{d}$ and, of course, ε taken as a small parameter.

We shall assume that $\alpha = \beta = \varepsilon t$ to recover Menyuk results. With this assumption, c is of the order ot ε and $\dot{c}$ of the order of ε^2. As our analysis is of first order in ε, it is not usefull to keep track of the contribution of $\dot{c}$ and we may use formulae corresponding to the symmetric case ($c = 0$).

The first problem we shall address is to evaluate the region of phase space where we can expect the diffusion due to separatrix-crossing to take place.

In the libration domain, this is the region for which

$$J_3(\mu) \geq \min_{-1\leq\mu\leq 1} J_3^\star(\mu) \tag{21}$$

where we have taken $\mu = \cos 2\varepsilon t = \cos 2\lambda$. This corresponds to initial conditions such that the value h of the Hamiltonian (20) is larger than the root $h^\star$ of

$$\sqrt{b}\{(\alpha - 1)\mathbf{K}(\alpha) + \mathbf{E}(\alpha)\} = |a_1 - a_2|^{1/2} \tag{22}$$

where $\alpha = h^\star/2b$.

In the circulation domains D_i ($1 \leq i \leq 2$), this is the region for which

$$J_i(\mu) \leq \max_{-1\leq\mu\leq 1} J_i^\star(\mu) \tag{23}$$

and corresponds to the initial conditions such that the value of h of the Hamiltonian is smaller than the root $h^\star$ of

$$\sqrt{b/\beta}\,\mathbf{E}(\beta) = |a_1 + a_2|^{1/2} \tag{24}$$

where $\beta = 2b/h^\star$.

These regions are shown shaded in Fig. 22 on three different surfaces of section ($\mu = -1, 0, 1$) and for two different choices of the parameters a_i ($a_1 = a_2 = 0.5$, $a_1 = 3a_2 = 0.75$).

Remark that obviously the shape of the shaded domain depends upon the surface of section (the value of λ at which the section is taken) but, according to the definitions (21) and (23), the area of the shaded chaotic domain remains invariant from section to section.

Another point of interest concerns the diffusion coefficient associated with the chaotic motion, i.e. the coefficient $D(\bar{J})$ in the equation:

$$< (\Delta\bar{J}^2) >= D(\bar{J})t \tag{25}$$

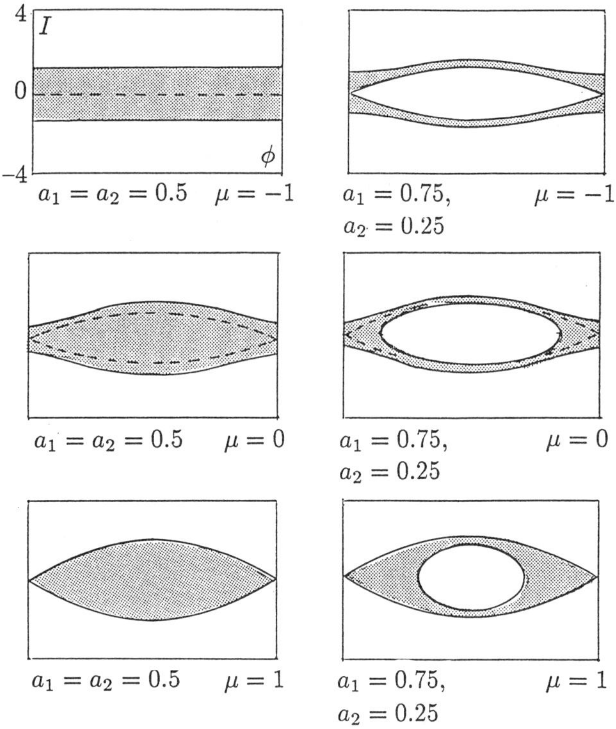

Fig. 22. Initial conditions which leads to a crossing of the separatrix.

which yields the root mean square change of $\bar{J}$ in its random walk due to separatrix crossing. Menyuk (1985) estimates it as scaling as ε^3. We shall see that we can give a more precise estimate for this quantity and show its dependence upon $\bar{J}$.

As the motion jumps from domain to domain and as the adiabatic invariant $\bar{J}$ has a different definition in each domain, we have first to decide in which of the domains D_i we shall estimate the variations $\Delta\bar{J}$. We choose the domain D_3 as we are sure that we always come back to it. Indeed, if the trajectory is initially in D_3, it will jump in domain D_1 or D_2 with equal probability and then come back to D_3 as shown in Fig. 10 for $C = 0$, and B alternatively negative and positive.

The first jump from D_3 to D_1 or D_2 leads to a value of $\bar{J}_i$ $(1 \le i \le 2)$ which, according to (II.92), is equal to

$$\bar{J}_i = J_i^\star(\mu_1) + \left(\frac{d\mu}{d\lambda}\right)_{\lambda_1} \left[\frac{\varepsilon}{\omega}\frac{\partial J_i^\star}{\partial\mu} G_{3i}(z_1,\mu)\right]_{\mu=\mu_1} \quad ; \tag{26}$$

we have taken into account that $g_i = g_3 = 0$ (see (III.25)). The quantity z_1 is a random variable (see Sect. II.6) the distribution of which is uniform in the unit interval $(0 \le z_1 \le 1)$. The quantity λ_1 is the pseudo crossing time from D_3 or D_2 (see (II.78)) and $\mu_1 = \cos 2\lambda_1$.

The jump back from D_1 (or D_2) to D_3 gives a new value to $\bar{J}_3$ equal to

$$\bar{J}_{3,\text{after}} = J_3^\star(\mu_2) + \left(\frac{d\mu}{d\lambda}\right)_{\lambda_2} \left[\frac{\varepsilon}{\omega}\frac{\partial J_3^\star}{\partial\mu} G_{i3}(z_2,\mu)\right]_{\mu=\mu_2} . \tag{27}$$

This time λ_2 and μ_2 refer to the pseudo crossing time of the second jump. The quantity z_2 is again a random variable the distribution of which is uniform in the unit interval and we shall assume that, it is independent of z_1. This assumption is equivalent to the "phase mixing" argument by which one derives that the phase of the system just before λ_2 (which is directly connected to z_2 - see Cary et al., 1986, for a more complete discussion of this fact) is so wild a function of the phase of the system just after λ_1 that these phases are essentially uncorrelated. The same argument justifies the use of "pseudo-random number generators" on a computer.

The definition of the pseudo-time of transition (see (II.78)) implies that

$$\bar{J}_i = J_2^\star(\mu_2) . \tag{28}$$

From (28) and (27), we can estimate the difference $\mu_2 - \mu_1$ which is a small quantity

$$\mu_2 - \mu_1 = \left(\frac{d\mu}{d\lambda}\right)_{\lambda_1} \left[\frac{\varepsilon}{\omega} G_{3i}(z_1,\mu)\right]_{\mu=\mu_1} . \tag{29}$$

From this estimate and from the fact that

$$\left(\frac{d\mu}{d\lambda}\right)_{\lambda_1} = -\left(\frac{d\mu}{d\lambda}\right)_{\lambda_2} , \tag{30}$$

because the direction of evolution of $J_3^\star$ is reversed between the two jumps, we conclude that the effect of the double jump from D_3 to D_1 or D_2 and then back to D_3 is given by

$$\begin{aligned} \Delta\bar{J}_3 &= \bar{J}_{3\text{after}} - \bar{J}_{3\text{before}} \\ &= \frac{\varepsilon}{\omega}\left(\frac{\partial J_3^\star}{\partial\lambda}\right)[G_{3i}(z_1,\mu) - G_{i3}(z_2,\mu)] \end{aligned} \tag{31}$$

where the right-hand member is evaluated at $\lambda = \lambda_1$, $\mu = \mu_1$.

The mean value of $\Delta\bar{J}_3$ is zero because the mean values of $G_{3i}(z_1)$ and $G_{i3}(z_2)$ are zero (see (II.96)) and the root mean square of $\Delta\bar{J}_3$ is given by

$$\sigma(\Delta J) = \sqrt{2/3}\left|\frac{2\dot{b}}{b}\right|_{\lambda_1} \tag{32}$$

(see (II.103) and (III.30)).

Taking into account the special form of b (see (19)) and the fact that the pseudo crossing time λ_1 is a function of the initial value of $\bar{J}_3$, we find, after some algebra, that

$$\sigma(\Delta J) = 2\varepsilon\sqrt{2/3}[J_M^\star - \bar{J}_3]^{1/2}[\bar{J}_3 - J_m^\star]^{1/2} \,/\, \bar{J}_3 \tag{33}$$

where $J_M^\star$ (resp. $J_m^\star$) is the maximum (resp. minimum) value of $J_3^\star$ given by

$$J_M^\star = \frac{8|a_1 + a_2|^{1/2}}{\pi} \quad , \quad J_m^\star = \frac{8|a_1 - a_2|^{1/2}}{\pi} \, . \tag{34}$$

As there are ε/π double transitions in the unit of time (because the full period of the separatrix sweeping is π/ε), the diffusion coefficient D in (25) is equal to

$$D(\bar{J}) = \frac{8}{3\pi}\varepsilon^3[J_M^\star - \bar{J}_3][\bar{J}_3 - J_m^\star] \,/\, \bar{J}_3^2 \, . \tag{35}$$

The diffusion coefficient is thus proportional to ε^3 as estimated by Menyuk (1985). Furthermore, it goes to zero close to the boundaries of the chaotic domain (the sticking phenomenon) as $\bar{J}_3$ is then close to either $J_m^\star$ of $J_M^\star$ and it reaches a maximum value of

$$\max(D(\bar{J})) = \varepsilon^3 \frac{8}{3\pi} \frac{(J_M^\star - J_m^\star)^2}{(J_M^\star + J_m^\star)^2} \tag{36}$$

for $\bar{J}_3 = (J_M^\star + J_m^\star)/2$.

We have used the words "chaotic motion" or "chaotic domain" in a loose way, from the phenomenological point of view usual in this context (see, for instance, Lichtenberg and Lieberman, 1983).

Following Escande (1987a, 1987b), we can, in this example, be somewhat more precise and exhibit in the "chaotic domain" an "homoclinic tangle" (Guckenheimer and Holmes, 1983) which can be shown to cover the domain with a very tight grid. We know that a dynamical system looks like a Smale-horseshoe close to such an homoclinic tangle by the Smale-Birkhoff homoclinic theorem (see, for instance Guckenheimer and Holmes, 1983). Of course, we cannot conclude from this that the full "chaotic domain" is ergodic (and probably it is not) but we can be sure that there is no large connex part of it which escapes to be ergodic. As pointed out by Escande (1987), the tight grid of the homoclinic tangle provides a strong skeleton to the chaotic domain. This is to be contrasted with the case of the "fast Hamiltonian chaos" (see, for instance, Chirikov, 1979; Lichtenberg and Lieberman, 1983; Escande, 1985).

Let us give a brief description of part of this homoclinic tangle. More detailed information are available in Escande (1987a, 1987b).

In the case where $c = 0$, the dynamical system described by (20) with b periodic as in (19) possesses trivially an unstable periodic orbit ($\phi = \pi$, $I = 0$, $\mu = \cos 2\varepsilon t$). The trace of this periodic orbit on the surface of section $\lambda = \tilde{\lambda}$ (as in Fig. 22) is a fixed point ($\phi = \pi$, $I = 0$) of the Poincaré's map.

To this periodic orbit are associated a 2-dimensional stable manifold (S) and a 2-dimensional unstable manifold (U) in the 3-dimensional space (ϕ, I, μ). The trace of these manifolds on the surface of section $\lambda = \tilde{\lambda}$ are one-dimensional curves (s) and (u), the stable and unstable manifolds of the Poincarés map.

In the vicinity of the periodic orbit ($\phi = \pi$, $I = 0$, $\mu = \cos 2\varepsilon t$), the manifold U follows closely the manifold $h = K(J, \lambda) = 0$ as we have seen in the Appendix 5.

Let us consider the intersection of the manifold U and the manifold $\phi = \pi$. It contains a piece of curve parametrized by λ_N:

$$\tilde{\lambda} < \lambda_N < \lambda_{\min} \quad , \quad h = h_N(\lambda_N) \sim 2\pi N \varepsilon \left(\frac{\partial J_3^\star}{\partial \lambda} \right)_{\lambda_N} \tag{37}$$

where N is of the order of $\varepsilon^{-1/3}$ and $\lambda_{\min}$ is the first value of λ following $\tilde{\lambda}$ such that $J_3^\star(\lambda)$ reaches its minimum value. The points on this piece of curve correspond to N complete traverses (vertex to vertex) after the main vertex as in Sect. II.7 and the function $h_N(\lambda_N)$ is defined in such a way that, after a translation $\psi = -2\pi N$, these points are on U very close to $h = 0$. The estimates of $h_N(\lambda_N)$ given in (37) is a consequence of (II.59).

Let us map the curve (37) in the plane $(\psi, \bar{J})$ at $\lambda = \tilde{\lambda}$. Using the adiabatic invariant theory, we have that the corresponding points are given by

$$\psi = \frac{1}{\varepsilon} \int_{\lambda_N}^{\tilde{\lambda}} \left(\frac{\partial K}{\partial J} \right) d\lambda \quad , \quad \bar{J} = J_3(h_N, \lambda_N) \, . \tag{38}$$

Hence the values of $\bar{J}$ range from (approximately) $J_3^\star(\tilde{\lambda})$ to (approximately) $J_3^\star(\lambda_{\min})$ which marks the boundary of the "chaotic domain" defined earlier.

Furthermore, the slope of this curve can be approximated by

$$\begin{aligned} \frac{\partial \bar{J}}{\partial \lambda_N} \Big/ \frac{\partial \psi}{\partial \lambda_N} &\sim \varepsilon \left\{ \left(\frac{\partial J^\star}{\partial \lambda} \right) \Big/ \left(\frac{\partial K}{\partial J} \right) \right\}_{\lambda = \lambda_N} \\ &\sim \frac{\varepsilon}{\log |he|^{-1}} \left(\frac{\partial J^\star}{\partial \lambda} \right)_{\lambda_N} \end{aligned} \tag{39}$$

which shows that this curve is a tightly wound spiral along the cylinder $(\psi, \bar{J})$.

Similar estimates are available for the stable manifold in the domain D_3 or for the stable and unstable manifold in the domains D_1 or D_2, leading to the picture shown in Fig. 23.

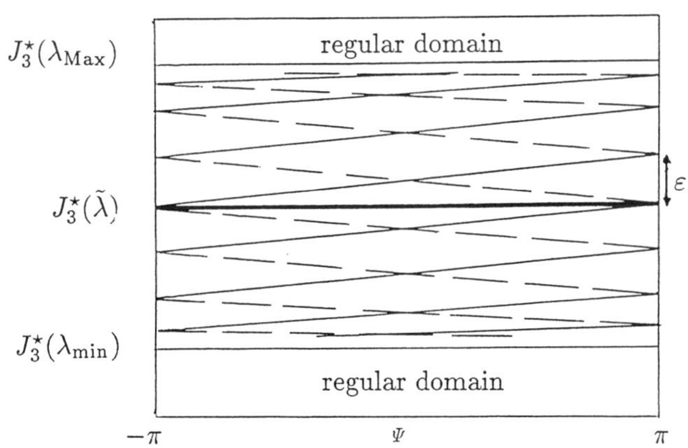

Fig. 23. Part of the stable (dashed line) and unstable (solid line) manifold in the plane $(\psi, \bar{J}, \lambda = \tilde{\lambda})$. The two manifolds cover the "chaotic domain" with $0(\varepsilon)$ losanges.

4. The Magnetic Bottle

If the adiabatic invariance received prominence because of its role in the early formulation of quantum mechanics, its importance in classical mechanics became first of significance for applications in connection with the magnetic momentum of gyration of a charged particle in a strong magnetic field. This was shown to be an invariant by Alfven (1950) in his investigation of cosmical rays. Very soon afterwards, its usefulness in the theoretical design of devices for controlling hot plasmas (Stellarators, Tokamaks, Mirror Machines, ...) was recognized (see, for instance, Kruskal, 1952; see also Freidberg, 1982 for a recent review).

We shall try in this section to suggest why the adiabatic invariance is so important in this context, without, of course, giving a full account of its technical applications. This would require by itself a complete review paper which we are not competent to write.

The motion of a particule of mass m and electric charge e in an electromagnetic field is controled by the Hamiltonian function:

$$H = \frac{1}{2m}||\mathbf{p} - \frac{e}{c}\mathbf{A}(\mathbf{x})||^2 + e\phi(\mathbf{x}) \tag{40}$$

where ϕ and $\mathbf{A}$ are respectively the electric potential and the magnetic vector potential of the field:

$$\mathbf{E} = -\text{grad}\ \phi \ , \tag{41}$$

$$\mathbf{B} = \text{rot}\ \mathbf{A} \ . \tag{42}$$

The vector $\mathbf{x}$ is the position vector of the particle and the vector $\mathbf{p}$ its momentum related to its velocity $\mathbf{V}$ by

$$\mathbf{p} = m\mathbf{V} + \frac{e}{c}\mathbf{A} \ . \tag{43}$$

Let us assume that the vector-potential and the magnetic field are given by

$$\mathbf{A} = B_0(1 + b(x_3))x_1\mathbf{e}_2 \ , \tag{44}$$

$$\mathbf{B} = B_0(1 + b)\mathbf{e}_3 - B_0 b' x_1 \mathbf{e}_1 \ , \tag{45}$$

where (x_1, x_2, x_3) are the cartesian coordinates of $\mathbf{x}$ in an orthonormal basis $(\mathbf{e}_1, \mathbf{e}_2, \mathbf{e}_3)$ and b' is the first derivative of b with respect to x_3.

This magnetic field describes a two-dimensional "magnetic bottle" (with two throats). The magnetic field is an almost constant field in the direction of $\mathbf{e}_3$ but slightly (if b' is small) modulated. The magnetic lines are given by

$$x_1 = \frac{x_1(0)}{1 + b(x_3)} \ , \quad x_2 = x_2(0) \tag{46}$$

and their shape is illustrated in Fig. 24 for $b = \beta^2 x_3^2$.

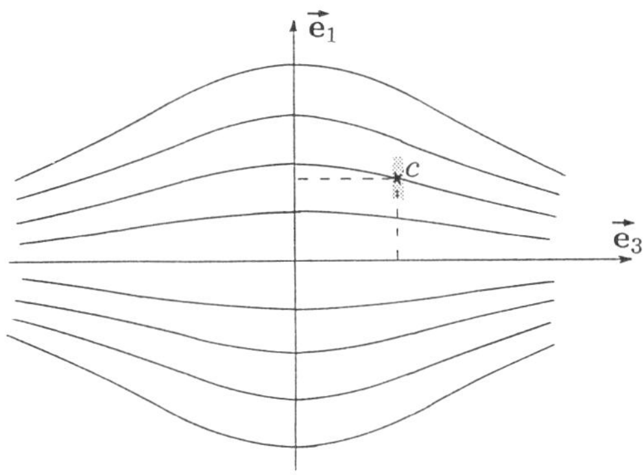

Fig. 24. The magnetic lines in the plane $(\mathbf{e}_1, \mathbf{e}_3)$. The motion of the particle is approximately a circle centered at the gyrocenter c. The gyrocenter itself moves slowly along the magnetic line bouncing back and forth in the "bottle".

It would have been more realistic to consider a magnetic field with cylindrical symmetry described for instance by:

$$\mathbf{A} = \frac{B_0}{2}(1+b)[x_1\mathbf{e}_2 - x_2\mathbf{e}_1] , \tag{47}$$

$$\mathbf{B} = B_0(1+b)\mathbf{e}_3 - \frac{B_0 b'}{2}[x_1\mathbf{e}_1 + x_2\mathbf{e}_2] \tag{48}$$

or a toroidal stellarator (Kovrizhnykh, 1984) but the geometry we have adopted will simplify the calculations without changing essentially the analysis.

Assuming at first no electric field, the Hamiltonian reads

$$H = \frac{1}{2m}\{p_1^2 + p_2^2 + [p_2 - \frac{eB_0}{c}(1+b)x_1]^2\} . \tag{49}$$

We introduce now the "guiding center" coordinates by means of the following canonical transformation

$$\begin{aligned} x_1 &= \tfrac{1}{1+b}Y_c + \tfrac{1}{(1+b)^{1/2}}y_g & p_1 &= \tfrac{eB_0}{c}(1+b)^{1/2}Y_g \\ x_2 &= y_c + \tfrac{1}{(1+b)^{1/2}}Y_g & p_2 &= \tfrac{eB_0}{c}Y_c \\ x_3 &= y_3 & p_3 &= \tfrac{eB_0}{c}\{Y_3 + \tfrac{b'}{2(1+b)}Y_g y_g + \tfrac{b'}{(1+b)^{3/2}}Y_c Y_g\} . \end{aligned} \tag{50}$$

The quantities (Y_g, Y_c, Y_3) are the momenta respectively conjugated to the variables (y_g, y_c, y_3). Geometrically speaking, $(Y_c/(1+b), y_c, y_3)$ are the coordinates of a point: the guiding center (or gyrocenter). Remark that the curves along which Y_c and y_c are constant are precisely the magnetic lines defined in (46). On the other hand, the quantities (y_g, Y_g) can be viewed as the coordinates (scaled by a factor $(1+b)^{1/2}$) of the moving particle in a frame centered at the guiding center.

This geometrical interpretation of the canonical transformation (50) may seem peculiar in the fact that both the variables (y_g, y_c) and their momenta (Y_g, Y_c) are interpreted in terms of position (respectively of the guiding center and of the particle). Actually, this is not that unusual. The Andoyer's variables (see, for instance, Deprit, 1967) used in the description of the Colombo's top (see Sect. III.3) have the same peculiarity. After all, the possibility of treating variables and momenta on the same foot is one of the advantages of Hamiltonian mechanics.

The new Hamiltonian function of the problem is:

$$K = \frac{c}{eB_0} H = \frac{eB_0}{2mc} \Big\{ (1+b)[y_g^2 + Y_g^2] + \Big[Y_3 + \frac{b' Y_g}{(1+b)^{3/2}} (Y_c + \frac{1}{2}(1+b)^{1/2} y_g) \Big]^2 \Big\} . \tag{51}$$

Let us choose the unit of time such that the gyrofrequency is unity:

$$\frac{eB_0}{mc} = 1 \tag{52}$$

and the unit of length such that the gyroradius (the norm of the vector (y_g, Y_g)) is of the order of unity. We assume that this unit of length is such that the scale on which the magnetic field changes significantly is large (say of the order of $1/\varepsilon^2$). With this assumption in mind, we introduce a scaling of the third dimension together with polar coordinates for the gyro-coordinates

$$\begin{aligned} y_g &= \sqrt{2G} \sin g \,, & Y_g &= \sqrt{2G} \cos g \,, \\ y_c &= y \,, & Y_c &= Y \,, \\ y_3 &= \tfrac{1}{\varepsilon} z \,, & Y_3 &= \varepsilon Z \,, \end{aligned} \tag{53}$$

which brings the Hamiltonian (51) under the form

$$K = (1+c)G + \frac{\varepsilon^2}{2} \Big[Z + \frac{\varepsilon c'}{(1+c)^{3/2}} \Big(Y + \frac{1}{2} \sqrt{2G(1+c)} \sin g \Big) \sqrt{2G} \cos g \Big]^2 \tag{54}$$

where c is a scaled version of the function b

$$b(y_3) = c(\varepsilon^2 y_3) = c(\varepsilon z) \,. \tag{55}$$

The function $1 + c(\cdot)$ and its derivatives are assumed to be of the order of unity in the domain of interest.

If we "freeze" the third coordinate by considering the function c as a constant, the Hamiltonian function (54) is actually a one-degree of freedom Hamiltonian expressed in its action-angle variables (g, G).

What makes the problem somewhat different from the other problems we have investigated so far is that the (hopefully slow) dependence upon the time is not direct but the result of its (slow) dependence upon a second-degree of freedom (z, Z). To investigate the motion of this second degree of freedom, we need some knowledge about the motion of the first one.

Hence the problem deviates from the narrow frame we have considered up to now and should be considered in the general frame of perturbation theory. We think that Delaunay (1867) deserves the credit of being the first to propose and actually use (in his celebrated mémoire on the theory of the motion of the Moon), changes of coordinates recursively defined in order to transform formally Hamiltonian functions such as (54) into an integrable form.

The idea was generalized by Poincaré (1899), implemented for a practical problem by von Zeipel (1915) and used later by various authors under the name of "von Zeipel method" (e.g. Brouwer, 1959) or "Lindstedt-Poincaré method" (e.g. Giacaglia, 1972). The idea was modified by Birkhoff (1927) to yield the "Birkhoff's normalization" and introduced by Krylov and Bogoliubov (1947) in the engineering (non Hamiltonian) context giving rise to the "averaging method". It was reformulated by Kruskal (1962) (and called the "nice variables method") in precisely the context we are now considering.

Whatever the method used to define the transformation, and we think that the more recent Lie Transforms method (Hori, 1966; Deprit, 1969) is the nicest of all, the result is that, for any integer N, we can define a canonical transformation from (g, G, z, Z) to $(\bar{g}, \bar{G}, \bar{z}, \bar{Z})$ such that, in the new (averaged) variables, the transformed Hamiltonian $\bar{K}$ depends upon $\bar{g}$ only through terms of the order of ε^{N+1} (see Sect. I.3):

$$\bar{K} = (1+c)\bar{G} + \frac{\varepsilon^2}{2}\bar{Z}^2 + \frac{\varepsilon^4}{2}\frac{c'^2\bar{G}}{(1+c)^3}[\bar{Y}^2 + \frac{\bar{G}}{8}(1+c)] + 0(\varepsilon^6) \,. \tag{56}$$

A first approximation of $\bar{K}$, the one which is explicitly written in (56) is, of course, the averaged value of K with respect to g.

We can now consider $\bar{G}$ as a constant. (It is an adiabatic invariant, its time derivative being of the order of ε^{N+1} – see Sect. I.4) and analyse (56) as a one-degree of freedom Hamiltonian in $(\bar{z}, \bar{Z})$. Let us restrict ourselves to a simple case where the function c is given by

$$c(x) = \frac{d}{2}x^2 \,. \tag{57}$$

Then the leading terms of (56) reproduce the harmonic oscillator

$$\bar{K} = \frac{\varepsilon^2}{2}[\bar{Z}^2 + (d^2\bar{G})\bar{z}^2] + 0(\varepsilon^4) \,, \tag{58}$$

the frequency of which is a function of $\bar{G}$, the (averaged) orbital magnetic momentum of the particle.

Hence, at least in a first approximation, the guiding center of the particle (coordinates: $Y/(1+b), y, z/\varepsilon$) bounces back and forth along a magnetic line (Y and y constant) between two "mirror points": $z = \pm z_M$ with

$$z_M = \frac{\bar{Z}(0)}{d}\frac{1}{\sqrt{\bar{G}}} \,. \tag{59}$$

Confinement of the plasma inside the magnetic bottle depends crucially upon the fact that z_M does not increase beyond a given bound on a very long time scale. Two things may happen: $\bar{Z}(0)$ may change due, for instance, to collisions between particles inside the plasma or $\bar{G}$ may change due also to collisions or to a default in the adiabatic invariance.

As we have just recalled, the invariance of $\bar{G}$ is only asymptotic. In the framework of the model just discussed for the magnetic bottle, Chirikov (1979) estimates the changes in $\bar{G}$ over a bounce period as proportional to

$$\Delta\bar{G} \sim \frac{1}{\varepsilon}\exp\left\{-\frac{2}{3\varepsilon}\right\} , \tag{60}$$

a quantity exponentially small with ε.

On the other hand, the model just discussed is, of course, only approximative. Fluctuations in the magnetic field or electric field may complicate the topology of the "frozen system" corresponding to (54) with $c = 0$. We have found this system to be just the harmonic oscillator with (g, G) as action variables. But fluctuations in the fields may introduce a separatrix in the phase space of the "frozen" system. The adiabatic invariance of the averaged $\bar{G}$ may then be in default at each crossing of the separatrix.

This effect has been investigated for instance by Dobrott and Greene (1971) in the case of the stellator in which a weak but short periodic poloidal magnetic field is superimposed on top of the main toroidal field (see also Kovrizhnykh, 1984) or by Aamodt (1971–1972) who considers short-wavelength fluctuations in the electric field due to collective modes in the plasma itself.

We shall discuss briefly this last application. Let us assume that superimposed on the magnetic field (45), there is a short-wavelength electric field in the direction perpendicular to the magnetic field and slightly modulated in the direction of the magnetic field:

$$\phi = \frac{m}{e}F(\varepsilon^2 x_3)\cos(kx_1) . \tag{61}$$

We introduce the "guiding center" coordinates as in (50) and a scaling of the third dimension:

$$y_3 = \frac{1}{\varepsilon}z \quad , \quad Y_3 = \varepsilon Z \tag{62}$$

to obtain

$$K = \frac{1}{2}(1+c)(Y_g^2 + y_g^2) + F(\varepsilon z)\cos\left[\frac{kY_c}{1+c} + \frac{ky_g}{(1+c)^{1/2}}\right] + 0(\varepsilon^2) . \tag{63}$$

As it can be seen in Fig. 25, the motion (y_g, Y_g) can be severely distorted by the electric field corresponding to ϕ. This does not preclude the application of the adiabatic invariant theory and the definition of "mirror points". Simply, the adiabatic invariant is no longer the (averaged) magnetic momentum $\bar{G}$ but a more complicated function and the mirror points are no longer given by (59).

Of course the mirror points may be much different for a trapped orbit (inside the loops in Fig. 25) than for un untrapped one (outside the loops). Also, we have to consider that the periodic jumps from one domain of the phase space of (y_g, Y_g) to another one generate a slow diffusion in the adiabatic invariant which may be much more important than the one estimated by Chirikov (see (60)).

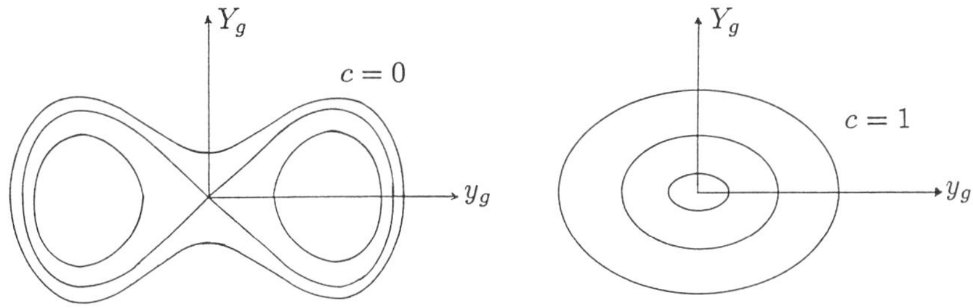

Fig. 25. Motion of the particle around its gyrocenter for the Hamiltonian (63) with $F = 2$, $k = 1$, $Y_c = 0$ and two particular values of c.

5. Orbit-Orbit Resonances in the Solar System

If two satellites orbiting the same planet (or two planets orbiting the Sun) have mean orbital angular velocities which have a ratio very near that of two (usually small) integers, the motions of the satellites are said to be commensurate and to define an orbital resonance. Orbital resonances amongst the satellites of the major planets are of interest because many more such resonances exist than can be accounted for by a random distribution of orbits. On the other hand, the distribution of Minor Planets (Asteroids) orbiting the Sun between Mars and Jupiter shows puzzling gaps (the Kirkwood's gaps) at the location of resonance with the orbit of Jupiter.

The study of the dynamical evolution of orbits in resonance has shown how very small dissipative forces may sculpt the Solar System to its present distribution of orbits and, at the same time, yield surprizingly tight constraints on the physics of tidal dissipation (see Yoder, 1979) or the nature of the primordial proto-solar nebula (see Henrard-Lemaître, 1983b).

The hypothesis of the formation of orbital resonances by the differential tidal expansion of initially randomly distributed orbits was first analysed by Goldreich (1965) and then in quick succession by Allan (1969–1970), Sinclair (1972–1974), Greenberg (1973) and Yoder (1973). The field has been reviewed by Peale (1976), Greenberg (1976) and very recently by Peale again (1986). One of the major accomplishments in this field is the explanation of the origin of the double resonance between the large satellites of Jupiter (the Galilean satellites) by Yoder (1979) and Yoder and Peale (1981).

Other types of non-conservative effects such as changes in the orbit of Jupiter (Torbett and Smoluchowski, 1980, 1982), drag forces (Gonczi et al., 1982; 1983) or the slow removal of the primitive proto-solar nebula (Henrard and Lemaître, 1983b; Lemaître, 1984, 1985), have been considered to explain the Kirkwood's gaps in the Asteroid belt. The interplay of resonances and the drag forces of the primitive nebula may also help to explain the formation of the planets by accretion of planetesimals (Stuart et al., 1985).

The basic mechanism (if not a detailed simulation) of all these phenomena may be understood with the help of the adiabatic invariant theory as we shall show in this section.

Let us consider the planar restricted three body describing the motion of a particle (a small satellite, an asteroid, a planetesimal, $\cdots$) in the gravitational field of a primary (the planet, the Sun, $\cdots$) and of a smaller secondary (another satellite or another planet, $\cdots$). The secondary is assumed to be on a circular orbit and the motion of the particle takes place in the plane of this orbit.

The problem is described by the Hamiltonian function

$$H = H_0(L) + mH_1(L, P; \lambda, p; \lambda') \tag{64}$$

where

$$H_0 = -\frac{\mu^2}{2L^2} \tag{65}$$

is the Hamiltonian of the two-body problem, particle-primary and μ is the gravitational mass of the primary. The perturbation mH_1 describes the interaction particle-secondary. It is proportional to the ratio, m, of the masses of the secondary and the primary which is assumed to be small (10^{-5} or smaller for Satellite/Planet and 10^{-3} for Jupiter/Sun). The function H_1 is a function of the Delaunay's variables describing the orbit of the particle:

$$\begin{aligned} \lambda &: \text{ mean longitude of the particle} \\ p = -g &\text{ where } g \text{ is the longitude of the pericenter of the particle} \end{aligned}$$

and their conjugate momenta

$$\begin{aligned} L &= \sqrt{\mu a} \\ P &= \sqrt{\mu a}(1 - \sqrt{1 - e^2}) \sim \frac{1}{2}\sqrt{\mu a}e^2 \end{aligned}$$

where a and e are respectively the semi-major axis and the eccentricity of the orbit of the particle. The Hamiltonian is also a function of

$$\lambda' : \text{mean longitude of the secondary}$$

which in this problem is simply a linear function of the time:

$$\lambda' = n't + \lambda_0' \, . \tag{66}$$

The perturbation H_1 is of course periodic in the angular variables λ, p, λ'. In a first approximation ($m = 0$), the two main longitudes (λ and λ') have finite frequencies of the same order of magnitude

$$\frac{d\lambda'}{dt} = n' \quad ; \quad \frac{d\lambda}{dt} = n = \frac{\mu^2}{L^3} = \sqrt{\frac{\mu}{a^3}} \, . \tag{67}$$

The corresponding periods will be called "short periods".

The classical theory of perturbation (see, for instance, Brouwer and Clemence, 1961; Arnol'd, 1978) shows that, to a very good approximation, the short periodic terms may be averaged out of the perturbation. Their only effect is to superimpose small (or the order of m) short periodic variations on top of a "mean" orbit.

Amongst the terms in H_1 which are not short periodic, we find the terms independent of λ and λ' (the so-called "secular terms") and possibly the "resonance terms" of the form:

$$F(L, P)\cos(i\lambda - j\lambda' + kp) \tag{68}$$

whenever the initial conditions (essentially the initial semi-major axis "a") is such that

$$in - jn' \simeq 0\,. \tag{69}$$

It happens that the coefficient $F(L, P)$ of such terms is of the order of $|i-j|$ in the eccentricity e. The eccentricities being usually small, the strongest resonances are the "first order" resonances such that $|i-j| = 1$.

Let us consider the internal 2/1 resonance for which $i = 1$ and $j = 2$. It is called internal because it implies that $n > n'$ and thus that $a < a'$; the orbit of the particle is inside the orbit of the secondary. External resonances (such that $a > a'$) or other first order resonances would be investigated in much the same way.

The most important term in the perturbation is then:

$$C(L)\sqrt{2P}\cos(\lambda - 2\lambda' + p) \tag{70}$$

where $C(L)$ may be taken as a constant in a first approximation. The Hamiltonian system (64) is then easily reduced to a one-degree of freedom system, which can be expressed in terms of the resonant angle s and its conjugated momentum S

$$s = \lambda - 2\lambda' + p \quad , \quad S = \frac{P}{\sqrt{\mu a}}\,. \tag{71}$$

The new Hamiltonian function reads (see, for instance, Peale, 1986):

$$K = \beta(S - S^\star)^2 + \varepsilon\sqrt{2S}\cos s \tag{72}$$

with

$$\beta = -\frac{3}{2}n \quad , \quad S^\star = \frac{n - 2n'}{3n} \quad , \quad \varepsilon = 0.75\ m \cdot n \tag{73}$$

which is just the "second fundamental model of resonance" we have investigated in Sect. III.3 (see (III.33)).

The scaled time and momentum proposed in (III.35) translate into

$$\tau = 0.595\ (m)^{1/3} nt \quad , \quad e = 0.794\ (m)^{1/3}\sqrt{2R} \tag{74}$$

which shows that they scale like the cubic root of the mass ratio m. The period of an orbit of the second fundamental model is large (of the order of $m^{-1/3}$).

The periodic excursions in eccentricity produced by the resonance are still small (of the order of $m^{1/3}$) but much larger than the periodic excursions produced by the short periodic terms (of the order of m).

The motion just described does not show any trend towards capture into resonance. Orbits which are in resonance remain in resonance and those which are not do not approach it. But, if we take into account the very small but dissipative forces produced by the tides raised upon the planet by the satellites, an evolutionary effect is produced.

Tides raised upon the planet by the satellites produce a periodic distorsion of the gravitational field of the planet. The principal effect upon the satellite is again small and periodic and may be neglected but the dissipation of energy due to the fact that the planet is not completely elastic does produce a systematic effect which can be modelized by (see, for instance, Goldreich, 1965):

$$\frac{1}{n_i^2}\frac{dn_i}{dt} = -\frac{27}{4}k_2 m_i \left(\frac{R_p}{a_i}\right)^5 \frac{1}{Q} \tag{75}$$

where n_i (resp. m_i, a_i) is the mean motion (resp. the mass ratio, the semi-major axis) of the i-th satellite and k_2 (the Love number, approximately equal to 0.34 for giant planets), R_p (the radius of the planet) and Q (the dissipation constant of the planet, of the order of 10^5 for giant planets) are constants pertaining to the planet.

The effects produced upon the parameters ε and β by the slow changes of the mean motions ($n = n_1$, $n' = n_2$) are neglectible but the effect produced upon the parameter $S^\star$ is relatively much larger. Indeed, $S^\star$ itself is small as it is obtained by subtracting two nearly identical quantities.

The rate of change of the mean motions being proportional to the fifth power of the distance to the planet, the inner satellite is usually pushed away from the planet faster than the outer one. The net effect is a slow increase of the parameter $S^\star$ and thus of δ (see (III.37)).

The effect upon the orbital evolution of the particle (the inner satellite) may be best visualized on what we have called the area index diagram (Henrard, Lemaître, 1983a).

For each value of δ, let us assign to each orbit of the second fundamental model a number A: the area index. For internal or external orbits, this number is just the area enclosed by the orbit (and thus 2π time the absolute value of the adiabatic invariant J). For resonance orbits, this number is defined as the area enclosed by the orbit plus the area enclosed by the smallest homoclinic orbit corresponding to the same value of δ. In this way, there is a one-to-one correspondance between the set of orbits and the set of positive real numbers. Indeed, the “largest” internal orbit is followed by the “smallest” (the equilibrium) resonance orbit and the “largest” resonance orbit is followed by the “smallest” external orbit.

Let us remark also that, far from the resonance (large absolute value of δ), the trajectories are almost circles (see Figs. 11 and 12 in Sect. III.3) and thus that the area enclosed by the orbit is directly proportional to the square of its radius and thus to the square of the eccentricity of the space trajectory of the particle

$$A \simeq 4.983\, m^{-2/3} e^2 \,. \tag{76}$$

If we plot the area index A of an orbit versus the value of δ to which it belongs, we obtain a two-dimensional diagram of possible orbits (see Fig. 26). The critical curves, giving, in function of δ, the area enclosed by the two homoclinic orbits, divide this diagram into three domains corresponding to internal, resonance and external orbits. The area enclosed by the two homoclinic orbits may be easily computed from (III.49).

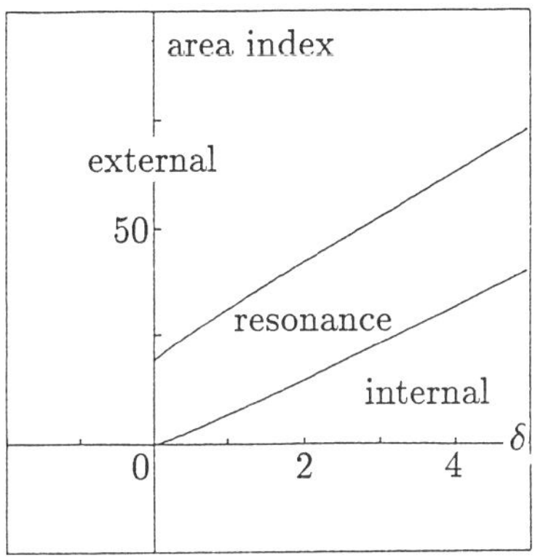

Fig. 26. The area index diagram. The two curves correspond respectively to the smallest (for the lower curve) and the largest (for the higher curve) homoclinic orbits.

Due to the fact that the adiabatic invariant is conserved, the evolutionary path forced by the change in the value of δ is easy to follow. In the internal and external domains, the area index is simply conserved. In the resonance domain, the path follows the boundary of the internal and resonance zone. Indeed, the difference between the area index and the area of the smallest homoclinic curve (which marks this boundary) is proportional to the adiabatic invariant and thus constant.

When the evolutionary path is forced to pass through a critical curve, on the boundaries of the zones, a jump may occur. The probability of such a jump is to be evaluated through the formula (III.51).

Typical examples of evolutionary path for increasing values of δ are shown in Fig. 27. Whenever the initial value of the area index (or of the eccentricity) is small enough, the capture into resonance is certain (see path (a) in Fig. 27).

When the initial value of the eccentricity is such that

$$e > 1.945\, m^{1/3} \tag{77}$$

(i.e. when A is larger than 6π), the evolutionary path meets the critical curve (see paths (b) and (c) in Fig. 27). It can then continue in the resonance zone (path (b)) with a probability which can be read from Fig. 15, Sect. III.3, or jumps to the internal domain (path (c)) accompanied with a diminution of the area index. Roughly speaking, a small (compared to $m^{1/3}$) starting eccentricity means a

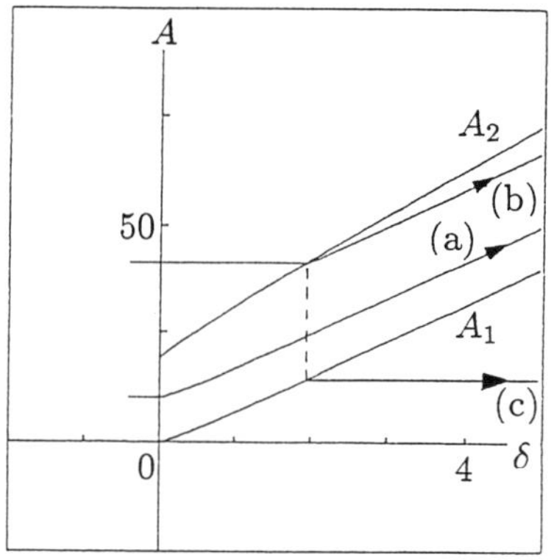

Fig. 27. Typical evolutionary paths in the area-index diagram for increasing values of δ.

certain or highly probable capture into resonance, a large eccentricity means an highly probable jump over the resonance with a damping of the eccentricity.

Applications of this scenario to particular planetary satellites can be found in the literature quoted at the beginning of this section. The recent review chapter by Peale (1986) in particular, does use the adiabatic invariant theory to evaluate the probabilities of capture into resonance and more generally to apply the basic mechanism we have outlined, to more realistic models.

Other types of small non-conservative effects may have been at work especially in the early stages of the Solar System and may have been instrumental in shaping it. As an illustration, we shall briefly describe the scenario proposed by Lemaître (Henrard and Lemaître, 1983b; Lemaître, 1984, 1985) to explain the Kirkwood's gaps.

Let us assume that an asteroid and Jupiter are orbiting aroung the Sun inside a rather thin and flat proto-solar nebula (the accretion disk). The principal gravitational effect of the disk is to make the mass of the Sun as seen by Jupiter and the mass of the Sun as seen by the asteroid slightly different. The motion of the asteroid may still be described by the Hamiltonian (72) but with slightly modified values of the coefficients β, $S^\star$ and ε.

The removal of the proto-solar nebula has for effect to bring the coefficients back to their present value and thus to induce a slow (if the removal of the proto-solar nebula is a smooth operation) variation of δ, the total value of which is essentially a function of the mass of this part of the proto-solar nebula which is contained between the orbit of the asteroid and the orbit of Jupiter. For the 2/1 resonance and for the type of proto-solar nebula proposed by Weidenschilling (1977), this variation of the parameter δ can be evaluated as

$$\Delta\delta = -116 M_D / M_\odot \tag{78}$$

where M_D is the mass of the disk between the two orbits and $M_\odot$ is the mass of the Sun. In order to obtain a value of $\Delta\delta$ equal to -3.5 which is suggested by the present state of the 2/1 Kirkwood's gap, the surface density of the disk between the two orbits should have been of the order of

$$\sigma \sim 4000\ gr/cm^2 , \tag{79}$$

a value somewhat larger but not incompatible with the estimate of Weidenschilling (1977) for the proto-solar nebula.

The effect of a decrease of δ on the evolutionary path of the orbit of an individual asteroid is illustrated in Fig. 28. An asteroid starting on an internal trajectory will, upon crossing the critical curve, jump over the resonance onto an external trajectory with probability one (Fig. 28, path (c)). A trajectory starting inside the resonance will leave it either without crossing the critical curve (Fig. 28, path (a)) or after crossing it (Fig. 28, path (b)) depending upon the initial value of its adiabatic invariant.

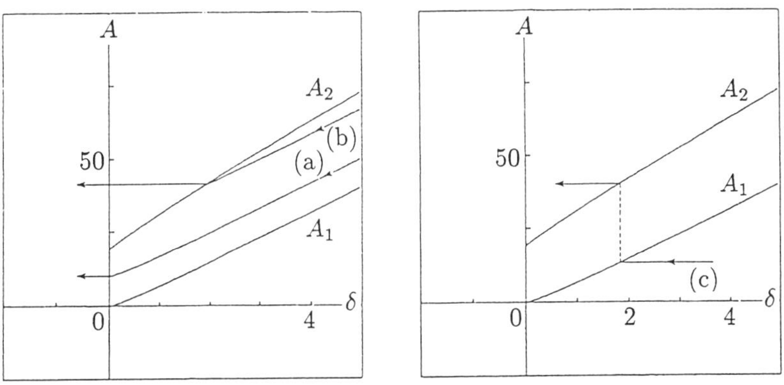

Fig. 28. Typical evolutionary path in the area index diagram for decreasing values of δ.

Knowing the evolutionary path of the orbit of each individual asteroid, we can describe the evolution of a swarm of asteroids. We have first to decide upon a more or less realistic original density distribution for the asteroids. Considering that objects in orbits with high eccentricity either are not formed or are soon removed by collisions with a planet, we consider a uniform density distribution in phase space modulated by a cut-off value in the eccentricity. The present distribution of non-resonant asteroids is close to this description.

A uniform distribution in phase space corresponds to a uniform distribution in the area index diagram. A cut-off value in the eccentricity corresponds to a cut-off value in the area index diagram. Outside the resonance zone, this cut-off value is independent of δ. Indeed, outside the resonance zone, the osculating eccentricity, the averaged eccentricity and the radius of the orbit in the phase diagram (see Figs. 11 and 12 in Sect. III.3) are almost the same. Inside the resonance zone, the relationship between eccentricity and area-index is more involved. If the cut-off value is chosen as 0.2 and applied to the mean eccentricity, it gives the density distribution reproduced in Fig. 29 (a).

A variation of δ of amplitude $\Delta\delta = -3.5$ applied to this density distribution yields the density distribution of Fig. 29 (b). It is characterized by a complete depletion of the resonance zone and by the formation of a region with higher area index (and thus eccentricity) just outside the resonance zone. The density of material in this region is slightly smaller than the original density.

These two features of the theoretical scenario can indeed be seen in the present distribution of the Asteroids close to the 2/1 gap as shown in Fig. 30.

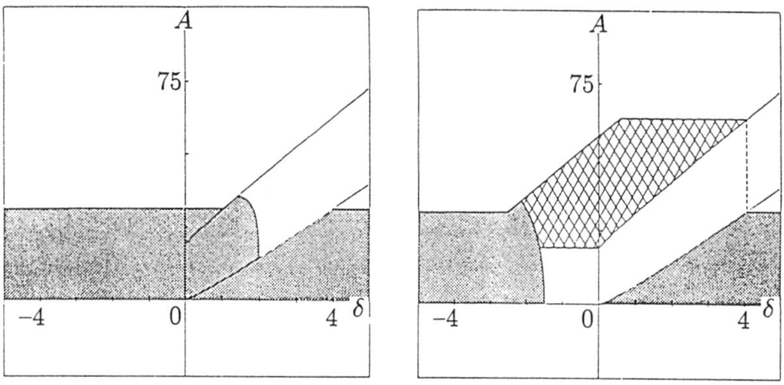

Fig. 29. Density distributions of asteroids: (a) Assumed density distribution at the origin, characterized by a cut-off at $e = 0.2$; (b) Subsequent distribution after a change of the parameter δ by -3.5.

The same mechanism can explain the shape of the other Kirkwood's gaps (resonance 3/1, 5/2, $\cdots$) and is thus a serious contender as the explanation of the 100 years old puzzle: where do the Kirkwood's gaps come from?

However the reader should be aware that another mechanism (Wisdom, 1983-1985) can explain the formation of the 3/1 gap. Whether it can explain also the main 2/1 gap is still uncertain.

6. Spin-Orbit Resonance in the Solar System

It has long been recognized that the rotation state of planets and satellites in the Solar System is far from being randomly distributed. The most obvious example being the Moon which shows always the same face to the Earth because its rotation rate is quite precisely synchroneous with its mean orbital angular velocity.

Part of this non-randomness should be attributed to the mechanism of formation of planets and satellites but part of it is clearly due to evolutionary processes. Amongst them, the interplay of resonance and small non-conservative effects plays a prominent role that we shall illustrate with two examples: the

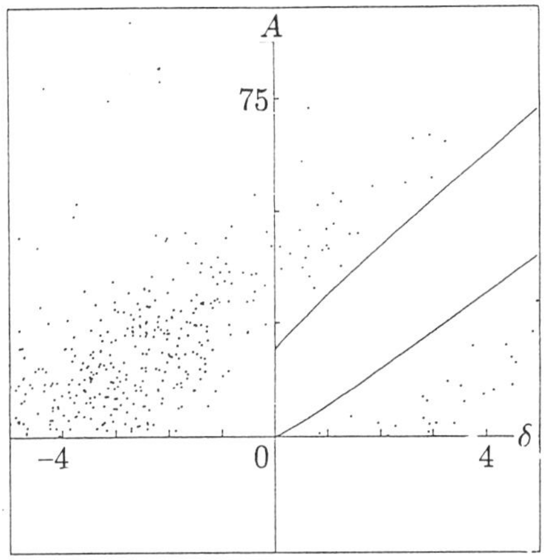

Fig. 30. Actual present distribution of the Asteroids in the vicinity of the 2/1 Kirkwood's gaps.

capture into resonance of the spin of Mercury and the history of the obliquity of Mars.

In 1965, radar observations revealed that Mercury's axial rotation has a sidereal period of about 59 days rather than the previously accepted value of 88 days (corotation value). Colombo (1965) pointed out that this observed spin period is almost exactly 2/3 of an orbital period. He suggested that a nearly uniform rotation at just 3/2 the mean orbital angular velocity might actually be stabilized if Mercury had a slight permanent equatorial asymmetry. Subsequent analysis (Colombo and Shapiro, 1966; Goldreich and Peale, 1966; Counselman and Shapiro, 1970; Burns, 1979) confirmed that such a spin-orbit resonance could be the result of an evolutionary process. We shall here give a brief summary of this analysis using specifically the concept of adiabatic invariance as in Henrard (1985).

Let us consider a planet (or satellite) the spin axis of which is normal to its orbital plane. The principal moments of inertia are A, B and C $(A < B < C)$ where C is the moment about the spin axis. We consider only the two torques which seem the most important. The torques exerted by the Sun (or the planet) on the permanent equatorial asymmetry and on the moving tidal bulge.

The torque exerted by the Sun on the permament asymmetry is given by (see e.g. Counselman and Shapiro, 1970):

$$T_p = -\frac{3}{2}\frac{GM}{r^3}(B-A)\sin(2\theta - 2f) \; . \tag{80}$$

The other external torque likely to be significant is the tidal lag torque. It is analogous to the asymmetry torque, the main difference being that the asymmetry in this case is not a permanent one but is due to the deformation of the non-rigid planet by the differential solar gravitational field. In an elastic

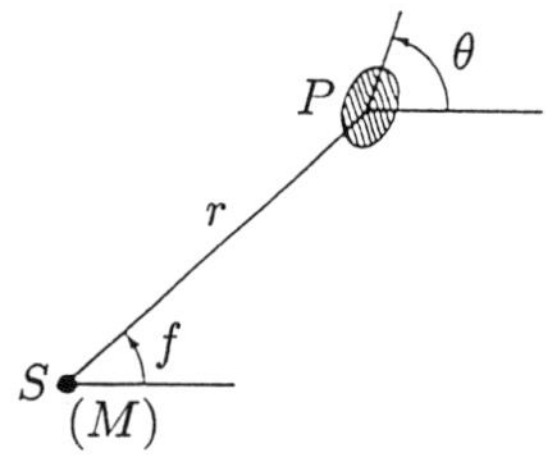

Fig. 31. Spin-Orbit geometry.

planet, the tidally induced bulge points towards the Sun so that the torque vanishes. But if there is internal dissipation, the motion of the tidal bulge lags behind the motion of the Planet-Sun direction.

The detailed mechanisms of tidal dissipation are uncertain, but it is possible to estimate the magnitude of the gross effect by

$$T_t = -\frac{3}{2}k_2\frac{GM^2R^5}{r^6}\sin 2\delta \tag{81}$$

where k_2 is the tidal Love number and R the planet's mean radius. The angle δ is the lag angle of the high tide. It is related to the tidal dissipation function Q, we have already met in (75), by $2\delta \sim Q^{-1}$. This angle δ is assumed to be small. It is not really known whether δ should be considered as constant or a function of the frequency or amplitude of the tide. We choose here a viscous model for which

$$\delta = 2(\dot\theta - \dot f)\Delta \tag{82}$$

is proportional to the velocity of the tide or equivalently for which the time lag of the tide is constant.

Since the two torques T_p and T_t are small, it follows that $\dot\theta$ changes very little over an orbital period. We may therefore average the torques over an orbital period holding $\dot\theta$ constant. Of course, if $2\dot\theta$ is close to kn (where k is an integer and n the orbital mean motion), the trigonometric line

$$\phi_k = 2\theta - k\ell \tag{83}$$

(where ℓ is the orbital mean anomaly) cannot be averaged out.

Hence close to a resonance, the mean value of the restoring torque T_p does not vanish

$$< T_p >_k = -\frac{3}{2}n^2(B-A)h_k(e)\sin\phi_k\ . \tag{84}$$

The function $h_k(e)$ of the orbital eccentricity is given approximately by

$$\begin{array}{ll} k = 2(\text{synchroneous case}) & h_2(e) = 1 - \frac{5}{2}e^2 + 0(e^4) \\ k = 3(\text{Mercury's case}) & h_3(e) = \frac{7}{2}e - \frac{123}{16}e^3 + 0(e^5) \\ k = 4 & h_4(e) = \frac{17}{2}e^2 - \frac{115}{64}e^4 + 0(e^6)\ . \end{array} \tag{85}$$

Likewise the tidal torque can be averaged out to yield

$$< T_t >_k \; = -\frac{C\beta}{2}\left[\dot{\phi}_k + \frac{k-2}{2}n - 7ne^2\right] + 0(e^4) \tag{86}$$

with

$$\beta = 6n^2 k_2 \left(\frac{mR^2}{C}\right)\left(\frac{M}{m}\right)\left(\frac{R}{a}\right)^3 \Delta\left(1 + \frac{13}{2}e^2\right) . \tag{87}$$

The averaged equation of motion for the resonance angle ϕ_k is thus

$$C\ddot{\theta} = \; < T_p > + < T_t >$$

or

$$\ddot{\phi} + \alpha \sin\phi + \beta\dot{\phi} + \gamma = 0 \tag{88}$$

with

$$\alpha = 3n^2\left(\frac{B-A}{C}\right) h_k(e) \quad , \quad \gamma = \frac{\beta}{2}[k - 2 - 14e^2]n . \tag{89}$$

In the case of Mercury, the values of the parameters may be roughly estimated as (we assume a unit of time such that $n = 1$):

$$\alpha \sim 2.10^{-4} \; , \quad 10^{-8} \leq \beta \leq 10^{-9} \; , \quad \gamma \sim \beta/4 \tag{90}$$

if we assume $e = 0.2$, $(B - A)/C = 1.10^{-4}$ and the tidal dissipation Q between 20 and 200.

The problem we are considering (which, by the way, is equivalent to the problem of the synchroneous motor – see Stoker, 1950, or Andronov et al., 1966) can thus be modelized as a forced pendulum the Hamiltonian of which is

$$H = \frac{1}{2}\left(I - \frac{\gamma}{\beta}\right)^2 - \alpha\cos\phi \tag{91}$$

plus a small linear dissipative force:

$$\dot{\phi} = \frac{\partial H}{\partial I} \quad , \quad \dot{I} = -\frac{\partial H}{\partial \phi} - \beta I . \tag{92}$$

We have seen in Sect. III.5 how to analyse such a system by using the adiabatic invariant theory. It amounts to consider the conservative system (91) with a slowly varying adiabatic invariant

$$J(t) = \exp\{-\beta t\} \cdot J(0) . \tag{93}$$

The analysis may then proceed as follows. Let us consider the "area" enclosed by the guiding trajectory in the phase space $(\phi, \dot{\phi})$. We have

$$\Gamma = \oint \dot{\phi}\, d\phi = \oint \left(I - \frac{\gamma}{\beta}\right) d\phi = 2\pi J - \frac{\gamma}{\beta}\oint d\phi . \tag{94}$$

In case of positive circulation ($\dot{\phi} > 0$),

$$\Gamma_1 = 2\pi e^{-\beta t} J_1(0) - 2\pi \frac{\gamma}{\beta} , \tag{95}$$

the area Γ_1 decreases and thus the spin is slowing down. The spin rate approaches (slowly) its value of resonance.

In case of libration,

$$\Gamma_3 = 2\pi e^{-\beta t} J_3(0) , \tag{96}$$

the area decreases exponentially and the trajectory approaches the stable equilibrium corresponding to exact resonance.

In case of negative circulation ($\dot{\phi} < 0$),

$$\Gamma_2 = 2\pi e^{-\beta t} J_2(0) + 2\pi \frac{\gamma}{\beta} , \tag{97}$$

two cases have to be considered.

If $8\sqrt{\alpha} > 2\pi(\gamma/\beta)$, then $J_2(0)$ has to be positive (because Γ_2 is larger than $8\sqrt{\alpha}$ which is the minimal "area" of a trajectory in the negative circulation regime). Hence Γ_2 decreases and the guiding trajectory approaches the resonance region. The spin rate actually increases.

If $8\sqrt{\alpha} < 2\pi(\gamma/\beta)$, there is one particular guiding trajectory for which $\Gamma_2 = 2\pi(\gamma/\beta)$ and thus $J_2(0) = 0$. The averaged tidal torque along this trajectory is zero and it does not evolve. Other trajectories either above it or below it evolve towards it. Indeed trajectories below it (resp. above it) have $J_2(0) > 0$ (resp. $J_2(0) < 0$) and thus their area Γ_2 decreases (resp. increases) with time.

Whenever the trajectory approaches the libration region coming from positive rotation, it may be captured into resonance (in the libration region) or jump over it. The probability that it is captured into resonance is given by

$$P_{\text{capt}} = 2 \Big/ \left[1 + \frac{\pi}{4} \frac{\gamma}{\beta\sqrt{\alpha}}\right] . \tag{98}$$

Indeed, according to Sect. III.5, we should take $b = \alpha \exp(2\beta t)$ and $c = \gamma \exp(\beta t)/\beta$ in the evaluation of the derivatives of the critical areas (see (III.26) and (III.27)), the ratio of which gives the probability of capture. The capture is, of course, certain when $8\sqrt{\alpha} > 2\pi(\gamma/\beta)$.

With the numerical value of the parameters α, β and γ given in (90), we find that the probability of capture of Mercury in the 3/2 resonance is about $10^{0}/_{0}$. If Mercury had not been captured by the resonance, it would have slowed down a little more to reach a spin state intermediary between the 3/2 resonance and the corotation (corresponding to the special trajectory for which $J_2(0) = 0$). The exact value of this spin rate depends mainly upon the ratio β/γ which in turns depends critically upon the model of tidal dissipation. With the viscous model we have used, $\beta/\gamma = 4$ and the critical rotation period is 64 days.

A second example of application of the adiabatic invariant theory to the spin motion of a planet concerns the history of the obliquity of Mars and has been developed by Ward et al. (1979). The possibility of large changes in the obliquity of Mars (the angle between its equator and its orbital plane) at some time in the past is an important piece in the debate concerning the climatic history of this planet and related questions regarding the origin of several of the planet's surface features. Indeed, changes in the obliquity mean changes in the insolation and thus in the climate. In the present climate on Mars, no water under its liquid form can exist. But several surface features suggest strongly that, at some time in the past, a fluid (most probably water) was indeed running on the surface and carving a primitive fluvial system.

Let us modelize the motion of the spin axis of Mars by the "Colombo's top" as in Sect. III.4. The precession of the orbital plane of the Sun around Mars (or of Mars around the Sun) is due to the perturbations of the other planets upon the orbit of Mars and evaluated as a result of the general planetary theory (e.g. Bretagnon, 1974). It is given as a sum of periodic terms, which being small can be treated independently. Two of them are of special interest because of their frequency. We shall investigate one of them (the term no. 2 in Bretagnon's classification).

It describes a precession of the orbital plane of Mars with an angular velocity of

$$\mu = 6.77 \text{ arc sec/year} = 0.9825\ 10^{-5} n \tag{99}$$

(where n is the orbital angular velocity of Mars). The angle between the normal to the plane of the orbit and the precession axis is $1.22\ 10^{-3}$ radians.

The frequency of this forced orbital precession term is close to the free spin precession frequency

$$\alpha = \frac{3}{2} \frac{n^2}{\varrho} \frac{C - A}{C} = 1.199\ 10^{-5} n$$

if we assume, for ϱ (the spin angular velocity of the planet), the value $669.585n$ and, for $(C - A)/C$ (the dynamical flattening of Mars), the value $5.3503\ 10^{-3}$ which are their present values.

With these values, the coefficients a and b (see (III.56)) are such that $a^{2/3} + b^{2/3}$ is smaller than one, so that a resonance zone appears in the phase space of the Colombo's top (Domain D_3 in Fig. 19, Sect. III.4).

However this resonance domain is so thin (because μ and α are not close enough) that the present value of the obliquity of Mars ($\arccos Z = 24^0 4$), put the spin axis of Mars well into the domain D_1 (see Fig. 19) so that it is not much affected by this resonance.

But the value of the dynamical flattening of Mars, $(C-A)/C$, may not always have been equal to its present value. Indeed part of the present value is certainly due to the big shield volcanoes in the Tharsis region close to the equator. Ward et al. (1979) estimate that prior to the Tharsis uplift, the dynamical flattening of Mars may have been smaller by $6^{0/0}$ to $7^{0/0}$. This would enlarge significantly the resonance domain (domain D_3 in Fig. 19) so much so that it could capture the spin axis of Mars.

Ward et al. (1979) envisage a dynamical history for Mars incorporating the following sequence of events. Differentiation with core formation and possibly large crater events decreasing the primeval α by $4^{0/0}$ to $5^{0/0}$ driving it through resonance. A combination of geophysical processes such as volcanic activity, impact cratering, mantle convection, polar caps evolution, etc., causing α to drift back and forth through resonance during a portion of Mars history. This situation was "turned off" by the Tharsis uplift which drove α away from the resonance and perhaps through resonance with another term in the precession of the orbital plane (term no. 26 in Bretagnon's list). Owing to subsequent partial isostatic compensation, α has drifted back to its present state between the two resonances.

The present value of the adiabatic invariant corresponds to the critical value $J_1^\star$ (see (III.69)) when the ratio α/μ reaches 1.132 (i.e. when $(C-A)/C = 4.963\ 10^{-3}$). Hence we can expect that, every time the dynamical flattening of Mars has reached this value in the past, the spin axis of Mars has jumped from one domain of the Colombo's top phase space to another one.

When the critical value is reached coming from domain D_1 or D_3 (α decreasing), the spin axis of Mars jumps into domain D_2. On the other hand, when the critical value is reached coming from domain D_2 (α increasing), there is a probability of 0.079 to be captured into resonance (domain D_3) against a probability of 0.921 to jump to domain D_1. These probabilities can be evaluated from formulae (III.72) as shown in Henrard (1987).

The effect upon the obliquity of Mars of such jumps can be described as follows.

The precession of the spin axis leads to a periodic (period of about 200 000 years) variation of the obliquity (arccos Z) as the spin axis follows a trajectory of the Colombo's top model. For a given value of α (or of the dynamical flattening of Mars), we can compute the maximum and minimum values of the obliquity given, of course, that the trajectory is known. But the trajectory is known when the value of the adiabatic invariant is known and this value is equal to the critical value at the time of the jump.

Hence the state of the rotation of Mars can be summarized in two diagrams (Figs. 32 and 33) giving the limits of the obliquity with respect to the value of α.

For values of α smaller than the critical value, one state only is possible with a mean value of the obliquity of about 31^0. For values of α larger than the critical value, two states are possible. One corresponding to the present state in domain D_1 (see Fig. 32) with a mean value of the obliquity of 24^04 and the other one in domain D_3 with much larger oscillations in obliquity (see Fig. 33). Whether one state or the other one is effective depends upon the history of Mars. At each transition with α increasing, we can only assign probabilities.

Even if the spin axis of Mars has never been captured in the resonance domain D_3, the jumps from D_1 to D_2 and back at each crossing of the critical value $\alpha/\mu = 1.132$ lead to changes in the mean obliquity of Mars large enough to cause a drastic change in the climate.

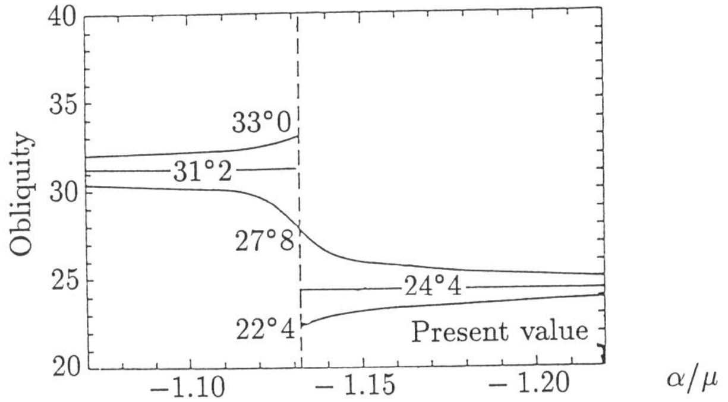

Fig. 32. Maximum and Minimum values of the obliquity of Mars versus α/μ. The values given for α/μ larger than the critical value correspond to trajectories in the domain D_1 (after Ward et al., 1979).

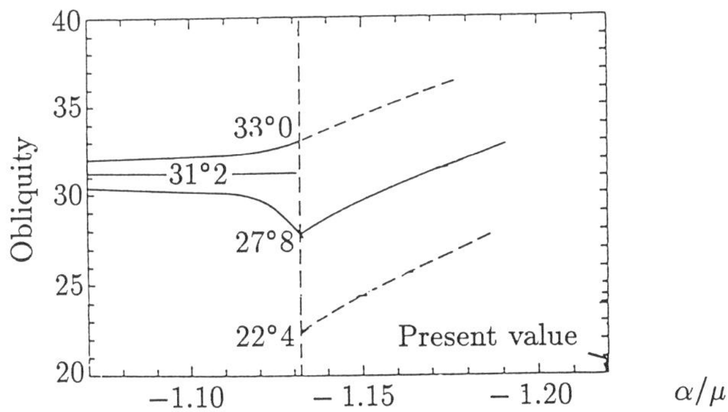

Fig. 33. Maximum and Minimum values of the obliquity of Mars versus α/μ. The values given for α/μ larger than the critical value correspond to trajectories in the domain D_3 (after Ward et al., 1979).

Appendix

Appendix 1: Variational Equations

Let $x = X'(t, x_0, \lambda)$ be the general solution of the dynamical system generated by the Hamiltonian function $H(x, \lambda)$. Each solution is periodic of period $T(x_0, \lambda)$.

Let us call $M(t)$ the principal matrix, solution of the variational equations

$$\frac{d}{dt} M = S \cdot H_{xx} \cdot M \quad \text{with } M(0) = I \tag{1}$$

where the Hessian matrix H_{xx} is evaluated along the solution $x = X'(t, x_0, \lambda)$. Whenever the initial conditions are function of a parameter (say J) the principal matrix enables us to compute the derivatives of X' with respect to J:

$$\frac{\partial X'}{\partial J} = M(t) \cdot \frac{\partial x_0}{\partial J} \, . \tag{2}$$

The principal matrix has also the basic property that the following row vector is constant

$$\left(\frac{\partial X'}{\partial t}\right)^{\tau} SM(t) = \text{ constant } = H_x(x_0) \tag{3}$$

This can be checked by differentiating the left-hand member of (3) with respect to time and verifying that the derivative is equal to zero: its value for $t = 0$ gives the right hand member.

The variational equations with respect to the parameter λ are given by

$$\frac{d}{dt}\left(\frac{\partial X'}{\partial \lambda}\right) = SH_{xx}\left(\frac{\partial X'}{\partial \lambda}\right) + SH_{x,\lambda} \, . \tag{4}$$

Its solution is:

$$\left(\frac{\partial X'}{\partial \lambda}\right) = M(t) \cdot \frac{\partial x_0}{\partial \lambda} + V(t) \tag{5}$$

where $V(t)$ is the particular solution of (4) corresponding to the initial conditions $V(0) = 0$.

Equation (5) together with (3) leads to

$$\left(\frac{\partial X'}{\partial t}\right)^{\tau} S\left(\frac{\partial X'}{\partial \lambda}\right) = H_x(x_0) \cdot \frac{\partial x_0}{\partial \lambda} + \left(\frac{\partial X'}{\partial t}\right)^{\tau} SV(t) \, . \tag{6}$$

Appendix 2: Fixing the Unstable Equilibrium and the Time Scale

Let us assume that the dynamical system described by the hamiltonian function $H(x, \lambda)$ possesses a non degenerate equilibrium $x^\star(\lambda)$; i.e. $x^\star(\lambda)$ is an isolated solution of

$$H_x(x^\star(\lambda), \lambda) = 0 \tag{7}$$

such that the Hessian matrix $H_{xx}(x^\star(\lambda), \lambda)$ is regular. For a sufficiently small ε there exists an holomorphic solution $x^{\star\star}(\lambda, \varepsilon)$ of

$$H_x(x^{\star\star}, \lambda) + \varepsilon S \frac{\partial x^{\star\star}}{\partial \lambda} = 0 \tag{8}$$

such that $x^{\star\star}(\lambda, 0) = x^\star(\lambda)$ (see Wasow, 1965).

Let us define the Hamiltonian function

$$M(x', \lambda) = H(x^{\star\star} + x', \lambda) + \varepsilon(x')^{\tau} S \frac{\partial x^{\star\star}}{\partial \lambda} \, . \tag{9}$$

The origin $x' = 0$ is a non-degenerate equilibrium of the system described by (9).

Now, if we consider the non-autonomous system described by $H(x, \varepsilon t)$ and apply the translation $x = x^{\star\star}(\varepsilon t, \varepsilon) + x'$, it will be described by the Hamiltonian

$$M(x', \varepsilon t) = H(x^{\star\star}(\varepsilon t, \varepsilon), \varepsilon t) + \varepsilon(x')^{\tau} S \frac{\partial x^{\star\star}}{\partial \lambda} \, . \tag{10}$$

Indeed the second term in the right-hand member of (10) is the remainder function of the (time-dependent) translation.

In chapter II, we substitute $M(x', \lambda)$ to $H(x, \lambda)$ in order to fix the unstable equilibrium. In doing so we do not change the dynamical system described by $H(x, \varepsilon t)$ but we do change the dynamical system described by $H(x, \lambda)$. In other words, we do no affect the "real problem" but we do distort slightly its "frozen approximation".

The functions which appear in the final results, $\omega(\lambda)$ (see (II.5)), $h_i^\star(\lambda)$ (see (II.21)), $J_i^\star(\lambda)$ (see (II.2)), $g_i(\lambda)$ (see (II.31)) are defined on the frozen approximation $M(x', \lambda)$. But we would rather compute them directly with $H(x, \lambda)$.

This can be done because those functions computed with $M(x', \lambda)$ or $H(x, \lambda)$ differ at most by terms of the order of ε and the relative accuracy of the final results is not as good (of the order of $\varepsilon^{1/3}$).

This is quite obvious for $\omega(\lambda)$ which is the eigenvalue of the Hessian matrix H_{xx} and for $J_i^\star(\lambda)$ and $h_i^\star(\lambda)$ which are functions related to the area enclosed by the homoclinic orbit or close-by orbits. This is less obvious for $\partial J_i^\star / \partial \lambda$ but it can be shown in the following way. We have

$$\frac{\partial}{\partial \lambda} J_i^\star(H) = - \int_{\Gamma_i} \frac{\partial H}{\partial \lambda} dt \, , \quad \frac{\partial}{\partial \lambda} J_i^\star(M) = - \int_{\Gamma_i} \frac{\partial M}{\partial \lambda} dt \tag{11}$$

according to (II.56). We then compute

$$\frac{\partial}{\partial \lambda} J_i^\star(M) = \frac{\partial}{\partial \lambda} J_i^\star(H) + \int_{\Gamma_i} (H_x)^\tau \frac{\partial x^{\star\star}}{\partial \lambda} dt + \varepsilon \int_{\Gamma_i} (x')^\tau S \frac{\partial^2 x^{\star\star}}{\partial \lambda^2} dt \, . \tag{12}$$

The second term vanishes because the curves Γ_i are closed:

$$\int_{\Gamma_i} (H_x)^\tau \frac{\partial x^{\star\star}}{\partial \lambda} dt = \int_{\Gamma_i} \left(\frac{\partial x}{\partial t} \right)^\tau S \frac{\partial x^{\star\star}}{\partial \lambda} dt = - \int_{\Gamma_i} \left(\frac{\partial x^{\star\star}}{\partial \lambda} \right)^\tau S dx = 0 \, . \tag{13}$$

The third term is of the order of ε as the integral is finite (x' goes exponentially to zero as t goes to infinity).

We have just shown that the equilibrium $x^\star(\lambda)$ can always be assumed to be at the origin. Let us go further and show that we can always assume that the eigenvalue $\omega(\lambda)$ of the matrix SH_{xx} can be made independent of λ.

This is obtained by scaling the Hamiltonian function and substituting to it the new Hamiltonian

$$N(x, \lambda) = \frac{\omega(\lambda_0)}{\omega(\lambda)} H(x, \lambda) \tag{14}$$

where $\lambda_0 = \varepsilon t_0$ is a fixed value of λ. This is equivalent to changing the independent variables from t to:

$$t' = t_0 + \int_{t_0}^{t} \frac{\omega(\lambda)}{\omega(\lambda_0)} dt \, . \tag{15}$$

This change of scale is valid as long as $\omega(\lambda)$ does not vanish. But our assumption that the unstable equilibrium is non-degenerate implies that $\omega(\lambda)$ is bounded away from zero.

The change of time variable (15) can be extended to the non-autonomous system obtained by the substitution $\lambda = \varepsilon t$. Of course the time transformation (15) will no longer be linear and this may be a nuisance in the interpretation of the results.

Actually this minor difficulty will be avoided. The time interval in which we are interested in will be restricted to the interval

$$|t' - t_0| \leq d_1 \varepsilon^{-1/3} . \tag{16}$$

In this interval, the original time t, and the new time t', cannot be desynchronized by more than

$$|t' - t| \leq d_2 \varepsilon^{1/3} . \tag{17}$$

As we have already mentioned, the relative accuracy of the final results is of the same order of magnitude so that t and t' can be interchanged without further loss of accuracy.

Appendix 3: Mean Value of $R_i(\psi_i, J_i, \lambda)$ $1 \leq i \leq 2$

Inside the disk of radius δ around the origin, the remainder function $R_i(\psi_i, J_i, \lambda)$ is the sum of the remainder function R_N of the transformation from (x) to (z) (see (II.4)) and the remainder function R' of the transformation from (z) to the (ψ_i, J_i) (see (II.13)). In order to make the notations less cumbersome, we shall drop in this appendix the subscript (i) in R_i, ψ_i, J_i etc.

We then have from (II.14)

$$R(\psi) = -\left(\frac{\partial \mathcal{J}}{\partial \lambda}\right)\psi + R_N(z_1, z_2) \quad \text{for } |\psi| \leq \psi_\delta = 1/2\left|\frac{\partial Z}{\partial J}\right| \log\frac{\delta^2}{|Z|} . \tag{18}$$

Outside the disk of radius δ the remainder function is:

$$R(\psi) = -\left(\frac{\partial \mathcal{J}}{\partial \lambda}\right)\psi + \tilde{R}(t, h, \lambda) \quad \text{for } |\psi| \geq \psi_\delta . \tag{19}$$

We evaluate $\tilde{R}$ in the following way. First from (I.65) we have

$$R(\psi) = R(\psi_\delta) + (\psi - \psi_\delta)\left\{\frac{\partial R}{\partial \psi}\right\}_{\psi=\psi_\delta} + \int_0^t \left(\frac{\partial X'}{\partial t}\right)^\tau SV(t)dt . \tag{20}$$

The origin of time $(t = 0)$ has been chosen at $\psi = \psi_\delta$ and we recall that $V(t)$ is the solution of the variational equations with respect to λ with zero initial conditions (see Appendix 1).

R and $\partial R/\partial \psi$ evaluated at $\psi = \psi_\delta$ can be computed from (18). We obtain

$$\tilde{R} = B_1(h) + B_2(h) \cdot t + \int_0^t \left(\frac{\partial X'}{\partial t}\right)^\tau SV(t)dt \tag{21}$$

with

$$\begin{aligned} B_1(h) &= R_N(z_1, z_2) \\ B_2(h) &= \left(\frac{\partial Z}{\partial h}\right)^{-1} \left\{\frac{\partial R_N}{\partial z_1} z_1 - \frac{\partial R_N}{\partial z_2} z_2\right\} \end{aligned} \tag{22}$$

evaluated at $z_1 = \pm\delta, z_2 = \pm Z/\delta$ (the upper sign correspond to D_2 and the lower sign to D_1).

From (19) and (20), we can evaluate the mean-value of $R(\psi)$

$$\frac{1}{2\pi}\int_{-\pi}^{\pi} R(\psi)\, d\psi = \frac{1}{2\pi}\int_{-\psi_\delta}^{\psi_\delta} R_N(z_1, z_2)\, d\psi + \frac{1}{T(h,\lambda)}\int_0^{T_\delta(h,\lambda)} \tilde{R}(t, h, \lambda)\, dt\ . \tag{23}$$

The first integral in the right-hand side of (23) is evaluated in the following way. The remainder function R_N is divided into three pieces

$$R_N(z_1, z_2) = ZB_3(Z) + z_1 B_4(Z, z_1) + z_2 B_5(Z, z_2) \tag{24}$$

where the functions B_j $(3 \leq j \leq 5)$ are analytical functions of their arguments. The first integral is then equal to:

$$\begin{aligned} \int_{-\psi_\delta}^{\psi_\delta} R_N\, d\psi &= \left(\frac{\partial \mathcal{J}}{\partial Z}\right)^{-1} \int_{\mp Z/\delta}^{\pm\delta} ZB_3 \frac{dz_1}{z_1} + B_4\, dz_1 - B_5\, dz_2 \\ &= \left(\frac{\partial \mathcal{J}}{\partial Z}\right)^{-1} \left\{B_6(Z,\lambda) + ZB_3(Z,\lambda)\log\frac{\delta^2}{|Z|}\right\}\ . \end{aligned} \tag{25}$$

The second integral in (23) is taken over a finite time interval (T_δ is defined in (II.23)) and can be evaluated numerically if need be. The integrand and the limit are analytical functions of h and λ. Hence the integral itself is an analytical function of those arguments or of Z and λ

$$B_8(Z,\lambda) = \int_0^{T_\delta} \tilde{R}(t)\, dt\ . \tag{26}$$

Collecting those results and using the expression (II.22) for the period $T(h, \lambda)$, we obtain

$$< R(\psi) > \; = \frac{1}{2\pi}\left(\frac{\partial \mathcal{J}}{\partial Z}\right)^{-1} \left\{B_6(Z,\lambda) + B_8(Z,\lambda) + ZB_3(Z,\lambda)\log\frac{\delta^2}{|Z|}\right\}\ . \tag{27}$$

For small values of Z (or h), we then have

$$< R(\psi) > \; = \left(\frac{\partial K}{\partial J}\right) \{-g(\lambda) + 0(h \log h^{-1})\} \tag{28}$$

where

$$g(\lambda) = \frac{-1}{2\pi\omega}\{B_6(0,\lambda) + B_8(0,\lambda)\}\ . \tag{29}$$

Appendix 4: Mean Value of $R_3(\psi_3, J_3, \lambda)$

In order to evaluate $R_3(\psi_3, J_3, \lambda)$, we shall compare it to $R_1(\psi_1, J_1, \lambda)$ and $R_2(\psi_2, J_2, \lambda)$ by estimating it on the two sub-intervals (see Fig. 34).

$$\begin{aligned} &(\mathrm{A}) \quad -\psi_\delta \leq \psi_3 \leq \psi^\star - \psi_\delta \quad \text{following the curve } \Gamma_1 \, , \\ &(\mathrm{B}) \; \psi^\star - \psi_\delta \leq \psi_3 \leq 2\pi - \psi_\delta \text{ following the curve } \Gamma_2 \, . \end{aligned} \tag{30}$$

The quantity $\psi^\star$ is the value of ψ_3 at the anti-apex (see Fig. 34) and ψ_δ is the value of ψ_3 where the orbit leaves the disk of radius δ around the origin:

$$\psi_\delta = \frac{1}{2} \left| \frac{\partial \mathcal{J}_3}{\partial Z} \right| \log \frac{\delta^2}{|Z|} \, . \tag{31}$$

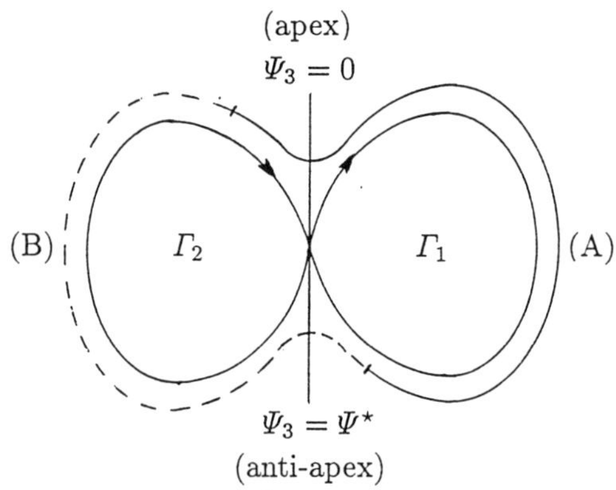

Fig. 34. The two subintervals (A) and (B) into which is divided a trajectory in the domain D_3.

On the interval (A), we can reproduce the argument of Appendix 3 and show that

$$R_3(\psi_3) = -\left(\frac{\partial \mathcal{J}_3}{\partial \lambda}\right) \psi_3 + R_N(z_1, z_2) \;\text{ for } |\psi_3| \leq \psi_\delta \, , \tag{32}$$

$$R_3(\psi_3) = -\left(\frac{\partial \mathcal{J}_3}{\partial \lambda}\right) \psi_3 + \tilde{R}_1(t, h, \lambda) \;\text{ for } \psi_\delta \leq \psi_3 \leq \psi^\star - \psi_\delta \, . \tag{33}$$

The function $\tilde{R}_1$ is the same as the one considered in Appendix 3. The only change is that it is now evaluated for positive values of h and not for negative values of h as in Appendix 3.

On the other hand, the "clocks" t, ψ_1, ψ_3 can be synchronized (at the apex) and are related by scale factors:

$$\psi_1 \left(\frac{\partial \mathcal{J}_1}{\partial Z}\right) = \left(\frac{\partial Z}{\partial h}\right)^{-1} t = \psi_3 \left(\frac{\partial \mathcal{J}_3}{\partial Z}\right) \tag{34}$$

so that (32) and (33) can be rewritten

$$R_3(\psi_3) = R_1(\psi_1) + \left(\frac{\partial \mathcal{J}_1}{\partial Z}\right)^{-1} \Delta'_{12} \cdot \psi_3 \quad \text{for } -\psi_\delta \le \psi \le \psi^\star - \psi_\delta \tag{35}$$

with

$$\Delta'_{12} = \frac{\partial \mathcal{J}_1}{\partial \lambda}\frac{\partial \mathcal{J}_2}{\partial Z} - \frac{\partial \mathcal{J}_2}{\partial \lambda}\frac{\partial \mathcal{J}_1}{\partial Z} . \tag{36}$$

Indeed, comparing (32) and (18), we find (35) with Δ'_{12} equal to

$$\Delta'_{12} = \frac{\partial \mathcal{J}_1}{\partial \lambda}\frac{\partial \mathcal{J}_3}{\partial Z} - \frac{\partial \mathcal{J}_3}{\partial \lambda}\frac{\partial \mathcal{J}_1}{\partial Z} \tag{37}$$

which is equivalent to (36) because $\mathcal{J}_3 = \mathcal{J}_1 + \mathcal{J}_2$.

Similarly, on the interval (B), $R_3(\psi_3)$ can be compared to $R_2(\psi_2)$ to yield:

$$R_3(\psi_3) = R_2(\psi_2) - \left(\frac{\partial \mathcal{J}_2}{\partial Z}\right)^{-1} \Delta'_{12}(\psi_3 - \psi^\star) + \{R_3(\psi^\star) - R_N(\psi^\star)\} . \tag{38}$$

The last term in (38) comes from the fact that, close to the anti-apex, a formula analogous to (35) gives the remainder function corresponding to initial conditions at the anti-apex. The correction to be made (the last term) in order to bring the initial conditions back to the apex is given by (I.17).

The value of $R_3(\psi^\star)$ can be evaluated by evaluating $R_3(\psi^\star - \psi_\delta)$ from (35) and then by using (18) to propagate this value to $\psi^\star$. We find

$$\begin{aligned} R_3(\psi^\star) &= R_N(\psi^\star) - R_1(2\pi) + \left(\frac{\partial \mathcal{J}_1}{\partial Z}\right)^{-1} \Delta'_{12}\psi^\star \\ &= R_N(\psi^\star) - R_1(0) + \left(\frac{\partial \mathcal{J}_2}{\partial Z}\right)^{-1} \Delta'_{12}(2\pi - \psi^\star) \end{aligned} \tag{39}$$

because R_1 is periodic and

$$\left(\frac{\partial \mathcal{J}_1}{\partial Z}\right)^{-1} \psi^\star = 2\pi \left(\frac{\partial \mathcal{J}_3}{\partial Z}\right)^{-1} = (2\pi - \psi^\star)\left(\frac{\partial \mathcal{J}_2}{\partial Z}\right)^{-1} \tag{40}$$

by (34) and $\mathcal{J}_3 = \mathcal{J}_1 + \mathcal{J}_2$.

Eventually, (38) reads

$$R_3(\psi_3) = R_2(\psi_2) + \left(\frac{\partial \mathcal{J}_2}{\partial Z}\right)^{-1} \Delta'_{12}(2\pi - \psi_3) - R_1(0) \quad \text{for } \psi^\star - \psi_\delta \le \psi \le 2\pi - \psi_\delta . \tag{41}$$

Using (35) and (41), we intregrate R_3 to find

$$\begin{aligned} \frac{1}{2\pi}\int_{-\psi_\delta}^{2\pi - \psi_\delta} R_3(\psi_3)\, d\psi_3 &= \pi \left(\frac{\partial \mathcal{J}_3}{\partial Z}\right)^{-1} \Delta'_{12} - \frac{1}{2\pi} R_1(0)(2\pi - \psi^\star) \\ &+ \left(\frac{\partial \mathcal{J}_3}{\partial Z}\right)^{-1} \left\{\left(\frac{\partial \mathcal{J}_1}{\partial Z}\right) < R_1 > + \left(\frac{\partial \mathcal{J}_2}{\partial Z}\right) < R_2 >\right\}. \end{aligned} \tag{42}$$

Taking into account (28) and the fact that $R_1(0) = 0(|Z|) = 0(h)$, we obtain

$$< R_3 >= \left(\frac{\partial K}{\partial J_3}\right)\{-g_3(\lambda) + \pi\Delta_{12} + 0(h\log h^{-1})\} \tag{43}$$

where we have put

$$g_3(\lambda) = g_1(\lambda) + g_2(\lambda)\ ,$$
$$\Delta_{12} = \frac{\partial J_1}{\partial\lambda}\frac{\partial J_2}{\partial h} - \frac{\partial J_2}{\partial\lambda}\frac{\partial J_1}{\partial h}\ . \tag{44}$$

Appendix 5: Estimation of the Trajectory Close to the Equilibrium

In the disk of radius δ around the origin, the Hamiltonian of the non-autonomous dynamical system is given by

$$H' = H_N(Z, \varepsilon t) + \varepsilon R_N(z_1, z_2, \varepsilon t) \tag{45}$$

where R_N (see (II.11) and (II.12)) is of the order two in the coordinates (z_1, z_2): $R_N = 0(||z||^2)$.

The corresponding differential equations are

$$\dot z_1 = \Omega(Z, \varepsilon t)z_1 + \varepsilon\frac{\partial R_N}{\partial z_2}\ , \quad \dot z_2 = -\Omega(Z, \varepsilon t)z_2 - \varepsilon\frac{\partial R_N}{\partial z_1} \tag{46}$$

where

$$\Omega(Z, \varepsilon t) = \frac{\partial H_N}{\partial Z}\ . \tag{47}$$

We shall write $z(t) = (z_1(t), z_2(t))$ the solution of (46) the initial conditions of which are (at the apex or anti-apex) $z(0) = (z_0, z_0)$. We shall compare it with the solution $\mathbf{u}^\star(t) = (\mathbf{u}_1^\star(t), \mathbf{u}_2^\star(t))$ of the system corresponding to $R_N = 0$. This last solution is easy to write. It is

$$\mathbf{u}_1^\star = z_0 e^{\mu}\ , \quad \mathbf{u}_2^\star = z_0 e^{-\mu} \tag{48}$$

with

$$\mu(t) = \int_0^t \Omega(z_0^2, \varepsilon t)\, dt\ .$$

We shall show that the difference $z(t) - \mathbf{u}^\star(t)$ remains of the order of ε as long as both solutions remain in the disk of radius δ. Of course, we shall have to impose some constraints on the initial condition z_0, namely

$$\delta\exp\{-1/c_1\varepsilon\} \le |z_0| \le \delta\max\left\{1/2, \sqrt{\omega_0/2c_2}\right\}\ . \tag{49}$$

The constants c_1 and c_2 are given by

$$c_1 = \max\left\{\frac{22c_3[15c_2 + 13\omega_0]}{\omega_0^2}, \frac{2}{\omega c_4}\right\} \tag{50}$$

where

$$\begin{aligned} c_2 &= \delta^2 \sup |\partial \Omega/\partial Z| , \\ c_3 &= \delta \sup |\partial^2 R_N/\partial z_i \partial z_j| , \\ c_4 &\text{ such that, for } |\lambda - \lambda_0| \le c_4 \text{ , we have } |\omega(\lambda) - \omega_0| \le 1/4\omega_0 . \end{aligned} \tag{51}$$

The suprema in (51) are taken on the disk of radius δ around the equilibrium and ω_0 in (50) and (51) is the value of $\omega(\lambda_0)$, the independent term of $\Omega(Z)$ evaluated at time $t = 0$.

We shall first evaluate the time T_0 at which the solution $\mathbf{u}^\star(t)$ (see (48)) reaches the boundary of the disk. It is given by

$$\mu(T_0) = \log \frac{\delta}{z_0} . \tag{52}$$

Let

$$T_1 = \frac{2}{\omega_0 c_1 \varepsilon} \le \frac{c_4}{\varepsilon} . \tag{53}$$

On the interval $0 \le t \le T_1$, we have

$$|\Omega(z_0^2, \varepsilon t) - \omega_0| \le c_2 \frac{z_0^2}{\delta^2} + \frac{1}{4}\omega_0 \le \frac{1}{2}\omega_0 . \tag{54}$$

Hence the function $\mu(t)$ is increasing on this interval and such that

$$\frac{1}{2}\omega_0 \le \frac{d\mu}{dt} \le \frac{3}{2}\omega_0 . \tag{55}$$

The value of $\mu(T_1)$ is thus larger than $\frac{1}{2}\omega_0 T_1$

$$\mu(T_1) \ge \frac{1}{2}\omega_0 T_1 = \frac{1}{c_1 \varepsilon} \ge \log \frac{\delta}{z_0} \tag{56}$$

and there is a unique solution T_0 of (52) in the interval $[0, T_1]$

$$\varepsilon T_0 \le \varepsilon T_1 = \frac{2}{\omega_0 c_1} \le \frac{\omega_0}{11 c_3 [15 c_2 + 13\omega_0]} . \tag{57}$$

We use, in Section II.5, a better estimate of T_0 in case ω is assumed to be independent of λ. We develop this estimate here for further reference. We have

$$\mu(T_0) - \Omega_0 T_0 = \int_0^{T_0} \{\Omega(z_0^2, \varepsilon t) - \Omega(z_0^2, \lambda_0)\}\, dt \sim 0(\varepsilon z_0^2 T_0^2) \sim 0(\varepsilon h \log^2 h^{-1}) . \tag{58}$$

Hence, by (52), we have that

$$\left| T_0 - \frac{1}{\Omega_0} \log \frac{\delta}{z_0} \right| \sim 0(\varepsilon h \log^2 h^{-1}) . \tag{59}$$

The main argument of this appendix is based upon the principle of contracting maps.

Let us consider the set $\mathbf{F}$ of continuous functions $U(t) = (U_1(t), U_2(t))$ on the interval $0 \le t \le T_0$ and with initial conditions $U(0) = (z_0, z_0)$. The Tchebitchev norm

$$||U(t)|| = \max_{0 \le t \le T_0} \{|U_1(t)| + |U_2(t)|\} \tag{60}$$

makes it a complete space.

The solution of (46) for the given initial conditions is a fixed point of the operator $\mathbf{B}$ from $\mathbf{F}$ into $\mathbf{F}$

$$V = \mathbf{B}(U) \tag{61}$$

with

$$\begin{aligned} V_1 &= z_0 + \int_0^t \left\{ \Omega(U_1 U_2, \varepsilon t) U_1 + \frac{\partial R_N}{\partial z_2}(U_1, U_2, \varepsilon t) \right\} dt \,, \\ V_2 &= z_0 - \int_0^t \left\{ \Omega(U_1 U_2, \varepsilon t) U_2 + \frac{\partial R_N}{\partial z_1}(U_1, U_2, \varepsilon t) \right\} dt \,. \end{aligned} \tag{62}$$

Let us consider the ball of radius η

$$\eta = \frac{11}{\omega_0} c_3 \varepsilon \le \frac{\delta}{4} \tag{63}$$

around the function $\mathbf{u}^\star(t)$ (see (48)) in $\mathbf{F}$. In order to make η smaller than $\delta/4$ in (63), we have to assume that ε is small enough. Let us now estimate $||\mathbf{B}(U) - \mathbf{B}(\mathbf{u}^\star)||$ for U in the ball.

We find

$$|U_1 U_2 - z_0^2| \le \eta[\mathbf{u}_1^\star + \mathbf{u}_2^\star] + \eta^2 \le \frac{3}{2}\eta\delta \,, \tag{64}$$

$$|\Omega(U_1 U_2) - \Omega(z_0^2)| \le \max \left| \frac{\partial \Omega}{\partial Z} \right| \frac{3}{2}\eta\delta \le \frac{3}{2}\frac{\eta}{\delta} c_2 \,, \tag{65}$$

$$\begin{aligned} |\Omega(U_1 U_2) U_i - \Omega(z_0^2)\mathbf{u}_i^\star| &\le \frac{3}{2}\frac{\eta}{\delta} c_2 |U_i| + \max \left| \frac{d\mu}{dt} \right| \eta \\ &\le \frac{3}{2}\eta \left[\frac{5}{4} c_2 + \omega_0 \right] \,, \end{aligned} \tag{66}$$

$$\begin{aligned} \varepsilon \left| \frac{\partial R_N}{\partial z_k}(U_j) - \frac{\partial R_N}{\partial z_k}(\mathbf{u}_j^\star) \right| &\le 2 \sup \left| \frac{\partial^2 R_N}{\partial z_i \, \partial z_k} \right| \eta\varepsilon \\ &\le 2 c_3 \frac{\eta\varepsilon}{\delta} \,. \end{aligned} \tag{67}$$

Collecting those results, we obtain

$$\begin{aligned} ||\mathbf{B}(U) - \mathbf{B}(\mathbf{u}^\star)|| &\le 2T_0 \left\{ \frac{3}{2} \left[\frac{5}{4} c_2 + \omega_0 \right] + 2\frac{c_3 \varepsilon}{\delta} \right\} \eta \\ &\le \frac{T_0}{4} [15 c_2 + 13 \omega_0] \\ &\le \frac{1}{4} \,. \end{aligned} \tag{68}$$

Hence the operator $\mathbf{B}$ is contracting (with a factor of contraction less than 1/2) in the ball of radius η around the function $\mathbf{u}^\star$.

We estimate also that η is large enough, i.e.

$$\eta > 2||B(\mathbf{u}^\star) - \mathbf{u}^\star|| \, . \tag{69}$$

Indeed, if we call $w = B(\mathbf{u}^\star) - \mathbf{u}^\star$, we have

$$\begin{aligned} w_1 &= -z_0[e^\mu - 1] + \int_0^t \left\{ \frac{d\mu}{dt} z_0 e^\mu + \varepsilon \frac{\partial R_N}{\partial z_2} \right\} dt \, , \\ w_2 &= -z_0[e^{-\mu} - 1] - \int_0^t \left\{ \frac{d\mu}{dt} z_0 e^{-\mu} + \varepsilon \frac{\partial R_N}{\partial z_1} \right\} dt \, , \end{aligned} \tag{70}$$

from which it follows easily that

$$||w|| \le \frac{4\varepsilon}{\omega_0} \frac{c_3}{\delta} \{ z_0[e^\mu - 1] + z_0[e^{-\mu} - 1[\le \frac{5}{11}\eta \, . \tag{71}$$

For this last estimate, it is essential that R_N does not contain linear terms in (z_1, z_2) and thus that the unstable equilibrium has a position independent of the time (see Appendix 2).

Hence the operator $\mathbf{B}$ possesses, in the ball of radius η, one and only one fixed point to which converge the Cauchy series $\{\mathbf{B}^i \mathbf{u}^\star\}$. It is the solution of the differential system (46) and we have:

$$|z_1(t) - z_0 e^\mu| \le \frac{11 c_3}{\omega_0} \varepsilon \, , \quad |z_2(t) - z_0 e^{-\mu}| \le \frac{11 c_3}{\omega_0} \varepsilon \, . \tag{72}$$

This estimate is valid for z_0 in the interval

$$\delta \exp\left\{ -\frac{1}{c_1 \varepsilon} \right\} \le |z_0| \le \delta \max\left\{ 1/2, \sqrt{\omega_0 / 2c_2} \right\} \, . \tag{73}$$

The condition on z_0 may be interpreted in terms of $h_0 = H_N(z_0^2, \lambda_0)$, the value of h at time $t = 0$:

$$\delta^2 \omega(\lambda_0) \exp\left\{ -\frac{2}{c_1 \varepsilon} \right\} \le h_0 \le c_5 \tag{74}$$

where c_5 is a constant independent of ε.

Appendix 6: Computation of the True Time of Transit

We shall first evaluate $\Lambda_i = \lambda_0 - \lambda_N$ for one of the internal region D_i $(1 \le i \le 2)$. We shall use the notations and assumptions of Section II.7, and, as we did there, we shall drop the index (i) affecting the quantities J_i, $J_i^\star$, $h_i^\star$, g_i, $\cdots$. We shall also assume that ω does not depend upon λ (see Appendix 2).

The true time of transit Λ is evaluated by summing the individual increments (II.59)

$$|\Lambda| = \varepsilon \sum_{k=0}^{N-1} \left\{ \frac{1}{2\omega} \log \frac{h^\star}{|h_k|} + \frac{1}{2\omega} \log \frac{h^\star}{|h_{k+1}|} + 0(h_k, \varepsilon \log h_k^{-1}) \right\} . \tag{75}$$

The quantities $h^\star$ are supposed to be evaluated at $\lambda = \lambda_k$ or $\lambda = \lambda_{k+1}$, but they can all be evaluated at $\lambda = \lambda_0$ or better at $\lambda = \tau$ (the pseudo-crossing time) without increasing the error on the result.

Indeed, a first estimation of $\tau - \lambda_k$ gives its order of magnitude by considering that, at each apex A_k (except maybe at one of the last two A_1, A_0), the value of h_k is at least of the order of ε. Hence

$$\begin{aligned} \tau - \lambda_k &= 0(N\varepsilon \log \varepsilon^{-1}) && 2 \leq k \leq N \\ \tau - \lambda_k &= 0(\varepsilon \log h_m^{-1}, N\varepsilon \log \varepsilon^{-1}) && 0 \leq k \leq 1 \\ h_m &= \min(|h_0|, |h_1|) . \end{aligned} \tag{76}$$

On the other hand, the values of $|h_k|$ are related by (see (II.59))

$$|h_k| = kb - h_0 + 0(\varepsilon^2 \log h_m^{-1}, k\varepsilon^2 \log \varepsilon^{-1}) \tag{77}$$

where we define b as the absolute value of the estimate (II.59) for Δh, evaluated at $\lambda = \tau$:

$$b = 2\pi\varepsilon \left| \frac{\partial J^\star}{\partial \lambda} \right| . \tag{78}$$

Let us introduce also the notation

$$x = \frac{h_0}{b} \leq 1 . \tag{79}$$

Formula (75) then becomes

$$\begin{aligned} |\Lambda| = \frac{\varepsilon}{2\omega} \left\{ 2N \log \frac{h^\star}{b} - \log x - \log \prod_{k=1}^{N-1} (k - x) - \log \prod_{k=1}^{N} (k - x) \right\} \\ + 0(N\varepsilon^2 \log h_m^{-1}, N^2\varepsilon^2, N\varepsilon^2 \log \varepsilon^{-1}) . \end{aligned} \tag{80}$$

Introducing the following formula and approximation for the Gamma function (see Abramowitz and Stegun, 1965),

$$\begin{aligned} &\prod_{k=1}^{N} (k - x) = \frac{\Gamma(N + 1 - x)}{\Gamma(1 - x)} , \\ &\log \Gamma(y) = (y - 1/2) \log y - y + 1/2 \log 2\pi + 0(1/y) , \end{aligned} \tag{81}$$

we obtain

$$\begin{aligned} \left(\frac{\partial J^\star}{\partial \lambda} \right) \Lambda = \frac{-b}{2\pi\omega} \left\{ \log \frac{\Gamma(1 - x)}{\sqrt{2\pi x}} + x \log \frac{h^\star}{b} - \frac{h_N}{b} \log \frac{h^\star e}{|h_N|} \right\} \\ + 0\left(\frac{\varepsilon}{N}, N\varepsilon^2 \log h_m^{-1}, N^2\varepsilon^2, N\varepsilon^2 \log \varepsilon^{-1} \right) . \end{aligned} \tag{82}$$

We recall that ω, b and $h^\star$ are evaluated at $\lambda = \tau$.

The error term is minimal and of the order of $\varepsilon^{4/3} \log \varepsilon^{-1}$ if we choose

$$N \sim \varepsilon^{-1/3} , \quad \log h_m^{-1} \le \varepsilon^{-1/3} . \tag{83}$$

This means that we can take, for the quantity η defined in (II.60), the value announced in (II.61). We use this particular choice from (II.79) on.

For the evaluation of $\Lambda_3 = \lambda_M - \lambda_0$, the true time of transit in domain D_3, we have to consider passage by the apices A'_k and the corresponding anti-apices. Let us call h'_0 the value of h reached at the first anti-apex (after one traverse along the curve Γ_1):

$$h'_0 = h_0 + \Delta h_1 \ge 0 . \tag{84}$$

Summing the individual increment (II.59) and remembering that we have now two traverses (one along Γ_1 and the other along Γ_2) for a full trip from apex to apex, we obtain

$$\begin{aligned} |\Lambda_3| = \varepsilon \sum_{k=0}^{N-1} \left\{ \frac{1}{2\omega} \log \frac{h_1^\star}{h_k} + \frac{1}{2\omega} \log \frac{h_1^\star}{h'_k} + \frac{1}{2\omega} \log \frac{h_2^\star}{h'_k} + \frac{1}{2\omega} \log \frac{h_2^\star}{h_{k+1}} \right\} \\ + 0(\varepsilon^{4/3} \log \varepsilon^{-1}) \end{aligned} \tag{85}$$

where

$$h_k = h_0 + k b_3 , \tag{86}$$

$$b_3 = 2\pi\varepsilon \left| \frac{\partial J_3^\star}{\partial \lambda} \right| = |\Delta h_3| , \tag{87}$$

$$h'_k = h'_0 + k b_3 . \tag{88}$$

The quantities ω, $h_i^\star$, $\cdots$ are evaluated for each individual term at the corresponding apex or anti-apex but, because of (76), they can all be evaluated at $\lambda = \lambda_0$ or at $\lambda = \tau$ without increasing the error terms. The Gamma function may again be used to evaluate the sums of logarithms and, after some manipulations, we arrive at

$$\begin{aligned} \left(\frac{\partial J_3^\star}{\partial \lambda} \right) \Lambda = & -\frac{h_M}{\pi\omega} \log \frac{h_3^\star e}{h_M} - \frac{\varepsilon}{2\omega} \left[\frac{\partial J_1^\star}{\partial \lambda} \log \frac{h_2^\star}{h_M} - \frac{\partial J_2^\star}{\partial \lambda} \log \frac{h_1^\star}{h_M} \right] \operatorname{sgn} \left(\frac{\partial J_3^\star}{\partial \lambda} \right) \\ & + \frac{\varepsilon}{2\omega} \left[\frac{\partial J_1^\star}{\partial \lambda} \log \frac{h_2^\star}{b_3} - \frac{\partial J_2^\star}{\partial \lambda} \log \frac{h_1^\star}{b_3} \right] \operatorname{sgn} \left(\frac{\partial J_3^\star}{\partial \lambda} \right) \\ & - \frac{b_3}{2\pi\omega} \left[\log \left[\frac{\Gamma(y)\Gamma(1-\alpha+y)\sqrt{y}}{2\pi} \right] - 2y \log \frac{h_3^\star}{b_3} \right] \\ & + 0(\varepsilon^{4/3} \log \varepsilon^{-1}) \end{aligned} \tag{89}$$

where

$$\alpha = \frac{\partial J_2^\star}{\partial \lambda} \Big/ \frac{\partial J_3^\star}{\partial \lambda} \ge 0 , \quad y = \frac{h_0}{b_3} \ge 0 .$$

Note added in Proof: Y. Elskens has indicated to us that the error term in (82) may be improved in the sense that the error term $N\varepsilon^2 \log h_m^{-1}$ may be replaced by $\varepsilon^2 \log h_m^{-1}$. It follows that the condition on $\log h_m^{-1}$ (see (83)) may be replaced by $\log h_m^{-1} \leq \varepsilon^{-2/3}$. The price to pay for this improvement is to base the recurrence defining h_k (see (77)) not on h_0 but on $h'_0 = h_2 - 2b$. The error terms $\varepsilon^2 \log h_m^{-1}$ in (77) appear then only for h_0 and h_1 and not for h_k $(k \geq 2)$. Of course the variable x in (82) is no longer h_0/b but h'_0/b.

Appendix 7: The Diffusion Parameter in Non-Symmetric Cases

The function $\Delta_2(i,j)$ corresponding to a jump from domain D_i to domain D_j, has been defined in (II.92). It can also be written as:

$$\Delta_2 - < \Delta_2 > = \frac{\varepsilon}{\omega}\left(\frac{\partial J_i^\star}{\partial \lambda}\right)\left\{\frac{1}{2}(1-2z)K_{i,j} + F_{i,j} - < F_{i,j} >\right\} \tag{90}$$

where the variable z is now always defined in the interval $[0,1]$. The constant $K_{i,j}$ is given by:

$$K_{i,j} = (2\delta_j^3 - 1)(1+\delta_i^3)\mathrm{Pr}(i,j)\log\frac{h_i^\star}{b_i} + (2\delta_i^3 - 1)(1+\delta_j^3)\mathrm{Pr}(j,i)\log\frac{h_j^\star}{b_j} \tag{91}$$

where the δ_m^n are the usual Kronecker symbols (equal to 1 for $m = n$ and to zero for $m \neq n$). The $\mathrm{Pr}(i,j)$ are the probability functions of a jump from domain D_i to domain D_j (see (II.72).

When one of the domains D_i or D_j is D_3, the functions $F_{i,j}$ are defined by (with $\alpha = b_k/b_3$)

$$\begin{aligned} F_{k,3} = F_{3,k} &= \log\left\{\frac{\Gamma(\alpha-\alpha z)\Gamma(1-\alpha z)\Gamma(z)}{(2\pi)^{3/2}}\right\} \text{ for } \alpha < 1\,, \\ F_{k,3} = F_{3,k} &= \log\left\{\frac{\Gamma(\alpha-z)\Gamma(1-z)\Gamma(\alpha^{-1}z)}{(2\pi)^{3/2}}\right\} \text{ for } \alpha > 1\,, \end{aligned} \tag{92}$$

and, when neither of the domain is D_3, we have (with $\alpha = \min\{b_i/b_j, b_j/b_i\}$)

$$F_{i,j} = \log\left\{\frac{\Gamma(1-\alpha z)\Gamma(1-z)}{2\pi\sqrt{\alpha}z}\right\}\,. \tag{93}$$

The diffusion parameter, the root-mean-square of Δ_2, is given by

$$\sigma_{i,j}^2 = \left(\frac{b_j}{2\pi\omega}\right)^2\left\{\frac{1}{12}K_{i,j}^2 + K_{i,j}M_{i,j}(\alpha) + N_{i,j}(\alpha)\right\} \tag{94}$$

with

$$M_{i,j}(\alpha) = \int_0^1 (1-2z)[F_{i,j} - < F_{i,j} >]\,dz\,, \tag{95}$$

$$N_{i,j} = \int_0^1 [F_{i,j} - < F_{i,j} >]^2\,dz\,. \tag{96}$$

The integrals $< F_{i,j} >$, $M_{i,j}$ and $N_{i,j}$ have been evaluated numerically and the results are presented in Figs. 35 and 36.

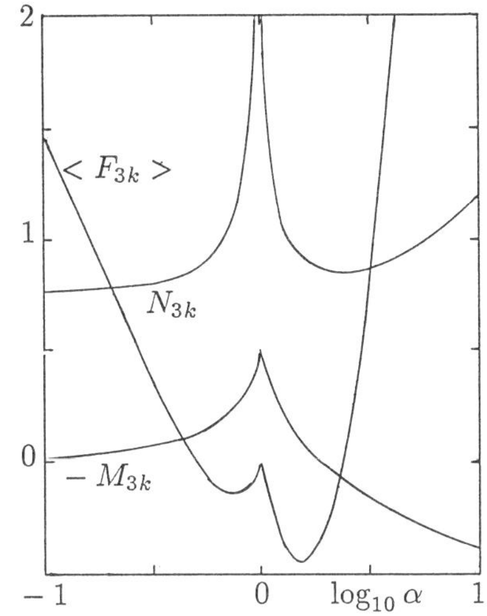

Fig. 35. The functions $< F_{i,j} >$, $M_{i,j}$, $N_{i,j}$ in the case where one of the indices i or j is equal to 3.

An analytical approximation of these integrals can be obtained by replacing, in the expression of the function $F_{i,j}$, the Gamma function $\Gamma(x)$ by $1/x$. This approximation is rather good in the interval $0 \leq x \leq 1$. In this case, we find

$$F'_{i,j} - < F'_{i,j} > = -\log\{z(1-z)(1-\beta z)\} - 3 + \frac{\beta - 1}{\beta}\log(1-\beta) \tag{97}$$

with

$$\beta = \min\{\alpha, \alpha^{-1}\} \leq 1\,. \tag{98}$$

The corresponding integrals are given by

$$M'_{i,j} = -\frac{1}{\beta}\left\{1 - \frac{\beta}{2} + \frac{1-\beta}{\beta}\log(1-\beta)\right\}\,, \tag{99}$$

$$\begin{aligned} N'_{i,j} &= 1 - \frac{\pi^2}{3} + -4\frac{1-\beta}{\beta}\log(1-\beta) - \frac{1-\beta}{\beta^2}\log^2(1-\beta) \\ &\quad + 2\sum_{j\geq 1}\frac{\beta^j}{j(j+1)}\left\{\frac{1}{j+1} + \sum_{k=1}^{j+1}\frac{1}{k}\right\}\,. \end{aligned} \tag{100}$$

To give an idea of the accuracy of the analytical approximation so obtained, we display in Figs. 37 and 38 the differences $< F'_{i,j} > - < F_{i,j} >$, $M_{i,j} - M'_{i,j}$ and $N_{i,j} - N'_{i,j}$.

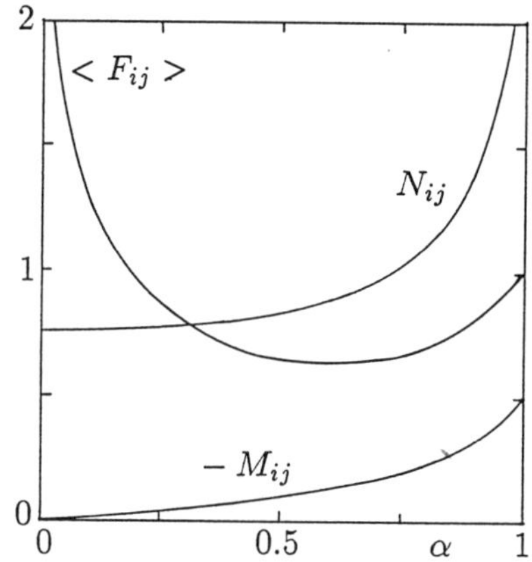

Fig. 36. The functions $< F_{i,j} >$, $M_{i,j}$, $N_{i,j}$ in the case where neither of the indices i or j is equal to 3.

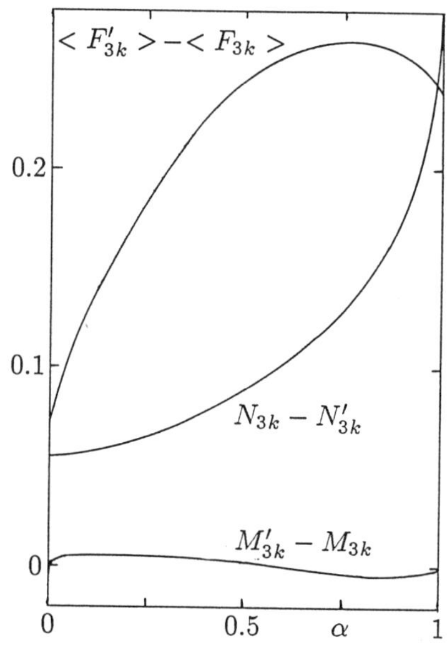

Fig. 37. The differences $< F'_{i,j} > - < F_{i,j} >$, $M_{i,j} - M'_{i,j}$ and $N_{i,j} - N'_{i,j}$ in the case where one of the indices i or j is equal to 3.

Appendix 8: Remarks on the Paper "On the Generalization of a Theorem of A. Liapounoff", by J. Moser (Comm. P. Appl. Math. 9, 257–271, 1958)

In this paper, J. Moser establishes the existence and some properties of a four-dimensional invariant manifold in the neighborhood of the equilibrium of an analytical Hamiltonian system.

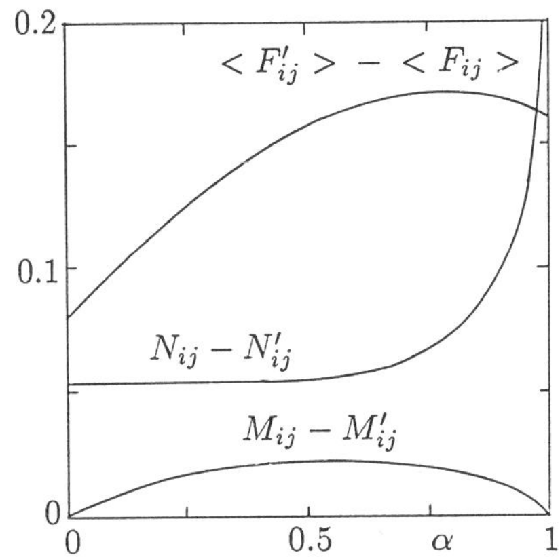

Fig. 38. The differences $< F'_{i,j} > - < F_{i,j} >$, $M_{i,j} - M'_{i,j}$ and $N_{i,j} - N'_{i,j}$ in the case where neither of the indices i or j is equal to 3.

When the Hamiltonian system has two degrees of freedom, this result describes the general solution of the system. We consider this particular case in our remarks.

Let the Hamiltonian function of the variables x_1, x_2 and their momenta x_3, x_4

$$H(x_1, x_2, x_3, x_4) \tag{101}$$

be real analytical in a neighborhood of the origin ($x_1 = x_2 = x_3 = x_4 = 0$) which is assumed to be an equilibrium (grad$H = 0$ at the origin). Let us assume further that the eigenvalues of SH_{xx} evaluated at the origin (S is the principal symplectic matrix and H_{xx} the Hessian of H) are

$$\alpha_1 \ , \ \alpha_2 \ , \ -\alpha_1 \ , \ -\alpha_2 \tag{102}$$

with α_1, α_2 independent over the reals. Hence as the complex conjugated of the eigenvalues are also eigenvalues, we have either that α_1 is the complex conjugated of α_2 (or of $-\alpha_2$) or that one of the α_1, α_2 is real and the other purely imaginary.

Moser recalls the existence of the Birkhoff normal form, i.e. that there exist formal power series

$$x_i = X_i(z_1, z_2, z_3, z_4) \tag{103}$$

which are formally canonical transformations, and which transform the Hamiltonian function (101) into the formal power series

$$\tilde{H}(z_1, z_2, z_3, z_4) = \alpha_1 Z_1 + \alpha_2 Z_2 + \tilde{H}_2(Z_1, Z_2) \ , \tag{104}$$

with

$$Z_1 = z_1 z_3 \ , \quad Z_2 = z_2 z_4 \ . \tag{105}$$

The $\tilde{H}_2$ is a formal power series starting with second degree terms in Z_1, Z_2. It follows that Z_1 and Z_2 are formal first integrals of the system, the formal solution of which is easy to produce.

Furthermore, Moser establishes the existence of formal power series

$$x_i = X_i'(y_1, y_2, y_3, y_4) \tag{106}$$

which are not necessarily canonical but which transform the differential equations associated with (101) into

$$\dot{y}_i = a_i(Y_1, Y_2)y_i \,, \quad \dot{y}_{i+2} = -a_i(Y_1, Y_2)y_{i+2} \tag{107}$$

with

$$Y_1 = y_1 y_3 \,, \quad Y_2 = y_2 y_4 \,. \tag{108}$$

The normalizing transformation (106) is not unique but the set of normalizing transformations form a group, each element of which is related to the other ones by the formulae

$$y_i' = y_i G_i(Y_1, Y_2) \quad 1 \leq i \leq 4 \,, \tag{109}$$

where the G_i are formal power series and $G_i(0,0) = 1$. Of course, the canonical transformations (103) form a subgroup of this group.

The essential part of Moser's paper is to show that there exists a subgroup of the group of normalizing transformations which is not only formal but analytical as the corresponding power series (106) actually converge.

Moser does not show however that the intersection of the subgroup of canonical normalizing transformations and the subgroup of converging normalizing transformations is not empty. The interest of the question did not escape him but apparently he left it for a subsequent paper which was never published (or we did not find!). He states on page 266 that this question "will not be decided in this paper".

It is this question that we would like to settle in this appendix.

Let the transformation (106) be an analytical normalizing transformation and the transformation (103) a canonical normalizing transformation. As they are both normalizing, they are related by

$$z_i = y_i G_i(Y_1, Y_2) \tag{110}$$

where the G_i are formal power series. Furthermore, as the transformation (103) is canonical, we have that formally

$$(y_i G_i; y_j G_j) = (X_i'; X_j') \,, \quad 1 \leq i < j \leq 4 \,. \tag{111}$$

the Poisson's brackets $(-;-)$ being taken with respect to the coordinates (y_i).

The right-hand members of (111) are known analytical functions of the (y_i). The equations themselves can be looked upon as partial differential equations in the functions G_i. We know that there exist formal solutions to them and we would like to show that they are analytical solutions as well. If we can find an analytical solution, this will show that the formal series (103) are actually convergent as they can be obtained by substituting the inverse of (110) into (106).

We shall first find a solution to a subset of the equations (111). Let us consider

$$\begin{aligned}(y_1G_1;y_3G_3) &= G_1G_3 + Y_1\frac{\partial(G_1G_3)}{\partial Y_1} &= (X_1';X_3')\,,\\(y_2G_2;y_4G_4) &= G_2G_4 + Y_2\frac{\partial(G_2G_4)}{\partial Y_2} &= (X_2';X_4')\,,\\(y_1G_1;y_2G_2) &= y_1y_2[G_1\tfrac{\partial G_2}{\partial Y_1} - G_2\tfrac{\partial G_1}{\partial Y_2}] &= (X_1';X_2')\,.\end{aligned} \tag{112}$$

These equations possess a formal solution. We conclude that $(X_1';X_2')$ is divisible by y_1y_2 and that $(X_1';X_3')$, $(X_2';X_4')$ and $(X_1';X_2')/y_1y_2$ are functions of Y_1, Y_2 alone. If we take now $G_1^\star = 1$, it is easy to find, by integration, an analytical solution of (112) for $G_2^\star$, $G_3^\star$, $G_4^\star$.

Let us take now, as new normalizing coordinates, non-necessarily canonical but analytical, the coordinates

$$y_i' = y_iG_i^\star(Y_1,Y_2) \tag{113}$$

with the $G_i^\star$ just defined. The new series (106) (we shall note them again as X_i') are now such that

$$(X_1';X_3') = (X_2';X_4') = 1\,,\quad (X_1';X_2') = 0\,. \tag{114}$$

It remains to show that, by construction, we must have also that

$$(X_1';X_4') = (X_2';X_3') = (X_3';X_4') = 0 \tag{115}$$

in order to obtain that the transformation is indeed canonical.

Considering again the set of equations (112) with the new coordinates y_i' (which we denote again y_i in order to simplify the notations), we observe that the only formal solution possible is such that

$$G_1G_3 = G_2G_4 = 1\,,\quad G_1\frac{\partial G_2}{\partial Y_1} = G_2\frac{\partial G_1}{\partial Y_2} \tag{116}$$

because the righ-hand members of (112) are now one or zero according to (114).

The remaining set of equations (111) can be written:

$$\begin{aligned}y_1y_4\left\{G_1\frac{\partial G_4}{\partial Y_1} + G_4\frac{\partial G_1}{\partial Y_2}\right\} &= (X_1';X_4')\,,\\y_2y_3\left\{G_2\frac{\partial G_3}{\partial Y_2} + G_3\frac{\partial G_2}{\partial Y_1}\right\} &= (X_2';X_4')\,,\\y_3y_4\left\{G_4\frac{\partial G_3}{\partial Y_2} - G_3\frac{\partial G_4}{\partial Y_1}\right\} &= (X_3';X_4')\,,\end{aligned} \tag{117}$$

and we know that there exists a formal solution of them which verifies (116). Inserting (116) into (117), we find that the left-hand members vanish. Hence the right-hand members are analytical functions with vanishing formal expansions. They are thus the zero functions.

References

Aamodt, R.E., 1971, Particle containment in mirror traps in the presence of fluctuating electric fields, Phys. Rev. Lett. 27, 135–138

Aamodt, R.E., 1972, Mirror containment with low-frequency short-wavelength fluctuations, Phys. of Fluids 15, 512–514

Abramowitz M. and Stegun, I.A., 1965, Handbook of Mathematical functions, Dover P.I., New York

Alfven, H., 1950, Cosmical Electrodynamics, Clarendon Press, Oxford

Allan, R.R., 1969, Evolution of mimas-tethys commensurability, Astron. J. 74, 497–508

Allan, R.R., 1970, On the evolution of commensurabilities between natural satellites, Symposia Mathematica 3, 75–96

Andronov, A.A., Vitt, A.A. and Khaikin, S.E., 1966, Theory of Oscillators, Addison-Wesley, Reading, Mass.

Arnol'd, V.I., 1964, Small denominators and problems of stability of motion in classical and celestial mechanics, Russian Math. Survey 18, 85–191

Arnol'd, V.I., 1978, Mathematical Methods of Classical Mechanics, Springer-Verlag

Arnol'd, V.I., 1980, Chapitres Supplémentaires de la Théorie des Equations Différentielles Ordinaires, Mir, Moscou)

Arnol'd, V.I. and Avez, A., 1968, Ergodic Problems of Classical Mechanics, W.A. Benjamin Inc.

Best, R.W.B., 1968, On the motion of charged particles in a slightly damped sinusoidal potential wave, Physica 40, 182–196

Birkhoff, G., 1927, Dynamical Systems, Am. Math. Soc. Coll. Pub. IX

Bogoliubov, N.M. and Mitropolsky, 1961, Asymptotic Methods in the Theory of Non-linear Oscillations, Hindustan P.C.

Borderies, N., 1980, La Rotation de Mars. Thesis, University of Toulouse

Borderies, N. and Goldreich, P., 1984, A simple derivation of capture probabilities for the $j + 1/j$ and $j + 2/j$ orbit-orbit resonance problems, Celestial Mechanics 32, 127–136

Born, H. and Fock, V., 1928, Beweis des Adiabatensatzes, Z. Physik 51, 165–180

Bretagnon, P., 1974, Termes à longues périodes dans le système solaire, Astron. Astrophys. 30, 141–154

Brouwer, D., 1959, Solution of the problem of an artificial satellite without drag, Astron. J. 64, 378

Brouwer, D. and Clemence, G.M., 1961, Methods of Celestial Mechanics, Academic Press, N.Y.

Burns, T.J., 1979, On the rotation of Mercury, Celestial Mechanics 19, 297–313

Cary, J.R., Escande, D.F. and Tennyson, J.L., 1984, Institute for Fusion Studies, Report no. IFRS-155

Cary, J.R., Escande, D.F. and Tennyson, J.L., 1986, Adiabatic invariant change due to separatrix crossing, Physical Review A34, 4256–4275

Cassini, G.D., 1693, Traité de l'Origine et des Progrès de l'Astronomie, Paris

Chirikov, B.V., 1959, The passage of a nonlinear oscillating system through resonance, Sov. Phys. Doklady 4, 390–394

Chirikov, B.V., 1979, A universal instability of many-dimensional oscillator systems, Physics Reports 52, 263–379

Colombo, G., 1965, Rotation period of the planet Mercury, Nature 208, 575

Colombo, G., 1966, Cassini's second and third laws, Astron. J. 71, 891–896

Colombo, G. and Shapiro, I.I., 1966, The rotation of the planet Mercury, Astrophys. J. 145, 296–307

Counselman C.C. and Shapiro, I.I., 1970, Spin-orbit resonance of Mercury, Symposia Mathematica, 3, 121–169
Deprit, A., 1967, Free rotation of a ridig body studied in the phase plane, Am. J. of Phys. 35, 424–428
Deprit, A., 1969, Canonical transformations depending on a small parameter, Celestial Mechanics 1, 12–30
Deprit, A. and Henrard, J., 1969, Construction of orbits asymptotic to a periodic orbit, Astron. J. 74, 308–318
Delaunay, C., 1867, Théorie du mouvement de la Lune, Mém. Acad. Sci. Paris 29
Dobrott, D. and Greene, J.M., 1971, Probability of trapping-state transitions in a toroidal device, Phys. of Fluids 14, 1525–1531
Dommanget, J., 1963, Recherches sur l'évolution des étoiles doubles, Ann. Observatoire Royal de Belgique, III, 9, 7–92
Ehrenfest, P., 1916, Adiabatische Invarianten und Quantentheorie, Ann. d. Phys. 51, 327
Escande, D.F., 1985, Change of adiabatic invariant at separatrix crossing: Application to slow Hamiltonian chaos, in "Advances in Nonlinear Dynamics and Stochastic Processes" (R. Livi and A. Politi eds.), World Scientific Singapore, 67–79
Escande, D.F., 1985, Stochasticity in classical Hamiltonian systems: universal aspects, Physics Reports 121, 165–261
Escande, D.F., 1987a, Slow Hamiltonian Chaos in "Advances in Nonlinear Dynamics and Stochastic Processes II" (Paladin and Vulpiani eds). World Scientific Singapore
Escande, D.F., 1987b., Hamiltonian Chaos and adiabaticity. Proc. Int. Workshop Kiev. World Scientific
Freidberg, J.P., 1982, Ideal magnetohydrodynamic theory of magnetic fusion systems, Rev. of Modern Physics 54, 801–902
Gardner, C.S., 1959, Adiabatic invariants of periodic classical systems, Phys. Rev. 115, 791–794
Giacaglia, G.E.O., 1972, Perturbation Methods in Non-Linear Systems, Springer-Verlag
Giorgilli, A. and Galgani, L., 1985, Rigorous estimates for the series expansions of Hamiltonian perturbation theory, Celestial Mechanics 37, 95–112
Goldreich, P., 1965, An explanation of the frequent occurrence of commensurable mean motions in the Solar System, M.N.R.A.S. 130, 159–181
Goldreich, P., 1986, Final spin states of planets and satellites, Astron. J. 71, 1–7
Goldreich, P. and Peale, S., 1966, Spin-orbit coupling in the Solar System, Astron. J. 71, 425–437
Goldreich, P. and Toomre, A., 1969, Some remarks on polar wandering, J. of Geophys. Res. 74, 2555–2567
Gonczi, R., Froeschlé, Ch. and Froeschlé, Cl., 1982, Poynting-Robertson drag on orbital resonance, Icarus 51, 633–654
Gonczi, R., Froeschlé, Ch. and Froeschlé, Cl., 1983, Trapping time of resonant orbits in presence of Poynting-Robertson drag, in "Dynamical Trapping and Evolution in the Solar System" (Markellos V.V. and Koza˘ Y. eds.), Reidel
Gradshteyn, I.S. and Ryzhic, I.M., 1965, Tables of Integrals, Series and Products Academic Press
Greenberg, R., 1973, Evolution of satellite resonances by tidal dissipation, Astron. J. 78, 338–346
Greenberg, R., 1977, Orbit-orbit resonances in the solar system: varieties and similarities, Vistas in Astronomy 21, 209–239
Guckenheimer, J. and Holmes Ph., 1983, Nonlinear Oscillations, Dynamical Systems and Bifurcations of Vector Fields, Springer-Verlag
Hagihara, Y., 1972, Celestial Mechanics, II: Perturbation Theory. MIT Press

Hannay, J.H., 1986, Accuracy loss of action invariance in adiabatic change of a one degree of freedom Hamiltonian, J. Phys. A, 19, 1067–1072
Helwig, G., 1955, Z. Naturforsch. 10A, 508
Henrard, J., 1970a, On a perturbation theory using Lie transforms, Celestial Mechanics 1, 111–222
Henrard, J., 1970b, Perturbation technique in the theory of nonlinear oscillations and in Celestial Mechanics, Boeing Scient. Res. Lab. Techn. Note 051
Henrard, J., 1974, Virtual singularities in the Artificial satellite theory, Celestial Mechanics 10, 437–449
Henrard, J., 1982a, Capture into resonance: An extension of the use of the adiabatic invariants, Celestial Mechanics 27, 3–22
Henrard, J., 1982b, The adiabatic Invariant: Its use in Celestial Mechanics in applications of modern dynamics (V. Szebehely ed.) pp. 153–171, Reidel, Dordrecht
Henrard, J., 1983, On Brown's conjecture, Celestial Mechanics 31, 115–122
Henrard, J., 1985, Spin-orbit resonance and the adiabatic invariant, in "Resonances in the motion of Planets" (Ferras-Mello and Sessin eds.), Univ. Saõ Paulo
Henrard, J. and Lemaître, A., 1983a, A second fundamental model for resonance, Celestial Mechanics 30, 197–218
Henrard, J. and Lemaître, A., 1983b, A mechanism of formation for the Kirkwood Gaps, Icarus 55, 482–494
Henrard, J. and Lemaître, A., 1986a, A perturbative treatment of the 2/1 Jovian resonance, Icarus 69, 266–279
Henrard, J. and Lemaître, A., 1986b, A perturbation method for problems with two critical arguments, Celestial Mechanics 39, 213–238
Henrard, J. and Murigande, Ch., 1987, Colombo's top, Celestial Mechanics 40, 345–366
Henrard, J. and Roels, J., 1974, Equivalence for Lie transforms, Celestial Mechanics 10, 497–512
Hori, G.I., 1966, Theory of general perturbation with unspecified canonical variables, Pub. Astron. Soc. of Japan 18, 287–296
Jeans, J.H., 1924, Cosmogenic problems associated with a secular decrease of mass., M.N.R.A.S. 85, 2
Jeans, J.H., 1924, The effect of varying Mass on a binary system, M.N.R.A.S. 85, 912
Jefferys, W.H., 1968, Perturbation theory for strongly perturbed dynamical systems, Astron. J. 73, 522–527
Kirchgraber, U. and Stiefel, E., 1978, Methoden der analytischen Störungsrechnung und ihre Anwendungen, B.G. Teubner, Stuttgart
Kneser, H., 1924, Die adiabatisshe Invarianz des Phasenintegrals bei einem Freiheitsgrad, Math. Ann. 91, 156–160
Kovrizhnykh, L.M., 1984, Progress in stellarator theory, Plasma Phys. 26, 195–207
Krilov, N. and Bogoliubov, N., 1947, Introduction to non-linear mechanics, Annals of Mathematics Studies, II, Princeton U.P.
Kruskal, M., 1952, U.S. Atomic Energy Commission Report N40-998 (PM-S-5)
Kruskal, M., 1957, Rendiconti del Tezzo Congresso Internazionale sui Fenomeni d'Ionizzazione nei Gas tenuto a Venezia, Societa Italiana di Fisica, Milan
Kruskal, M., 1962, Asymptotic theory of Hamiltonian and other systems with all solutions nearly periodic, J. of Math. Phys. 3, 806–828
Kulsrud, R., 1957, Adiabatic Invariant of the Harmonic Oscillator, Phys. Rev. 106, 205–207
Langevin, P. and De Broglie, M., 1912, La théorie du rayonnement et les quanta, Report on the meeting at the Institute Solvay at Brussels, Gauthier Villars, Paris (p. 450)
Lemaître, A., 1984, High-order resonances in the restricted three body problem, Celestial Mechanics 32, 109–126

Lemaître, A., 1984a, Analysis of a simple mechanism to deplete the Kirkwood Gaps, in Dynamical Trapping and Evolution in the Solar system (V.V. Markellos and Y. Kozai eds) Reidel
Lemaître, A., 1984b, L'origine des lacunes de Kirkwood. Thesis, University of Namur
Lemaître, A., 1985, The formation of the Kirkwood gaps in the Asteroid Belt, Celestial Mechanics 34, 329–341
Lemaître, A. and Henrard, J., 1988, Chaotic Motions in the 2/1 Resonance, in preparation
Lenard, A., 1959, Adiabatic Invariance to all orders, Annals of Physics 6, 261–276
Leung, A. and Meyer, K., 1975, Adiabatic Invariants for Linear Hamiltonian Systems, J. of Dif. Equ. 17, 33–43
Lichtenberg, A.J. and Lieberman, M.A., 1983, Regular and Stochastic Motion, Springer-Verlag
Littlejohn, R.G., 1979, A guiding center Hamiltonian: A new approach, J. Math. Phys. 20, 2445–2458
Littlejohn, R.G., 1983, Variational principles of guiding center motion, J. Plasma Physics 29, 111–125
Littlewood, J.E., 1963, Lorentz's pendulum problem, Ann. Physics 21, 233–242
Menyuk, C.R., 1985, Particle motion in the field of a modulated wave, Phys. Rev. A 31, 3282–3290
Message, P.J., 1966, On nearly-commensurable periods in the restricted problem of three bodies, Proceedings IAU, Symposium 25, 197–222
Meyer, R.E., 1976, Adiabatic variation – Part V: Nonlinear near-periodic oscillator, ZAMP 27, 181–195
Meyer, R.E., 1980, Exponential asymptotics, SIAM Review 22, 213–224
Moser, J., 1958, On a generalization of a theorem of A. Liapounoff, Comm. Pure. Applied. Math. XI, 257–271
Nayfeh, A.H., 1973, Perturbations Methods, J. Wiley and Sons
Neishtadt, A.I., 1975, Passage through a separatrix in a resonance problem with a slowly-varying parameter, Prikl. Matem. Mekhun 39, 621–632
Neishtadt, A.I., 1986, Change in adiabatic invariant at a separatrix, Sov. J. Plasma Phys. 12, 568–573 (Fiz. Plazung 12, 992–1001)
Peggs, S.G. and Talman, R.M., 1986, Nonlinear problems in accelerator physics, Ann. Rev. Nucl. Part. Sci. 36, 287–325
Peale, S.J., 1969, Generalized Cassini's laws, Astron. J. 74, 483–489
Peale, S.J., 1974, Possible histories of the obliquity of Mercury, Astron. J. 79, 722–744
Peale, S.J., 1976, Orbital resonances in the Solar system. Ann. Rev. Astron. Astrophys. 14, 215–246
Peale, S.J., 1986, Dynamical evolution of natural satellites: Some examples and consequences, A chapter of the book Natural Satellites (Burns J.A. and Matthews M.S. eds.), Univ. of Arizona Press
Poincaré, H., 1899, Les méthodes nouvelles de la mécanique céleste, Gauthier-Villars
Poincaré, H., 1902, Sur les Planètes du type d'Hecube, Bull. Astron. 19, 289–310
Poincaré, H., 1911, Leçons sur les hypothèses cosmogoniques, Paris
Rüssmann, H., 1964, Über das Verhalten analytischer Hamiltonscher Differentialgleichungen in der Nähe einer Gleichgewichtslösung, Math. Ann. 154, 285–300
Schubart, J., 1966, Special cases of the restricted problem of three bodies, Proceedings IAU Symposium 25, 187–193
Sessin, W., 1981, Thesis, University of Saõ Paulo
Sessin, W. and Ferraz-Mello, S., 1984, Motion of two planets with period commensurable in the ratio 2/1, Celestial Mechanics 32, 307–332
Siegel, C.L. and Moser, J.K., 1971, Lectures on Celestial Mechanics, Springer-Verlag

Sinclair, A.T., 1972, On the origin of the commensurabilities amongst the satellites of Saturn, Mon. Not. Roy. Astr. Soc. 160, 169–187

Sinclair, A.T., 1974, On the origin of the commensurabilities amongst the satellites of Saturn II, Mon. Not. Roy. Astr. Soc. 166, 165–179

Stengle, G., 1977, Asymptotic estimates for the adiabatic invariance of a simple oscillator, SIAM J. Math. Anal. 8, 640–651

Stuart, J., Weidenschilling, S.J. and Davis, D.R., 1985, Orbital resonances in the solar nebula. implications for planetary accretion, Icarus 62, 16–29

Stern, D.P., 1971, Classical adiabatic perturbation theory, J. of Math. Phys. 12, 2231–2242

Stoker, J.J., 1950, Nonlinear vibrations, Interscience, New York

Tennyson, J., 1979, In nonlinear dynamics and the beam-beam interaction, A.I.P. Conference Proceedings 57, 158

Tennyson, J.L., Cary, J.R. and Escande, D.F., 1986, Change of the adiabatic invariant due to separatrix crossing, Phys. Rev. Letters 56, 2117–2120

Timofeev, A.V., 1978, On the constancy of the adiabatic invariant when the nature of the motion changes, Sov. Phys. JETP 48, 656–659

Torbett, M. and Smoluchowski, R., 1980, Sweeping of the Jovian resonances and the evolution of the asteroids, Icarus 44, 722–729

Torbett, M. and Smoluchowski, R., 1982, Motion of the Jovian Commensurability resonances and the character of the celestial mechanics in the asteroids zone: Implications for kinematics and structure, Astron. Astrophys. 110, 43–49

Urabe, M., 1954, Infinitesimal deformation of the periodic solution of the second kind and its application to the equation of a pendulum, J. Sci. Hiroshima Univ. A18, 183

Urabe, M., 1955, The least upper bound of a damping coefficient ensuring the existence of a periodic motion of a pendulum under constant torque, J. Sci. Hiroshima Univ. A18, 379

Ward, W.R., 1975, Tidal friction and generalized Cassini's laws in the Solar System, Astron. J. 80, 64–70

Ward, W.R., Burns, J.A. and Toon, O.B., 1979, Past obliquity oscillations of Mars: the role of the Tharsis Uplift, J. of Geophys. Res. 84, 243–258

Wasow, W., 1976, Adiabatic invariants and the asymptotic theory of ordinary linear differential equations, SIAM-AMS Proceedings 10, 131–144

Wasow, W., 1965, Asymptotic Expansions for Ordinary Differential Equations, J. Wiley and Son, 1965 (p. 128)

Wasow, W., 1973, Adiabatic invariance of a simple oscillator, SIAM J. Math. Anal. 4, 78–88

Weidenschilling, S.J., 1977, The distribution of mass in the planetary system and solar nebula, Astrophys. Space Sci. 51, 153–158

Whittaker, E.T., 1916, Proc. Roy. Soc. Edinburgh 37, 95

Whittaker, E.T., 1927, Treatise on the Analytical Dynamics of Particles and Rigid Bodies, 3rd. Edition

Wintner, A., 1941, The Analytical Foundations of Celestial Mechanics, Princeton U.P. (p. 24)

Wisdom, J., 1983, Chaotic behaviour and the origin of the 3/1 Kirkwood gap, Icarus 56, 51–74

Wisdom, J., 1985, A perturbative treatment of motion near the 3/1 Jovian commensurability, Icarus 63, 272–289

Yoder, C.F., 1973, On the establishment and evolution of orbit-orbit resonances, Thesis, University of California, Santa-Barbara

Yoder, C.F., 1979a, Diagrammatic theory of transition of Pendulum like systems, Celestial Mechanics 19, 3–29

Yoder, C.F., 1979b, How tidal heating in Io drives the Galilean orbital resonance locks, Nature 279, 747–750

Yoder, C.F. and Peale, S.J., 1981, The tides of Io, Icarus 47, 1–35

Zeipel, H. von, 1915, Recherche sur le mouvement des petites planètes, Arkiv Math. Astro. Fys. 11, 1–58

List of Contributors

H. S. Dumas
University of Cincinnati
Department of Mathematical Sciences
Cincinnati, OH 45221–0025
USA

Chr. Genecand
Goldauerstrasse 10
8006 Zürich
Switzerland

J. Henrard
Facultés Universitaires de Namur
Département de Mathématique
Rempart de la Vierge, 8
5000 Namur
Belgium

J. Komorník
Univerzita Komenského
Matematicko-fyzikálna fakulta
Mlynská dolina
842 15 Bratislava
Slovak Republic

PHYSICS